Studies in Systems, Decision and Control

Volume 336

The series “Studies in Systems, Decision and Control” (SSDC) covers both new developments and advances, as well as the state of the art, in the various areas of broadly perceived systems, decision making and control–quickly, up to date and with a high quality. The intent is to cover the theory, applications, and perspectives on the state of the art and future developments relevant to systems, decision making, control, complex processes and related areas, as embedded in the fields of engineering, computer science, physics, economics, social and life sciences, as well as the paradigms and methodologies behind them. The series contains monographs, textbooks, lecture notes and edited volumes in systems, decision making and control spanning the areas of Cyber-Physical Systems, Autonomous Systems, Sensor Networks, Control Systems, Energy Systems, Automotive Systems, Biological Systems, Vehicular Networking and Connected Vehicles, Aerospace Systems, Automation, Manufacturing, Smart Grids, Nonlinear Systems, Power Systems, Robotics, Social Systems, Economic Systems and other. Of particular value to both the contributors and the readership are the short publication timeframe and the world-wide distribution and exposure which enable both a wide and rapid dissemination of research output.

Indexed by SCOPUS, DBLP, WTI Frankfurt eG, zbMATH, SCImago.

All books published in the series are submitted for consideration in Web of Science.

More information about this series at http://www.springer.com/series/13304

Jitesh J. Thakkar

Multi-Criteria Decision Making

Springer

Jitesh J. Thakkar
National Rail and Transportation Institute
(NRTI)
Vadodara, Gujarat, India

ISSN 2198-4182 ISSN 2198-4190 (electronic)
Studies in Systems, Decision and Control
ISBN 978-981-33-4744-1 ISBN 978-981-33-4745-8 (eBook)
https://doi.org/10.1007/978-981-33-4745-8

This Springer imprint is published by the registered company Springer Nature Singapore Pte Ltd.
The registered company address is: 152 Beach Road, #21-01/04 Gateway East, Singapore 189721, Singapore

Dedicated to

my Guru
Prof. S. G. Deshmukh

Preface

The book covers the domain of multi-criteria decision making. This topic has gained significant attention of both researchers and practitioners for enhancing their decision making in real-life situation. The last two decades have witnessed the growth of numerous application of multi-criteria decision making (MCDM) in all sorts of areas such as engineering, business, management, humanities, psychology, law and others. These techniques involve the decision makers in the process and value their subjective and objective inputs. As the decision makers are involved in the model building and analysis stage, they gain greater confidence in implementing the recommendations drawn from such techniques. This book intends to benefit larger audience in academia and industry both for adopting such techniques. The book includes a total of 21 chapters. This covers 17 multi-criteria decision-making techniques like AHP, ANP, TOPSIS, ELECTRE, VIKOR, MOORA, LINMAP, SWARA, ARAS, PROMETHEE, etc., with its detailed applications for interesting problems. The book also includes three chapters on integrated applications of MCDM techniques. The application of the problems is demonstrated for a hypothetical data set. The analysis for the selected hypothetical problems is just for the purpose of illustrating various MCDM techniques, and results do not resemble the reality. The book is designed to help the potential researchers and students to develop the following competencies.

- Select an appropriate technique for the specificity of the given problem.
- Develop a case description for MCDM problem.
- Formulate MCDM problem and define the structure in detail.
- Implement a step-by-step procedure of MCDM technique and appreciate intricacies involved in each step.
- Develop meaningful inferences and recommendations useful to industry and academia.

Vadodara, India
February 2021

Jitesh J. Thakkar

Acknowledgements

It gives me an immense pleasure to deliver the book on *Multi-Criteria Decision Making (MCDM)* to the students. The book will benefit the students/researchers in engineering, management, law and social science field. The book will immensely benefit the industry professionals in solving real-life problems with MCDM techniques.

The field of multi-criteria decision making (MCDM) has received the contributions from many scholars, and hence, these individuals and scholars have contributed to the development of this book.

The discussion on the topic and growing research applications of MCDM has helped to shape this book.

I have always received inspiration and energy for executing academic projects from my teacher and Guru Prof. S. G. Deshmukh. I express my deep gratitude for his valuable inputs in improving the quality of this book. This book could not have attained its present form, both in content and in presentation, without his active interest, direction and help. I feel proud by mentioning that whatever I am today, it is because of his inspiration, affection and personal care. Professor Deshmukh has always inspired me for academic excellence and contributions.

I acknowledge the support of my students in coding and verification of the models. I thank my colleagues and researchers at IIT Kharagpur for providing valuable inputs and raising thoughtful queries.

My father—Jayprakashbhai Thakkar—and mother—Late Ushaben Thakkar—have provided a constant moral support and motivation for this work.

I deeply express my love and affection for my wife, Amee, daughter, Prachi, and son, Harshit, for giving me freedom and moral support in timely completion of this book.

I am thankful to the publisher Springer Nature Singapore Pte Ltd. for supporting my book proposal.

Vadodara, India
February 2021

Dr. Jitesh J. Thakkar

Contents

About the Author

Dr. Jitesh J. Thakkar is a Professor at National Rail and Transportation Institute (NRTI), India. He served as faculty at Department of Industrial and Systems Engineering, Indian Institute of Technology (IIT) Kharagpur for 10 years. He holds a Ph.D in Supply Chain Management from IIT Delhi, M.Tech. in Industrial Engineering from IIT Delhi and Bachelor's of Engineering in Mechanical Engineering Sardar Patel University, India. His areas of interests include transportation management, logistics and supply chain management, project management, service operations management, Six Sigma, lean manufacturing, optimization, statistical modeling and system dynamics. He has supervised 14 Ph.D. and 80 B.Tech./M.Tech. projects at IIT Kharagpur. He has a H-index 26 and 70 research papers published in SCI/SCOPUS listed journals. The research papers are published in high impact factor, A* & A grade journals of ABDC (Australian Business Deans Council) such as International Journal of Production Economics, Transportation Research (Part-E), International Journal of Production Research, Computers and Industrial Engineering, Production Planning and Control, Expert Systems with Applications, Journal of Cleaner Production. He has delivered over 100 invited sessions at various reputed platforms and contributed to journals of national and international repute.

Chapter 1
Introduction

1.1 Background

Decision making is a systematic and formal process which involves four steps: (i) identification of the problem; (ii) deriving the preferences; (iii) evaluation of alternatives; and (iv) identification of the best alternatives. The domain of decision making involves three kinds of analysis:

1. Descriptive analysis: It extensive involves the decision maker in deriving the inferences from the graphical or tabular presentation of the data/information.
2. Prescriptive analysis: It uses the methods to enabler decision makers for an improved analysis.
3. Normative analysis: It deals with the problems which extensively demand the involvement of decision maker in problem-solving through evaluation of set of arguments trying to present right and wrong about the situation. It demands cross-examination of the opinions and a consensus need to be sought in arriving at a specific decision.

The MCDM techniques discussed in this book are primarily fall into the domain of descriptive and normative analysis. Typically, the techniques in descriptive and normative analysis are considered within the broader domain of operations research (OR).

The purpose of operations research is to enable a quantitative analysis of solving the complex problems in organizational, national or societal domain. OR models are constructed with the help of fundamentals of mathematics, statistics and probability theory to develop optimal solutions. The approaches such as simulation and multi-criteria decision making primarily do not look at the optimal solution but intends to analyse various alternatives and scenarios for "what-if" analysis. Typically, OR modelling adopts a simple approach for solving the problems. At the first instance, it tries to break the problem into meaningful and manageable components and then solves it using well-defined theories of mathematics.

J. J. Thakkar, *Multi-Criteria Decision Making*, Studies in Systems, Decision and Control 336, https://doi.org/10.1007/978-981-33-4745-8_1

The history of decision science has reported significant advancements from operational researchers, behavioural scientists and mathematicians for structuring and solving a complex problem. A decision maker is expected to examine the set of potential and feasible alternatives and identify the optimal or most preferred one. The most important requirement for a decision maker is to identify the problem with all the relevant information from internal and external environment of the organization and derive an appropriate structure of the problem. A well-structured problem helps the decision maker and his team to develop greater insights into the intricacies of the problem and specifically the factors that may have an influence on the decision. Typically, the process of evaluating the alternatives and identifying the most preferred one is known as screening. This results into the identification of a smaller and manageable set of alternatives which may further be examined in detail to select the most appropriate one. The screening of alternatives may be done using various approaches such as Pareto-optimality, sequential approach, distance-based approach, trade-off based on weightages and others.

Decision makers have to apply highly intuitive approach when they deal with single criterion problems. This is quite simple as one has to select an alternative with a very high preference. The process of decision making becomes complex when it demands an evaluation of multiple criteria which are conflicting in nature and have different weightages. This demands the development of methods which can accommodate the trade-offs among the criteria and alternative and real-life issues and concerns of the decision makers. A typical decision-making process may be visualized as presented in Fig. 1.1.

The problems involving the evaluation of multiple criteria and alternatives, typically, are considered as a part of multi-criteria decision making (MCDM) domain. The problems in this area can be broadly classified into two categories: (i) multiple attribute decision making (MADM) which considers limited number of predetermined alternatives and discrete preference ratings and (ii) multiple objective decision making (MODM) which accommodates design and planning related problems for deriving an optimal solution of set of goals for set of interactions prevailing among the constrains. The process of MCDM analysis is governed by three steps:

(1) Identification of the relevant criteria and alternatives from the existing theory and practice.
(2) Assignment of numerical values to criteria to indicate their relative importance and to quantify the impacts of the alternatives on these criteria.
(3) Using a formal mathematical procedure for analysing numerical values to determine the ranking (priorities) of the alternatives.

Multiple criteria decision making (MCDM) is also known as multi-criteria decision analysis (MCDA). The methodologies in the domain of multi-criteria decision making are considered as subset of operations research. We are constantly involved in evaluating the trade-offs among conflicting criteria in our daily personal and professional life. Usually, such decisions are taken based on past experiences and intuitions of the person. For example, you want to optimize your financial portfolio by selecting

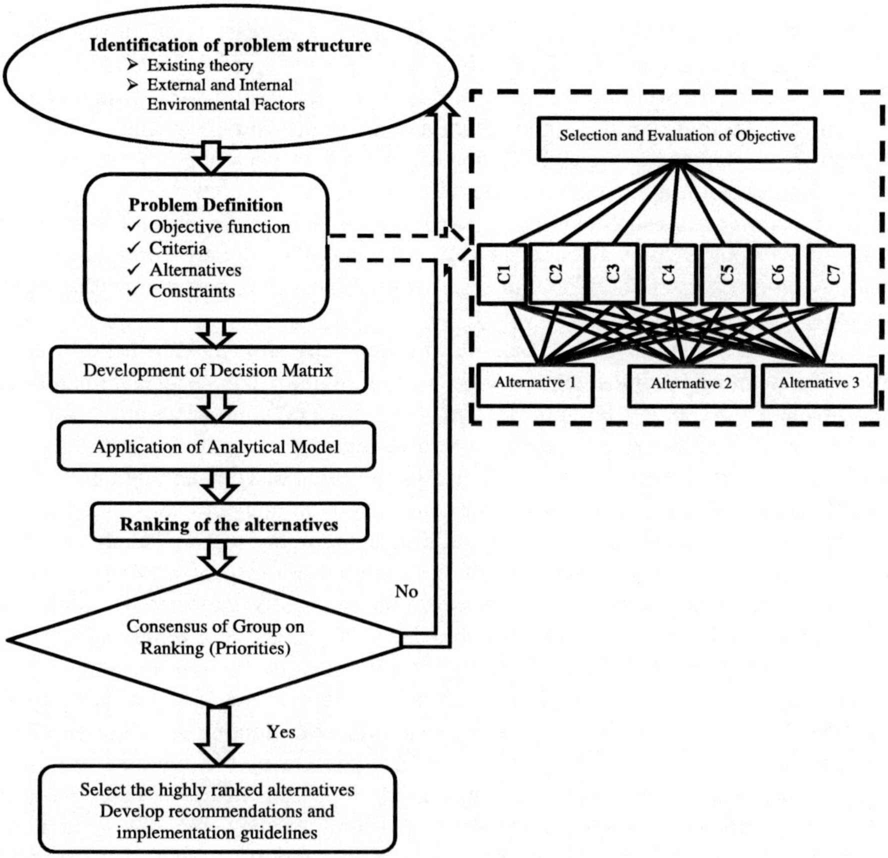

Fig. 1.1 Typical decision making process

various investment options. You want to maximize your daily productivity by prioritizing the various activities to be carried out. You want to prioritize your social and professional interactions and hence ensure an effective utilization of time and energy. We may think about many such examples in our day-to-day life where the situation demands trade-offs among various conflicting criteria. The criteria may be tangible in nature or intangible. For example, a marketing manager wants to launch an aggressive marketing campaign and considers increase in sales, sales effort, publicity and advertisement cost, commitment of the sales people, brand and image as some of the key criteria. We can very well see that the first three criteria (increase in sales, sales effort, publicity and advertisement cost) are tangible in nature whereas the other two criteria (commitment of the sales people, brand and image) are intangible in nature. A close examination may reveal that even the criteria may have a conflicting nature and hence setting a trade-off for a particular objective may be not easy to perform using individual intuitions. Another example in health care may help to appreciate the need of multi-criteria decision making better. For example, a patient is looking for a treatment which accommodates four criteria: quality of the treatment, cost, convenience and recovery period. One may see that the criteria like cost, quality and convenience are quite conflicting in nature and hence needs to be evaluated with an appropriate systematic approach. Further, if a patient is also looking for the market reputation of healthcare service provider or hospital, then the convenience may have to be compromised as such facility is usually available in urban cities. This kind of complex analysis enables a deeper investigation and reveal many important information that can help an individual or organization to execute their plan with a right strategy.

In last three decades, the field of multi-criteria decision making has received extensive attention in almost all the fields such as engineering, science, humanities, management, psychology, law, politics and others. MCDM approaches are extremely useful for the problem situation where policy makers need to decide priorities by resolving key conflicting objectives. Typically, the situations demanding the evaluation of such conflicting trade-offs do not demand any unique optimal solution but it looks for various preferences that a model can generate for "what-if" analysis. This helps the decision makers to address the problem more holistically by considering various environmental and external factors which are usually neglected in traditional OR approaches. MCDM solutions help the decision makers to examine the various scenarios and practical utility of the proposed solutions in a quite flexible manner.

In many of the real-life situations, decision makers do not look for an optimal solution but wish to investigate the set of non-dominated solutions. A non-dominated solution is the one which does not permit the deviation from it to any other solution without sacrificing in at least one criterion. The selection of set of potential and promising solutions from a non-dominated set is complex and demands a systematic approach. MCDM analysis ensures that the proposed solution is acceptable to a group of people or organization who will be affected through this. Also, the trade-offs evaluated through a critical analysis in MCDM techniques ensure that the adopted solutions will have an overall positive impact on the system and will not try to optimize on part of the system at the cost of another. For these reasons, MCDM has seen a phenomenal growth in different application areas, especially as new methods develop and as old methods improve.

1.2 Single-Criteria Versus Multi-criteria Decision Making

The problem structured may be governed by single criteria like maximization of profit or minimization of cost. There are well-established optimization procedures like linear programming, integer programming, nonlinear optimization, etc. in operations research to deal with such class of the problems. The real-life situation demands the evaluation of set of alternatives against multiple criteria and typically structuring it as multi-criteria decision-making problem. A comprehensive summary of single criteria versus multi-criteria is presented in Table 1.1.

Table 1.1 Single-criteria versus multi-criteria decision making

Characteristics	Single-criteria decision making	Multi-criteria decision making
Definition	Decision problem is characterized by single criteria or a single aggregate measure like cost	A decision problem may have multiple criteria like cost, profit, performance, etc.
Problem Structure	✓ Objective function: single criterion ✓ Constraints: indicates the requirements on the alternatives ✓ Depending on the characteristics of the optimization problem, different optimization techniques such as linear programming, nonlinear programming, discrete optimization can be used	It is characterized by a finite number of criteria and feasible alternatives

1.3 Applications of MCDM in Personal Versus Professional Life

Multi-criteria decision making has received an attention in all the facets of life. The problems are structured and solved for variety of the problems in personal and professional life. To appreciate, the holistic nature and broader utility of MCDM approach, the application of MCDM in personal and professional life for variety of the problems is presented in Tables 1.2 and 1.3.

1.4 Elements and Formulation of MCDM Problem

MCDM techniques are employed with a key focus to find adequate trade-offs among conflicting tangible and intangible criteria. For example, if we consider a small problem like purchasing a car for a family. The decision maker (an individual or family) has to evaluate many trade-offs among conflicting criteria such as: brand, type of car (Luxury or Economy), space, ergonomics, fuel efficiency, maintenance requirement, serviceability, colour, resale value. It is likely possible that a family may end up with many good options. Having arrived at this list, now how do they decide the one which can closely satisfy their requirements? This is where the use of MCDM helps to reach to the most acceptable solution on a consensus basis. The MCDM model consists of various elements, depending on the nature of the decision problems. Figure 1.2 depicts the elements which are generally found in the decision space.

A typical MCDM model is a systematic approach of evaluating, prioritizing and selecting the most favourable alternative from a set of available ones. The decision maker makes the final decision by considering both inter-attribute and intra-attribute

Table 1.2 Examples of MCDM in personal life

Goal	Possible criteria	Possible alternatives
Decide best school	• Distance • Reputation • Cost • Teacher	School names
Select best teacher	• Educational background • Subject knowledge • Motivation and enthusiasm • Self-Discipline	List of candidates
Determine Ph.D. topic	• Novelty • Usefulness to society • Availability of existing literature • Research cost	List of PhD topics
Finding best apartment	• Price • Down payment • Distance from shops/bus stand/metro station/school • Neighbour's friendliness	List of apartments under consideration
Buy house	• Price • Locality • Space • Nearness to market, hospital or airport	Types of apartments
Outsourcing decision—decide whether manufacture or produce in-house	• Availability of supplier • Proximity to supplier base • History of supplier • Cost of outsourcing	Outsource or Produce In-house
Vacation destination	• Entertainment • Facilities • Accommodation cost • Travel cost	List of places
Selecting hotel	• Ambience • Food quality • Price level • Noise level • Cleanliness • Rating	List of hotels
Selection of mobile phone	• Price • Availability • Storage • Camera • Looks • Serviceability	List of mobile phones

(continued)

Table 1.2 (continued)

Goal	Possible criteria	Possible alternatives
Campus recruitment	• Communication Skill • Knowledge • Personality • Academic performance	List of candidates

comparisons with necessary trade-offs. Mathematically, a typical MCDM problem with "m" alternatives and "n" criteria is modelled by the matrix as given in Fig. 1.3.

where $xR_{ij}R \in [0, 1]$ represents the level of importance of alternative aR_iR, $i = 1, \ldots, m$ with respect to criterion cR_jR, $j = 1, \ldots, n$, with 0 interpreted as "no satisfaction" and 1 corresponding to "complete satisfaction". The weight vector $w \in [0, 1]P^nP$, $\Sigma\ wR_jR = 1$ having its generic component wR_jR, $j = 1, \ldots, n$ represents the relatives importance of criterion cR_jR in terms of its weight.

MCDM approach evaluates the criteria using an aggregation function f: $[0, 1]P^nP \rightarrow [0, 1]$, which maps vectors of criteria values xR_iR, $i = 1, \ldots, m$ to the [0, 1] interval and satisfies, for all $x, y \in [0, 1]\ P^nP$. The resulting score indicates the attractiveness of the associated alternative, with the common understanding that 0 corresponds to "no preference" and 1 to "strongest preference". The scores derived through this helps to rank the alternatives and assist the decision maker in selecting the most preferred one.

1.5 Classification of MCDM Techniques

Multi-criteria decision making (MCDM) has two broad categories. This includes multi-attribute decision making (MADM) and multi-objective decision making (MODM). MADM deals with the evaluation of each alternative on the set of objectives to facilitate the selection. Attributes provide the basis for a comparison of the alternatives and consequently facilitate the selection.

✓ MADM approaches evaluate discrete alternatives. It covers a wide range of techniques for assessing the decision problems characterized by a large number of diverse attributes consisting of cardinal and ordinal data and preferences regarding the relative importance of the evaluating criteria. The relative importance of the evaluating criteria in terms of criteria weights in MADM though does not have any clear economic significance, but it plays an important role by providing the opportunity to represent the actual aspects of decision making (the preference structure). This class of MADM approaches is broadly covered into three categories: elementary methods, unique synthesizing methods, and outranking methods.

✓ MODM approaches evaluate the continuous alternatives and deals with multi-objective optimization problem. The method in MODM tries to evaluate the conflict among multiple objectives and search for the optimal solution.

Table 1.3 Examples of MCDM in professional life

Goal	Possible criteria	Possible alternatives
Supplier selection problem	• On time delivery • Quality and lead-time history • Technical Capability • Special factor	Number of suppliers
Vendor selection of a telecommunications system	• Cost – Capital expenditure – Operating expenditure • Quality – Technical – Operational – Vendor	Number of systems
Selection of knowledge management strategies	• Activating information • Improving performance • Promoting innovation – Top management Support – Communication – Attitude, values and adaptability of people – Incentives – Time – Cost	• Human-oriented style • Dynamic style
Optimal relationship between buyer and seller	• Coercive power • Communication cost • Economic satisfaction • Non-economic satisfaction	Revenue categories • Under 50 million • 50–500 million • 500–5000 million • Above–5000 million
Selecting full-service advertising agencies	• Strategic planning ability – Marketing research – Whole planning – Business understanding • Media Ability – Media planning – Media buying • Creativity – Creative work that sells – Advertising awards • Service level – Service range – Personnel quality – Compatibility and timing • Cost Consciousness	Number of agencies
Selection of new warehouse	• Proximity to market • Overhead cost • Storage capacity • Capacity utilization • Operating cost	Various locations

(continued)

Table 1.3 (continued)

Goal	Possible criteria	Possible alternatives
Weapon selection	• Barrel • Recoil system • Hit precision • Weight • Usability in different climatic conditions and Maintenance • Modularity • Trigger system • Safety mechanism • Sight system • Magazine • Dust bolt or dust cover • Price	Alternatives
Foreign trade company	• Technological • Infrastructure • Project • Duration • Budget • After Sales • Support • Usability • Capabilities • Service Provider	Alternatives
Overall risk index in Indian petroleum supply chain	• Environment – Man-made disaster – Legal/political – Social – Economic crisis • Upstream – Inventory – Crude supply – Delay • Process – Manufacturing competitiveness – IT system failure – Quality – Cost • Downstream – Demand – Transportation/ logistics – Forecast • Organization structure – Merger/alliance – Intellectual property trust – Leadership	Risk index value

(continued)

Table 1.3 (continued)

Goal	Possible criteria	Possible alternatives
Prioritization of risks in supply chain	• Terror • Economic • Political • Cultural • Natural • Outsourcing risk • Supplier insolvency • Quality • Sudden risk in cost • Sudden fluctuation • Market changes • Competition changes • Forecasting error • Machine failure • Quality problems • Technological change	• Types of risks
Selection of a suitable manufacturing system	• Operational cost • Time required • Changes in organization size structure • Feasibility • Availability of resources • Risk of failure • Stress and fatigue • Job security • Increase in roles and responsibilities	• Modern manufacturing systems • Traditional manufacturing systems

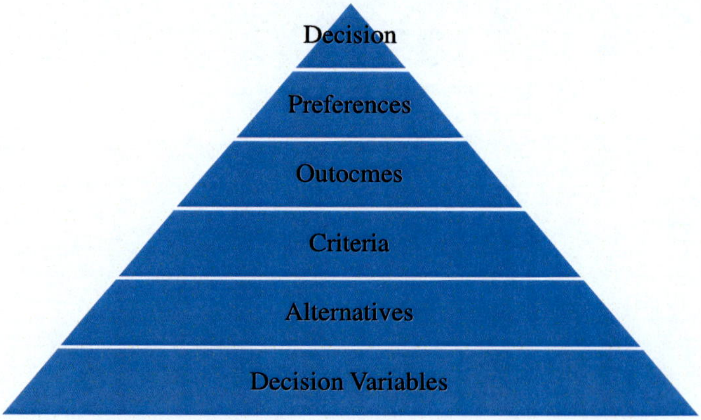

Fig. 1.2 Elements of MCDM Models

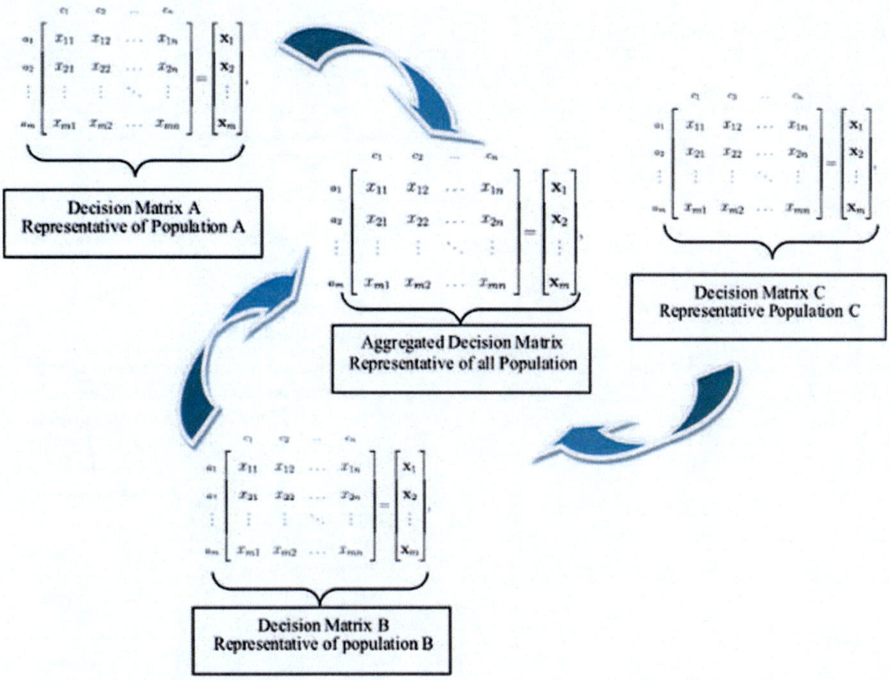

Fig. 1.3 Matrix formulation of MCDM Problem

Some of the most popular MADM tools are as following: analytic hierarchy process (AHP), analytic network process (ANP), technique for order preference by similarity to ideal solution (TOPSIS), Elimination and Choice Translating Reality (ELECTRE), VIsekriterijumska optimizacijai KOmpromisno Resenje (VIKOR), Preference Ranking Organization Method for Enrichment Evaluations (PROMETHEE), Simple Additive Weighting (SAW), Multi-Objective Optimization on basis of Ratio Analysis (MOORA), Complex Proportional Assessment (COPRAS), Complex Proportional ASsessment with Grey relations (COPRAS-G), Step-wise Weight Assessment Ratio Analysis (SWARA). The details of MADM and MODM techniques are given in Table 1.4.

1.6 Key Features of MCDM Techniques

An application of MCDM technique for a specific problem structure is governed by its capability to deal with the complexity of the problem. It is necessary for a decision maker to appreciate the key features of the MCDM technique before considering it as suitable for analysing the set of preferred alternatives. Table 1.5 depicts the key features of various MCDM techniques.

Table 1.4 Classification of MCDM approaches

		Type of method	Characteristic	Examples
MADM approaches	Category 1	Elementary methods	These are easy to use methods and analysis complexity is very low. The methods can be used by an individual decision maker without even the use of sophisticated computer software	Simple Additive Weightage (SAW)
	Category 2	Unique synthesizing methods	It can accommodate a complex problem structure and makes use of special mathematical and analytical approaches for analysing the set of alternatives with respect to underlying criteria. Typically, the methods in this category use the concept of ideal, anti-ideal and reference point to derive the ranking of the potential alternatives. The positive ideal or negative ideal is considered as the reference point solution and all other maximum values and minimum values are compared with this positive ideal and negative ideal, respectively, in the decision matrix	Technique for Order Preference by Similarity to Ideal Solution (TOPSIS) and Grey Relational Analysis (GRA)
	Category 3	Outranking methods	This class of MCDM approaches analyses the alternatives using series of pairwise comparisons with respect to criterion to derive an outranking relation indicating the degree of dominance of one alternative over other	Preference Ranking Organization METHod for Enrichment Evaluation (PROMETHEE) ELimination and Et Choice Translating REality (ELECTRE)

(continued)

Table 1.4 (continued)

		Type of method	Characteristic	Examples
MODM approaches			Evaluates the continuous alternatives and deals with multi-objective optimization problem	Linear programming Goal programming LINMAP Lexicographic

1.7 Real-Life Applications of MCDM Techniques

There are numerous successful applications of MCDM techniques in the various engineering and non-engineering domains. The techniques have extensively helped the practising managers in decision making and developing policy related guidelines for necessary resource allocation.

Bridge Construction

An application of MCDM technique can be employed for a bridge construction project in a city due to increased vehicle population and traffic load. This causes frequent traffic congestions and there is a need for an additional bridge. It is expected that the new bridge will meet three requirements: (i) reduce traffic congestion; (ii) reduce total distance to travel; and (iii) add to the elegance of the city. It is noted by the city planners that a new bridge is urgently needed as the service life of the old bridge is already over. The planners have also anticipated another important requirement as the proposed bridge needs to be constructed on a river. There is a need to keep the provision for spillway gates which are necessary for the bund (waterfront area) exist on the river. It is also necessary to identify and provide access points for the maintenance of the spillway gates. This problem situation can be considered as a fist case of MCDM modelling with a broader definition of the decision problem as "Select type of superstructure for proposed bridge".

Selection of a New Hub Airport

It primarily deals with the development and operation of hub-and-spoke transport networks. The numerous applications are reported in the operations research domain which mainly deals with the determination of the route structure and location of one and/or a few hubs, for maximizing the total network cost for a transport operator. If a real-life problem demands the consideration of criteria like delay in government approvals, perceived risk, etc. which might be difficult to accommodate in a typical OR model, then MCDM can be successfully employed to derive a holistic and policy-driven solution for such problem.

Software Selection

An educational institute wants to implement an ERP system with an objective to improve administrative accuracy, enhance knowledge sharing, improve transparency

Table 1.5 Key features of MCDM techniques

MCDM Technique/Chapter	Nature of relationship		Key features
	Linear	Cause and Effect	
SAW (Chap. 2)	✓		• Involves simple calculation • Intuitive to decision makers • The results may not reflect the real situation and do not sound logical
AHP (Chap. 3)	✓		• It accommodates a linear hierarchical structure of criteria and sub-criteria • Difficulty in accommodating interdependence between criteria and alternatives. This leads to inconsistencies between judgment and ranking criteria; rank reversal
ANP (Chap. 4)		✓	• Independence among elements is not required • Prediction is accurate because priorities are improved by feedback • Time consuming • Uncertainty is not supported • Hard to convince decision making
TOPSIS (Chap. 5)	✓		• A very simple Euclidean distance-based approach because of standardized steps regardless of the number of attributes • Does not accommodate the correlation of attributes • Difficulty in ensuring consistency of judgments
ELECTRE (Chap. 6)	✓		• Accommodates uncertainty and vagueness present in the judgements • Difficult to interpret the results as the ranking is based on the strength of the relationships and it is difficult interpret the complex relationships present among the criteria
PROMETHEE (Chap. 7)	✓		• It is an easy to use technique with standard steps • It does not demand an assumption to be satisfied that criteria are proportionate • Does not provide a clear method by which to assign weights

(continued)

Table 1.5 (continued)

MCDM Technique/Chapter	Nature of relationship		Key features
	Linear	Cause and Effect	
VIKOR (Chap. 8)	✓		• It is based on a calculation of positive- and negative-ideal solutions and proposes a compromise solution with advantage rate
DEMATEL (Chap. 9)		✓	• It helps decision maker to visualize complex causal relationships between criteria employing matrices and digraphs • It converts cause-effect relationship into a structural model for developing strategy maps
Fuzzy Integral and gray relation (Chap. 10)	✓		• Fuzzy integral has the ability to demonstrate the interaction between criteria, ranging from substitutive to multiplicative effects • Gray system approach allows for systematic analysis of relationships and model construction of a system in the condition of uncertainty and incomplete data • Grey relation model is an evaluation method for determining the relationships between individual attributes of the target system and screening out the attributes which has significant impact on the operational objectives of a system
MOORA (Chap. 11)	✓		• The strength of this method lies in the use of separate mathematical models to benefit the non-benefit and use of graphical qualitative criteria of the decision matrix. This enhances the discriminatory power of decision making as the chance of losing information is very small
LINMAP (Chap. 12)	✓		• The LINMAP method employs decision maker's pairwise comparisons of alternatives to compute best compromise alternative having shortest distance to the positive-ideal solution (PIS) in terms of a weighted Euclidean distance measure
COPRAS (Chap. 13)	✓		• The aggregation procedure used for ranking the criteria is comparatively more complex • It has an advantage that there is no need to transform cost to benefit type criteria

(continued)

Table 1.5 (continued)

MCDM Technique/Chapter	Nature of relationship		Key features
	Linear	Cause and Effect	
ARAS/ARM (Chap. 14)	✓		• It uses a utility function value to determine the complex relative efficiency of a feasible alternative which is directly proportional to the relative effect of values and weights of the main criteria considered in a problem
WASPAS (Chap. 15)	✓		• It is a unique combination of weighted sum model (WSM) and weighted product model (WPM)
SWARA (Chap. 16)	✓		• It estimates experts' opinion about importance ratio of the criteria in the process of their weights determination • This method has a significant utility for select real-life problems where priorities are defined based on policies of companies or countries and there aren't any needs for evaluation to rank criteria
GTA (Chap. 17)		✓	• It is originated from the fields of 45Tmathematics and sociology • It has been applied in the various fields such as engineering, management, law, psychology and others to examine a predefined model • The model is represented with regions of interest (called "nodes" or "vertices") and their connections (called "edges")
ISM (Chap. 18)		✓	• It identifies the complex relationships among the variables or specific issues under investigation • It enables the decision maker to develop a holistic and interpretive view about his understanding and belief and also to enhance his knowledge through interrelationships discovered by the method

in the system, expedite approval process, etc. There are various leading software companies like TCS, Infosys, JDA, etc. which provides such kind of customized solutions. The real issues in the purpose of standard or customized software from a leading IT company are to see the trade-off between cost of purchase, cost of annual maintenance contract, user-friendliness, ease in customizing modules, process reengineering efforts and in-house technical expertise needed. It is evident from the literature that ERP implementation cost is always much higher than the purchase cost. Also it has been noticed that more than 50% ERP implementation projects have experienced sever failure in improving the overall efficiency of the organization. This is primarily because of a wrong selection of the ERP package. Under this situation, selection of ERP package in the light of various conflicting criteria becomes a perfect candidate for MCDM analysis.

Ranking environmental risks caused by petrochemical industries

A petrochemical industry has always experienced a very high frequency of environmental hazards and accidents. This has motivated the experts in this industry to prioritize risk in the light of various hazard factors for developing an appropriate risk mitigation plan. MCDM analysis helps to identify the priorities of the critical risks by conducting multiple pairwise comparisons. This relative assessment of the risks based on the subjective opinion of the experts provides greater insights into the severity of risk and hence the development of suitable risk mitigation plans. The risks in petrochemical industry may be identified using experts' opinions based on three criteria: "impact intensity", "occurrence probability", and the "extent of contamination dispersion in the environment". The shortlisted environmental risks can be prioritized using a suitable MCDM technique.

Construction Industry

The construction portfolio is highly vertical and horizontal in nature which involves hundreds of the work packages and each work package contains number of activities and sub-activities. This kind of complex and uncertain environment demands prioritization of various issues such as investment decisions, supplier selection, contract rating, risks, etc. and hence can be considered as an appropriate context for MCDM analysis.

Application in manufacturing industries

A manufacturing industry is characterized by high market competition, work pressures and a competitive environment. It is really challenging for the managers to meet market expectations about price, quality and availability of the products. A highly competitive and dynamic structure of the market does not provide them a much of the space in decision making. A trial and error approach may prove to be fatal and it is extremely important to prioritize supply chain and resource utilization decisions to improve market competitiveness. The areas such as facility location, supplier selection, inventory management, risk management, outsourcing, knowledge management are extensively modelled with MCDM techniques for many real-life industries.

Evaluating vehicle telematics system

This is a new and upcoming area which employs the advanced integration of communications, information and vehicle technologies, Vehicle Telematics Systems (VTS), for satisfying consumers' needs with respect to automobile movement. This technology makes the vehicle a multifunction mobile-services platform. For example, cars are not only designed for transportation, but also extend value-added services such as navigation, safety, security, information, communications and entertainment. This helps drivers or passengers to contact a call centre via VTS to access aspired/desired services and information online. Therefore, VTS increases both the utility/functions and the safety of driving. It is really a great challenge to develop an optimal VTS system which closely meets the customer requirements and hence this can be structured as MCDM problem. This would identify and prioritize set of customers' requirements for the next e-era generation VTS.

1.8 Why MCDM Techniques Are More Popular in Industry Practitioners?

MCDM has gained significant attention of practising managers for solving various real-life problems due to its potential to evaluate various conflicting alternatives (may be called choice, strategy, policy, scenario, etc.) on set of criteria. The process helps the managers to select the best/suitable alternative without compromising with their subjective and intuitive insights. The alternatives evaluated with such process may further be evaluated for greater insights through "what-if" analysis. Typically, MCDM analysis exhibits select unique characteristics such as: (i) accommodates multiple non-commensurable and conflicting criteria; (ii) can handle different units of measurement among the criteria; and (iii) can include the evaluation of different alternatives. It has a great advantage in evaluating the intangible factors like brand, image, risk, difficulty level, etc. which may have a great influence on final decision. This can be seen as the most significant reason for the popularity of MCDM approaches in industry practitioners.

In last two decades, there is a phenomenal growth in the development of various MCDM techniques. This book reports the details of total 17 MCDM techniques with illustrative applications. The availability of various MCDM techniques and their ability to accommodate complexity of the various industry problems as well as their potential to present the results in the format as desired by the industry practitioners has given a large-scale popularity to these techniques. Industry managers have the options to choose an appropriate MCDM method which suits to their problem environment by evaluating their relative advantages and disadvantages.

In the recent time, numerous applications for real-life problems are reported where two or more MCDM techniques are successfully integrated to capture and analyse the problem complexity in greater detail. It has also been observed that successful integration of MCDM techniques with OR tools (e.g., AHP and Goal Programming,

TOPSIS and Integer Programming) have significantly improved the discriminatory power of traditional OR-based decision making. Typically, such integrations are widely seen for the decision making in the areas such as: evaluation of project investment decisions, advanced manufacturing systems, layout design, and also in other engineering, social science and management issues.

In real-life situation, practising managers and policy makers are many a times more interested in understanding the relative importance of various potential alternatives and not only obtaining the one best alternative. For example, a research organization like DST (Department of Science & Technology) has received multiple research proposals for funding. It is necessary for a funding organization to evaluate the relative importance of such project proposals in light of the key objectives/ problems declared by the Nation. MCDM has a great role to play in helping the experts for funding such projects which has a large-scale implication for the development and growth of the Nation. A real-life system such as any local or federal government, industry, or business activity involves, in one way or the other, the evaluation of a set of alternatives in terms of a set of decision criteria and hence MCDM techniques have gained the special attention of researchers and decision makers.

Irrespective of the types and nature of problems, MCDM has an ultimate goal: "to help a Decision-Maker (DM) to find the 'most preferred' solution for his/her decision problem". Decision maker tries to obtain the most preferred solution called as non-dominated solution when the criteria and the decision alternatives are assumed to be known. It is not possible for a decision maker to compare all possible solutions to the final solution at the moment of making final choice. Therefore, MCDM methods are always based on more or less restrictive implicit or explicit assumptions concerning the DM's preference structure and behaviour. Any non-dominated solution can be regarded as a rational choice, if there is no information about the DM's preferences.

MCDM process is executed with an objective to enable decision making in the presence of multiple objectives. A decision maker is expected to accommodate both tangible and intangible criteria. The objectives are usually conflicting in nature, and therefore, the solution is primarily driven by the preferences of the decision maker. Typically, MCDM methodology involves decision makers from different functions in order to minimize the bias in obtaining the final priorities of the alternatives. This way, MCDM approach accommodates conflicting criteria and points of view, within a well-structured framework. The recent trends demonstrate an extensive use of MCDM techniques in the variety of areas such as integrated manufacturing systems, evaluations of technology investment, water and agriculture management, health care, IT Services, energy planning.

1.9 Historical Milestones in MCDM

The practice of solving multi-criteria decision making is as old as the history of mankind. A human being had always intuitively evaluated the trade-offs among conflicting criteria to arrive at the best alternative in personal or professional life.

There is a well-known story of King Solomon (during 1011–931 BC). He was the first who demonstrated the use and importance of MCDM in solving select critical problems. His wisdom-based intuitive and iterative approach has extensively helped in making many importance decisions for the development of Israel. He applied the process of MCDM successfully in many development works like infrastructure development, solving military, economic and social problems, etc.

Moral Algebra

The approach adopted by Benjamin Franklin in solving multi-criteria decision-making problem is highly noted as "Moral Algebra". He has worked with a simple paper system to decide how to tackle important issues. He explained his procedure to his friend Joseph Priestly as he would summarize the various arguments which are favouring the problem under consideration on the one side of the paper and the arguments opposing on the other side. The arguments on both the sides of the paper having relatively equal importance would be crossed out. At the end of this process, he would examine the argument which is left out on either side of the paper and would consider as a strong candidate to support the decision. His approach was clearly an indication of MCDM problem-solving process.

Voting System

The Marquis de Condorcet (whose name was Marie Jean Antoine Nicolas Caritat 1743–1794) has developed Condorcet's paradox for addressing voting problem. This approach starts with an understanding that majority preferences may be intransitive, even though individual preferences are perfectly transitive. The another notable contribution developed in 1785 was the Condorcet voting method. The method was successfully used for holding fair elections with more than two candidates. According to Condorcet, if one person is to be elected from n candidates, then the person elected would have to win in a head-to-head contest each of the other n-1 candidates. Subsequently, Jean-Charles de Borda (1733–1799), a French mathematician and political scientist, developed a system of ranking candidates by allocating them points on the basis of their rank. Both the researchers—Condorcet and de Borda—have significantly contributed in the area of "outranking methods" in the era of 1960s in France.

Pareto-Optimality

The economist Vilfredo Pareto (1848–1923) is regarded as the first researcher who has formally contributed in the area of MCDM. In his 1906 publication, he reported a key observation that 20% of the population of Italy owned 80% of the property. This was later generalized by Joseph M. Juran as the Pareto principle (also termed the 80–20 rule). Pareto was the first (or at least one of the first) to mathematically study the aggregation of conflicting criteria into a single composite index. He is also considered as the pioneer who introduced the concept of efficiency which became known as Pareto-optimality. The concept has gained significant attention in solving some of the problems in economics, negotiation science and modern MCDM theory.

A Pareto-optimal allocation of resources is achieved when it is not possible to make anyone better off without making at least one other person worse off.

Indifference Curves and Edgeworth Box

The foundation of utility theory was established by Francis Edgeworth (1845–1926) and introduced the concept of indifference curve. He has extensively contributed in the field of distribution of resources and his contribution became famous as "Edgeworth box". The original idea of Edgeworth was reworked by Pareto in 1906. Both the researchers have created profound effect in the areas of economics, negotiation science and modern MCDM.

A summary of key milestones in the history of MCDM is summarized in Table 1.6.

1.10 Organization of This Book

This book is structured to help the readers to appreciate the use of various multi-criteria decision-making techniques for real-life applications. There are various ways to see the advancements in the domain of MCDM techniques. There are techniques like ISM, DEMATEL, and GTA which accommodates the interrelationships among the criteria while majority of the other techniques like AHP, LINMAP, VIKOR, etc. primarily focuses on linear relationships. The techniques like ELECTRE, PROMETHEE and TOPSIS are based on outranking of criteria and may be considered as the techniques falling under one group. SAW is the most elementary technique and its use is adopted in other advanced MCDM techniques. Last two decades have contributed significantly in the development of various MCDM techniques such as VIKOR, DEMATEL, LINMAP, SWARA, WASPAS and this can be considered as recent advances in the domain of MCDM techniques. There are some techniques which shares the common features in terms of analysing the criteria and alternatives. For example, TOPSIS and VIKOR both the techniques use the concept of positive- and negative-ideal solutions in ranking the alternatives. Similarly, GTA and ISM can be seen as the techniques which predominantly help the decision makers to visualize the solution in a graphical format. The techniques such as PROMETHEE and ELECTRE share the similarities in prioritizing the alternatives through outranking process. The technique like LINMAP integrates the properties of select techniques like AHP and TOPSIS/VIKOR to combine the use of pairwise comparison with Euclidean distance measure (positive and negative ideal). The technique like WASPAS is a unique combination of weighted sum model (WSM) and weighted product model (WPM) and can be seen as using the properties of SAW method.

Table 1.6 Historical timeline of MCDM

Year	Author	MCDM Technique Developed	Key feature of the technique
1706–1790	Benjamin Franklin	–	Simple paper system for deciding important issues
1951	When Kuhn and Tucker	Nonlinear programming	• For finding out the optimal solution • Used to solve the multi-objective problems
1955	Charnes, Cooper, and Ferguson	Goal programming	• Extension of linear **programming** which evaluates multiple, normally conflicting objective measures • Provides the best satisfying solution to determine the required resources to achieve a desired set of objectives
mid-1960s	Bernard Roy	ELECTRE	Methods construct a set of outranking decisions, or decisions that should be considered as best
1973	Zionts Jyrki Wallenius	Zionts-Wallenius interactive method	Solving the linear programming problem with multi-objective problems
Late 1970s	Zionts Jyrki Wallenius Pekka Korhonen	Decision support system	Interactive multiple objective mathematical programming
1970s	Thomas Saaty	AHP and ANP	AHP and the ANP are used to measure intangibles using human judgment. AHP/ANP is the most powerful synthesis methodologies for combining judgment and data to effectively rank options and predict outcomes
1976	Ralph Keeney and Howard Raiffa	Multi-attribute value theory (including utility theory)	It considers the concept of utility and value and gained significant attention in MCDM problems
1977	Daniel Kahneman and late Amos Tversky	Behavioural decision theory	It accommodates courses of action which closely reflects the decision maker's beliefs and values and how a decision maker incorporates this into their decisions

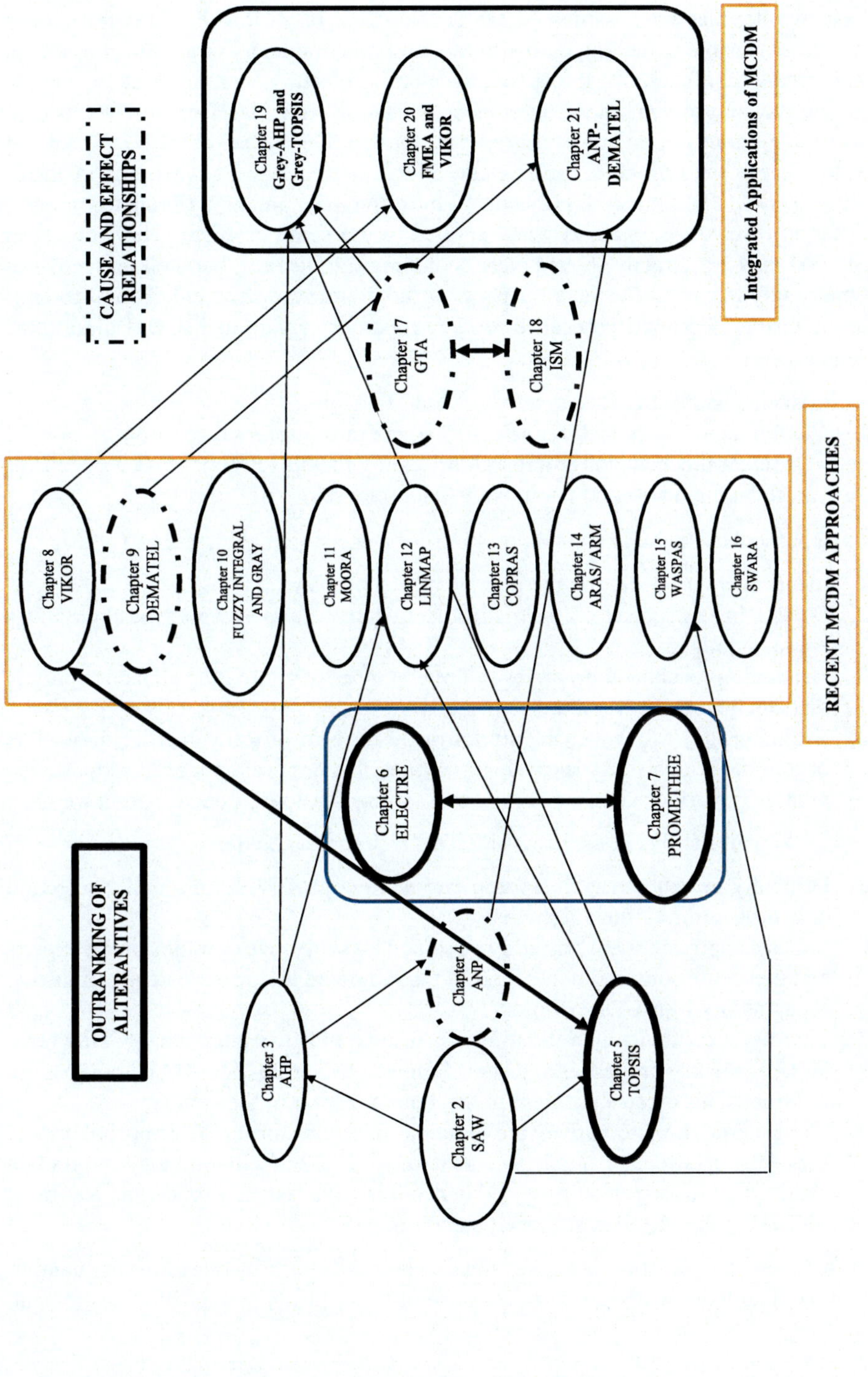
OUTRANKING OF ALTERANTIVES
CAUSE AND EFFECT RELATIONSHIPS
RECENT MCDM APPROACHES
Integrated Applications of MCDM
Chapter 2 SAW
Chapter 3 AHP
Chapter 4 ANP
Chapter 5 TOPSIS
Chapter 6 ELECTRE
Chapter 7 PROMETHEE
Chapter 8 VIKOR
Chapter 9 DEMATEL
Chapter 10 FUZZY INTEGRAL AND GRAY
Chapter 11 MOORA
Chapter 12 LINMAP
Chapter 13 COPRAS
Chapter 14 ARAS/ ARM
Chapter 15 WASPAS
Chapter 16 SWARA
Chapter 17 GTA
Chapter 18 ISM
Chapter 19 Grey-AHP and Grey-TOPSIS
Chapter 20 FMEA and VIKOR
Chapter 21 ANP-DEMATEL

1.11 Summary of Key Points on MCDM

Last two decades have witnessed the development of various MCDM techniques which are capable of dealing with different level of complexity of the select problem. This creates confusion for the decision maker in selecting an appropriate technique for the typical problem characteristics he is dealing with. Each method has its own strength, weaknesses and possibilities to be applied. This causes phenomenon known as the inconsistent problem ranking caused by different MCDM methods. A major criticism of MCDM methods is that due to the differences among different techniques, different results are obtained when applied to the same problem. The weights in MCDM may not directly show a clear economic significance, but their use provides opportunity to model the actual aspects of the preference structure. The difference in the results is primarily because of key properties of various MCDM algorithms such as:

✓ Mechanism of MCDM in terms of use of weights;
✓ Difference in approach for selecting of the most preferred solution;
✓ Different mechanisms employed for scaling the objectives;
✓ Introducing additional parameters that affect solution.

In particular, the main steps of multi-criteria decision making are as follows:

1. Identifying and declaring the main goal of a problem;
2. Developing a structure for the main objectives or criteria by which the alternatives are to be judged;
3. Generating feasible alternatives (a finite number of alternative plans or options) that can be implemented to achieve goals;
4. Evaluating an impact of each criterion on the decision-making function or weights of criteria. A decision maker should express his/her preferences in terms of the relative importance of criteria, and one approach is to introduce criteria weights.

The key limitations in the use of MCDM technique include:

1. Difficulty in structuring an appropriate hierarchy will lead to a faulty selection or prioritization of the alternatives.
2. MCDM approaches are primarily governed by subjective opinions of the experts and their intuitions and hence bias in the judgment and inconsistency limits the acceptability of MCDM results.
3. There is a difficulty in validating the results with the alternate MCDM technique as the key properties and steps followed by various MCDM techniques are different. This creates problem in the generalization of the results.
4. The criteria and alternatives are dynamic in nature for a real-life problem and hence the priorities or ranking generated by MCDM technique may not remain valid for a longer-period time and hence it is advisable to support the results of MCDM with a detailed sensitivity analysis.

An alternative in multi-criteria evaluation is usually described in terms of quantitative and qualitative criteria. The criteria usually have different units of measurement.

It is necessary to normalize the decision matrix comprising criteria with different units of measurement to obtain a comparable scale of the criteria values. There are numerous procedures available for normalization.

Selected Bibliography and Further Reading

1. Triantaphyllou, E. (2000). Multi-criteria decision making methods. In *Multi-criteria decision making methods: A comparative study* (pp. 5–21). Boston, MA: Springer.
2. Ho, W., Xu, X., & Dey, P. K. (2010). Multi-criteria decision making approaches for supplier evaluation and selection: A literature review. *European Journal of Operational Research, 202*(1), 16–24.
3. Wang, J. J., Jing, Y. Y., Zhang, C. F., & Zhao, J. H. (2009). Review on multi-criteria decision analysis aid in sustainable energy decision-making. *Renewable and Sustainable Energy Reviews, 13*(9), 2263–2278.
4. Kahraman, C. (Ed.) (2008). *Fuzzy multi-criteria decision making: theory and applications with recent developments* (Vol. 16). Berlin: Springer Science & Business Media.
5. Wang, G., Huang, S. H., & Dismukes, J. P. (2004). Product-driven supply chain selection using integrated multi-criteria decision-making methodology. *International Journal of Production Economics, 91*(1), 1–15.
6. Sanguanduan, N., & Nititvattananon, V. (2011). Strategic decision making for urban water reuse application: A case from Thailand. *Desalination, 268,* 141–149.
7. Supriyasilp, T., Pongput, K., & Boonyasirikul, T. (2009). Hydropower development priority using MCDM method. *Energy Policy, 37,* 1866–1875.

Chapter 2
Simple Additive Weightage (SAW)

The first application of SAW was reported for a portfolio selection problem. It is also known as a weighted linear combination or scoring method. It is a very simple and highly preferred method used for a less complex problem environment. It calculates the priorities for each alternative by multiplying the scaled value given to the alternative with the weights of relative importance assigned by the decision makers to the attribute and finally summing up all the products for each criterion. The method is based on the weighted average.

2.1 Step-by-Step Procedure of SAW

The procedural steps involved in SAW are simple and easy to follow. The summary of steps is given below.

Step 1	Construct a pairwise comparison matrix for criteria, and according to the importance of criteria assign a score to that criteria
Step 2	Finding the weighting sum matrix by multiplying pairwise comparison matrix and priority vector, comparison matrix is its column total and priority vector is its row averages
Step 3	Find consistency index by using the formulae $CI = (k - n)/(n - 1)$ K = average value of all the elements after they are divided by respective priority vector element n = no. of elements
Step 4	Calculate consistency ratio CR as follows CR = CI/RI The value of RI is to be obtained from the standard table proposed by T.L. Saaty. This depends upon number of criteria
Step 5	Calculate a decision matrix ($m * n$) that includes m personnel and n criteria and calculate the normalized decision matrix for positive criteria

(continued)

J. J. Thakkar, *Multi-Criteria Decision Making*, Studies in Systems, Decision and Control 336, https://doi.org/10.1007/978-981-33-4745-8_2

(continued)

Step 6	Multiply the normalized weight of each criterion with respective criteria weightage
Step 7	Obtain the sum total of each raw representing various alternatives and decide the priority of alternatives

2.2 Illustrative Applications of SAW

Application 1: Prioritization of hotels

The customer satisfaction is derived from the perceived fairness of exchange situations. Customers perceive higher equity when they receive distinct advantages from the service provider, and good conflict handling policy, characterized by distributive, procedural justice and interactional justice (fairness). The problem considered here attempts to prioritize various hotels for set of key criteria

Step 1: Obtaining the decision matrix for set of criteria and types of hotels

C1: Location of the hotel in Hyderabad.
C2: Price in Rs.
C3: Rating of hotel.
C4: Type of hotel.
C5: Amenities.
Types of hotels.
P1: Ellaa hotel in Gachibowli.
P2: The Manohar hotel in Begumpet.
P3: Eaglewood hotel in Gachibowli.
P4: Radisson Blu Plaza hotel in Banjara hills.
P5: Dream Valley Resorts in Shamshabad.
P6: Aalankrita Resorts and spa in Ameerpet.
See Table 2.1.
Weightages: C1 = 0.1, C2 = 0.3, C3 = 0.1, C4 = 0.2, C5 = 0.3
(C2 is cost criteria and remaining are benefit criteria).

Table 2.1 Decision matrix for service recovery attributes in hotel industry

	C1	C2	C3	C4	C5
P1	3.2	4700	4	5	5
P2	4.2	6000	3.4	5	5
P3	3.2	1375	4	3	3
P4	4.5	5680	4.3	5	5
P5	4.1	2600	4	3	3
P6	3.5	4100	4.4	4	5

Table 2.2 Normalized decision matrix

	C1	C2	C3	C4	C5
P1		3.418182	0.909091	1	1
P2	0.93333	4.363636	0.772727	1	1
P3	0.71111	1	0.909091	0.6	0.6
P4	1	4.130909	0.977273	1	1
P5	0.911111	1.890909	0.909091	0.6	0.6
P6	0.77778	2.981818	1	0.8	1

Step 2: Obtain the normalized decision matrix

$r_{ij} = x_{ij}/x_j*$, if the *j*th criterion is a benefit criterion.

$r_{ij} = x_j - /x_{ij}$, if the *j*th criterion is a cost criterion here only C2 has cost criteria. See Table 2.2.

Step 3: Using the weights for the different criteria obtain the weighted score for each alternative using the normalized decision matrix

The weighted scores.

P1 = (0.711111 * 0.1) + (3.418182 * 0.3) + (0.909091 * 0.1) + (1 * 0.2) + (1 * 0.3) = 1.687475.

P2 = (0.933333 * 0.1) + (4.363636 * 0.3) + (0.772727 * 0.1) + (1 * 0.2) + (1 * 0.3) = 1.979697.

P3 = (0.711111 * 0.1) + (1 * 0.3) + (0.909091 * 0.1) + (0.6 * 0.2) + (0.6 * 0.3) = 0.76202.

P4 = (1 * 0.1) + (4.130909 * 0.3) + (0.977273 * 0.1) + (1 * 0.2) + (1 * 0.3) = 1.937.

P5 = (0.911111 * 0.1) + (1.890909 * 0.3) + (0.909091 * 0.1) + (0.6 * 0.2) + (0.6 * 0.3) = 1.04929.

P6 = (0.777778 * 0.1) + (2.981818 * 0.3) + (1 * 0.1) + (0.8 * 0.2) + (1 * 0.3) = 1.232323.

Step 4: Based on the final scores, rank the alternatives

The rank is: **P2, P4, P1, P6, P5 and P3.**

Application 2: Prioritization of manufacturing systems

Some of the most important types of manufacturing systems are:

(i) Job production
(ii) Batch production
(iii) Mass or flow production
(iv) Projects

A production manager will have to choose most appropriate method for his enterprise. The final decision regarding any particular method of production is very much affected by the nature of the products and the quantity to be produced. In the problem we have taken these manufacturing systems taking several attributes that contribute a major role in choosing most appropriate method into considerations and have given rank to these manufacturing systems.

Step 1: Obtaining the decision matrix

Different types of manufacturing systems.

P1: Mass production.

P2: Batch production.

P3: Job shop.

P4: Project.

The factors in manufacturing systems.

X1: Capital investment.

X2: No of products to be produced.

X3: Flexibility.

X4: Variable cost.

X5: Labour content.

X6: Volume of units produced in year.

See Tables 2.3 and 2.4.

step 1: The decision matrix after converting intangibles to numbers

See Table 2.5.

Weightages: X1 = 0.3, X2 = 0.2, X3 = 0.1, X4 = 0.2, X5 = 0.3

Here X4 has cost criteria and remaining are benefit criteria.

Step 2: Obtain the normalized decision matrix

$r_{ij} = x_{ij}/x_j{}^*$, if the jth criterion is a benefit criterion.

Table 2.3 Decision matrix for manufacturing system selection problem

	X1	X2	X3	X4	X5
P1	Very high	Few (High)	Very low	Very low	Very low
P2	Medium	Several (Medium)	Medium	Medium	Medium
P3	Low	Many(Low)	High	High	High
P4	Very low	One (Low)	Very high	Very high	Very high

Table 2.4 Rating scale

Benefit attributes		Cost attributes	
Very low	1	Very low	9
Low	3	Low	7
Medium	5	Medium	5
High	7	High	3
Very High	9	Very high	1

Table 2.5 Decision matrix with numerical values

	X1	X2	X3	X4	X5
P1	9	5	3	7	3
P2	5	7	5	5	5
P3	3	9	7	3	7
P4	1	1	9	1	9

Table 2.6 Normalized decision matrix

	X1	X2	X3	X4	X5
P1	1	0.555556	0.333333	7	0.333333
P2	0.555556	0.777778	0.555556	5	0.555556
P3	0.333333	1		3	0.777778
P4	0.111111	0.111111	1	1	1

$r_{ij} = x_j - /x_{ij}$, if the *j*th criterion is a cost criterion here only X4 has cost criteria. See Table 2.6.

Step 3: Using the weights for the different criteria obtain the weighted score for each alternative using the normalized decision matrix

P1 = (1 * 0.3) + (0.555556 * 0.2) + (0.33333 * 0.1) + (7 * 0.2) + (0.3333 * 0.3) = 1.944444.

Likewise P2: 1.544444.

P3: 1.211111

P4: 0.655556

Step 4: Based on the final scores, rank the alternatives

Final Rank is: **P1, P2, P3 and P4.**

2.3 Advantages of SAW

- One of the advantages of this method is that it is a proportional linear transformation of the raw data which means that the relative order of magnitude of the standardized scores remains equal.
- For the solution of various problems, SAW is used as an important tool in integrated approaches along with sensitivity analysis, TOPSIS, statistical clustering.

2.4 Limitations of SAW

- It does not account for fuzziness in experts' judgements in the decision-making process.
- Its internal validity is affected by self-assessment bias.
- It demands all the criteria to be of maximizing nature. This is a drawback of this method, though minimizing criteria can be easily converted to the maximizing ones.
- It demands all criteria values to be positive.

SAW is the most simple and easy to use MCDM technique. The use of SAW is employed as an integral part of many MCDM techniques. SAW because of its simplicity is used for a very preliminary multi-criteria decision-making problems. A typical real-life problem is usually structured in hierarchy with multiple layers of criteria and sub-criteria. This demands a more sophisticated version of MCDM technique. The next chapter will discuss the key features and application of MCDM technique—Analytic Hierarchy Process (AHP). This technique accommodates multiple layers of linear hierarchy and evaluates the alternatives through multiple pairwise comparisons of criteria and alternatives.

Selected Bibliography and Further Reading

1. Chou, S. Y., Chang, Y. H., & Shen, C. Y. (2008). A fuzzy simple additive weighting system under group decision-making for facility location selection with objective/subjective attributes. *European Journal of Operational Research, 189*(1), 132–145.
2. Afshari, A., Mojahed, M., & Yusuff, R. M. (2010). Simple additive weighting approach to personnel selection problem. *International Journal of Innovation, Management and Technology, 1*(5), 511.
3. Abdullah, L., & Adawiyah, C. R. (2014). Simple additive weighting methods of multi criteria decision making and applications: A decade review. *International Journal of Information Processing and Management, 5*(1), 39.
4. Modarres, M., & Sadi-Nezhad, S. (2005). Fuzzy simple additive weighting method by preference ratio. *Intelligent Automation & Soft Computing, 11*(4), 235–244.
5. Kaliszewski, I., & Podkopaev, D. (2016). Simple additive weighting—A metamodel for multiple criteria decision analysis methods. *Expert Systems with Applications, 54,* 155–161.
6. Liu, S. M., Pan, S., Mi, Z. K., Meng, Q. M., & Xu, M. H. (2010, May). A simple additive weighting vertical handoff algorithm based on SINR and AHP for heterogeneous wireless networks. In *2010 International Conference on Intelligent Computation Technology and Automation* (pp. 347–350). New York: IEEE.

Chapter 3
Analytic Hierarchy Process (AHP)

3.1 Background

Both academia and industry have witnessed a huge application of Analytic Hierarchy Process (AHP). The technique has gained significant attention in the various domains such as machine selection, supplier selection, ambulance allocation, nurse and other resource prioritization in healthcare, prioritization of academic policies. The technique was developed in 1970 by Thomas L. Saaty is a well-known and widely popular MCDM technique. The technique accommodates linear hierarchical structure of the problems including criteria, sub-criteria and alternatives. It can include the inputs in both the forms like actual measurement of criteria such as cost, profit and also subjective assessment by experts that captures their satisfaction, opinions, beliefs and priorities related to any decision. AHP tries to identify the relative significance of each element of a decision by comparing each pair of alternatives at every hierarchy level. The technique thus examines multiple alternatives and provides the decision maker with relative priorities of the various considered alternatives. A decision maker has an ease in converting its subjective opinion into objective one while implementing AHP for a real-life problem. This gives a confidence to decision maker that his intuitions and experience are not overlooked in arriving to the final ranking of the alternatives. The simple structure and low mathematical complexity of the technique make it the most preferred one in various sectors, including engineering, health care, finance, public policies and business environments. In last two decades, researchers and practitioners have successfully implemented this technique for variety of decision making which involves "selection", "evaluation", "benefit–cost analysis", "allocations", "planning and development", "priority and ranking", "decision making", "forecasting", and "health and related fields". Because of its simplicity and flexibility in accommodating various conflicting criteria, this technique is successfully integrated with other approaches such as "integer programming", "quality function deployment (QFD)", "meta-heuristics", "SWOT analysis", and "data envelopment analysis (DEA)" for enhancing the power of decision making.

J. J. Thakkar, *Multi-Criteria Decision Making*, Studies in Systems, Decision and Control 336, https://doi.org/10.1007/978-981-33-4745-8_3

3.2 AHP: Basic Principles

AHP begins with the construction of a decision matrix representing relative significance of various "attributes" relative to each other. For example, from a customer perspective, what is the significance of cost of a laptop relative to its computing power (in terms of RAM and processor configurations). The decision makers are asked to judge the importance of cost in terms of "intensity of importance" like "very much more important", "more important", etc. as indicated in Table 3.1.

A very simple understating is used in rating the attributes. For example, if attribute A is "very much important" than attribute B then as per Table 3.1, it is rated at 7, and the importance of B to A is rated as 1/7. It is mandatory to carry out these pairwise comparisons with the help of experts' opinion for all the criteria. Subsequently, the relative weights of the factors are calculated for the problem under consideration. This helps a decision maker to understand the relative priorities of the alternatives and select the one or set of alternatives which ranks high in the list.

Finally, it is very much important to calculate a consistency ratio (CR) in AHP analysis. This ratio provides a measure of the consistency of the judgement considered through AHP compared to that of completely random samples of judgement. If the value of CR significantly exceeds 0.1, the judgement is deemed to be very close to being purely random which may require the entire process to be repeated or abandoned as futile. It is actually easier to achieve perfect consistency through calculations of remainder of judgement when a minimum required number of judgements have been performed. A typical AHP process is illustrated in Fig. 3.1.

Table 3.1 The Saaty rating scale

Intensity of importance	Definition	Explanation
1	Equal importance	Two factors contribute equally to the objective
3	Somewhat more important	Experience and judgement slightly favour one over the other
5	Much more important	Experience and judgement strongly favour one over the other
7	Very much more important	Experience and judgement very strongly favour one over the other. Its importance is demonstrated in practice
9	Absolutely more important	The evidence favouring one over the other is of the highest possible validity
2, 4, 6, 8	Intermediate values	When compromise is needed

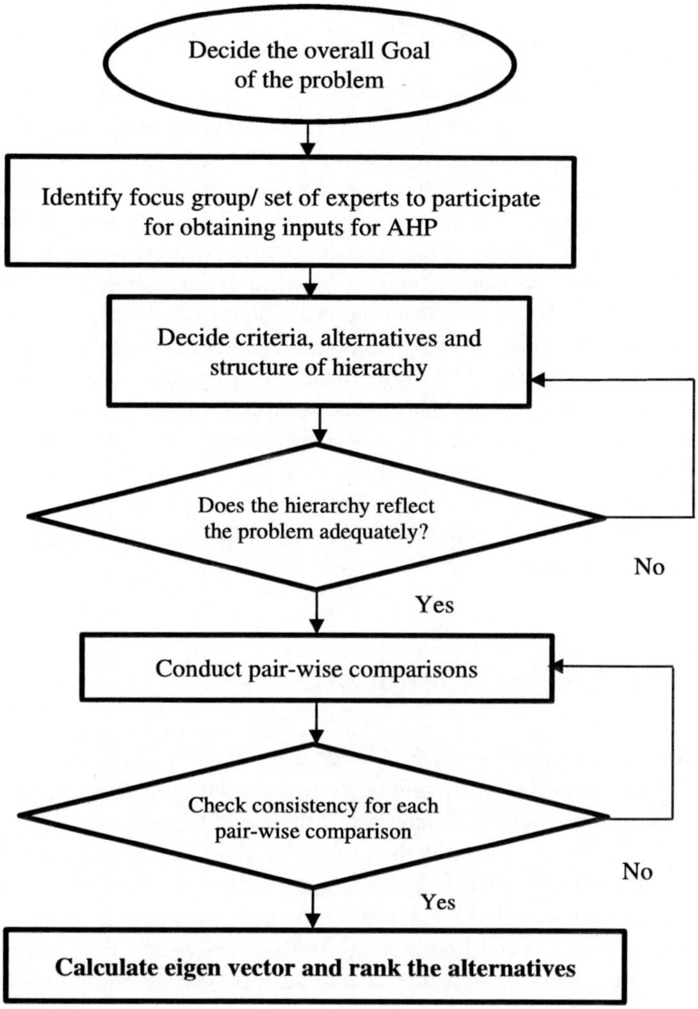

Fig. 3.1 AHP process

3.3 The AHP Theory

Consider n elements to be compared, $C_1, \ldots, C_n$ and denote the relative "weight" (or priority or significance) of C_i with respect to C_j by a_{ij} and form a square matrix $A = (a_{ij})$ of order n with the following conditions: $a_{ij} = 1/a_{ji}$, for $i \neq j$, and $a_{ij} = 1$, all $i = j$. The matrix formed is called the reciprocal matrix.

The weights are considered to be consistent if they exhibit transitivity, that is $a_{ik} = a_{ij}a_{jk}$ for all i, j, and k. Such a matrix might exist if the a_{ij} are calculated from exactly measured data. Then find a vector ω of order n such that $A\omega = \lambda\omega$. For such a matrix, ω is said to be an eigenvector (of order n) and λ is an eigenvalue. For a consistent matrix, $\lambda_{max} = n$.

The condition for consistent weights, $a_{ik} = a_{ij}a_{jk,}$ does not hold for matrices that are based on expert opinions or judgements as human judgements are more likely to be inconsistent. In such a case the ω vector satisfies the equation $A\omega = \lambda_{max}\omega$ and $\lambda_{max} \geq n$. The difference, if any, between λ_{max} and n indicates the degree of inconsistency involved in the judgement. If $\lambda_{max} = n$ indicates, that the judgements are consistent. A consistency index for judgements could be easily calculated using the formula: $(\lambda_{max} - n)/(n\text{-}1)$. This calculated index is required to be evaluated in comparison with purely random judgements. Saaty has calculated Consistency Indices for large samples of random matrices of increasing order against which the calculated judgement values can be compared. The comparison is done by calculating the consistency ratio which is a ratio of the consistency index based on the considered judgements and Index for the random judgement matrix. An acceptance value of 0.1 was suggested by Saaty, indicating that a consistency ratio value exceeding 0.1 may suggest highly unreliable judgement. Although, in actual practice, even CR value greater than 0.1 in certain circumstances are considered acceptable. A CR value of 0 indicates perfect consistency of judgement.

3.4 Step-by-Step Procedure of AHP

Calculation of eigenvector values can be done in using various methods. A very close approximation to the exact value can be obtained by calculating the product of elements in each row and calculating the nth root of these values. Normalization of the nth root values calculated is then normalized by dividing these values with the sum of all the nth root. Normalization ensures the calculated sum up to 1.0. In the matrix below (Table 3.2), the 4th root for the first row is 0.293, which is divided by 5.024 to give 0.058 as the first element in the eigenvector. For example, Table 3.2 gives a worked example in terms of four attributes to be compared which, for simplicity, we refer to as A, B, C and D.

The eigenvector of the relative importance or value of A, B, C and D is (0.058, 0.262, 0.454, 0.226). From the eigenvector values, we can infer that C is the element with the highest priority followed by elements, B and D, which have roughly similar

Table 3.2 Calculating eigenvector

	A	*B*	*C*	*D*	*n*th root of Product of values	Eigenvector
A	1	1/3	1/9	1/5	0.293	0.058
B	3	1	1	1	1.316	0.262
C	9	1	1	3	2.279	0.454
D	5	1	1/3	1	1.136	0.226
Totals					5.024	1.000

values and attribute A has the least priority. Once, the eigenvector values are calculated, we then calculate λ_{max} which is required in determining the consistency index and the consistency ratio. The values of judgement matrix are multiplied with the eigenvector values calculated, which results in a new vector. For example, the calculation for the first row will be as follows:

$$1 * 0.058 + 1/3 * 0.262 + 1/9 * 0.454 + 1/5 * 0.226 = 0.240$$

and similarly, for the other three rows we get 1.116, 1.916 and 0.928. The vector thus calculated, consisting of four elements, i.e. (0.240, 1.116, 1.916, 0.928) is, the product $A\omega$. From the theory of AHP we have that $A\omega = \lambda_{max}\omega$ and therefore we can obtain four estimation values of λ_{max}. These estimate values are obtained by simply dividing each element of (0.240, 1.116, 1.916, 0.928) with their respective eigenvector values. For example, for the first component we get 0.240/0.058 = 4.137 followed by 4.259, 4.22 and 4.11. We estimate λ_{max} by calculating the mean of these values which works out to be 4.18. A useful and simple check for calculation errors is provided by the fact that all these calculated λ_{max} values must be greater than n (i.e. 4 in the example case).

The consistency index for any matrix can be calculated using the formula $(\lambda_{max} - n)/(n - 1)$ and, since $n = 4$ for the above example matrix, the CI value turns out to be 0.060. Finally, based on the consistency index we calculate the consistency ratio by comparison with the CI values obtained for matrices based on completely random judgement values. Table 3.3, which is derived from Saaty's book on Analytic Hierarchy Process, provides the corresponding values of consistency for different order of the random matrix. The values in the below table represent average of Consistency Indices derived from a random samples of reciprocal matrices. A CR value below 10% indicates the adjustment is quite small in comparison with the actual calculated values of the eigenvector. However, a high CR value such as 85% would indicate that the pairwise judgements obtained are untrustworthy.

In the example provided above, the computed CR value is 0.0677(=0.060/0.90). Even though, Saaty suggests that a CR value greater than 0.1 indicates that the judgements are inconsistent, but consistency ratios slightly greater than 0.1 need to be accepted in certain cases. In the above example, we get a value below 0.1, so we accept the judgement as being fairly consistent. Typically, AHP involves large number of pairwise comparisons and its important to check the consistency in experts' opinion for each of the matrix of pairwise comparison. If the determined value of consistency index does not qualify the standard, then experts are advised to repeat the pairwise comparisons among the criteria or alternatives till the value of consistency index comes in the desirable range.

In summary, the consistency checking procedure for AHP involves five steps as follows:

1. Multiply each column of the pairwise comparison matrix by the corresponding weight.

Table 3.3 Saaty's table for CR calculation

Order of the random matrix	1	2	3	4	5	6	7	8	9	10	11	12	13	14	15
Random Index (RI)	0.00	0.00	0.58	0.90	1.12	1.24	1.32	1.41	1.45	1.49	1.51	1.48	1.56	1.57	1.59

$$A.C = \begin{bmatrix} a_{11} & a_{12} & \ldots & a_{1m} \\ a_{21} & a_{22} & \ldots & a_{2m} \\ \ldots & \ldots & \ldots & \ldots \\ a_{m1} & a_{m2} & & a_{mm} \end{bmatrix} \begin{bmatrix} c_1 \\ c_2 \\ \ldots \\ c_m \end{bmatrix} = \begin{bmatrix} x_1 \\ x_2 \\ \ldots \\ x_m \end{bmatrix}$$

2. Calculate the eigenvector λ by the following equation

$$\begin{bmatrix} \lambda_1 \\ \lambda_2 \\ \lambda_3 \\ \ldots \\ \ldots \\ \lambda_n \end{bmatrix} = \begin{bmatrix} x_1/c_1 \\ x_2/c_2 \\ x_3/c_3 \\ \ldots \\ \ldots \\ x_n/c_n \end{bmatrix}$$

3. Compute the average of the values, denote it by λ_{avg} which is maximum eigenvalue of the pairwise comparison matrix.

$$\lambda_{\text{avg}} = \left| \frac{\sum \lambda_i}{n} \right|$$

4. Calculate CI—the consistency index

$$\left| \frac{\lambda_{\text{avg}} - n}{n - 1} \right|$$

5. If this ratio (CI/RI) is very large (Saaty suggests >0.10), then we are not consistent enough and the best thing to do is go back and revise the comparisons.

3.5 Illustrative Examples on AHP

Application 1: AHP for prioritizing Humanitarian SC Challenges

A consultancy company, GMDN Solutions Pvt Ltd, received a consultancy contract from MHA, GOI for the development of "India Disaster Recovery Framework" (IDRF) in collaboration with United Nations Development Program (UNDP). Under this contract the firm has to provide a list of the challenges/issues prevailing in humanitarian supply chain along with their prioritization, to MHA, GOI. These data provided for IDRF will be used in National Disaster Management Plan (NDMP) also. The scope of work for GMDN includes the study of humanitarian supply

Table 3.4 HSCM challenges

1	Distribution planning
2	Information and communication system
3	Sourcing and supplier management
4	Supply chain coordination and integration
5	Transportation

Table 3.5 HSCM criteria

1	Reliability
2	Effectiveness
3	Ease to monitoring
4	Leanness
5	Controllability
6	Flexibility
7	Agility
8	Risk controllability

chain in India and finalizes at least five challenges/issues which has to be prioritized. GMDN has set up a team to dwell upon the best multiple criteria decision model. Later it has decided to use Analytical Hierarchical Process (AHP) for prioritizing issues/challenges of humanitarian supply chain. After brainstorming, the team has arrived to the following challenges/issues of HSCM. Challenges/issues in humanitarian supply chain in India are tabulated in Table 3.4.

The prioritizing of these challenges requires few criterions based on which the issues are to be compared pairwise and AHP is to be applied. The team did extensive surveys for NDMP and IDRF documents and finalized eight criteria against which the challenges/issues of HSCM have to be prioritized. The criterions are summarized in Table 3.5.

These challenges and criterions are then used in AHP technique to obtain the desired result by GMDN Solutions Pvt Ltd team.

Problem structure

The company has divided the problem into 3 levels (in Table 3.6) viz. objective, criteria, issues/challenges as shown in the table below. A group which is having knowledge in the field of humanitarian supply chain and its challenges, is formed.

The company has started the process of AHP technique and the team started to rank the criterions based on the following scale (Table 3.7).

The definition of the criterions and challenges is defined in Table 3.9, which provides details of criterions selected in terms of challenges of humanitarian supply chain. The relation of alternatives and their relationship with the criteria can be well understood. Table 3.8 called consistency ratio table derived from Saaty's book is used to calculate consistency ratio (CR).

Table 3.6 Three level problem structure

Level 1: *Objective*	Prioritizing of humanitarian supply chain issues							
Level 2: *Criteria*	Reliability (R)	Effectiveness (E)	Ease of Monitoring ((EoM))	Leanness (L)	Controllability (C)	Flexibility (F)	Agility (A)	Risk Controllability (RC)
Level 3: *issues*	Distribution planning (DP) Information and communication system (ICS) Sourcing and supplier management (SSM) Supply chain coordination and integration (SCCI) Transportation, model choice & routing (TMR)							

Table 3.7 Scale adopted for pairwise comparison of HSCM problem

Numerical value	Preference level
1	Equally preferred
2	Equally to moderately preferred
3	Moderately preferred
4	Moderately to strongly preferred
5	Strongly preferred
6	Strongly to very strongly preferred
7	Very strongly preferred
8	Very strongly to extremely preferred
9	Extremely preferred

Step-by-Step Application of the AHP for HSCM

Initially, pairwise comparisons were done to criterions selected (Table 3.9) to determine which criteria get the maximum point. Here the values of rows below the diagonal (in Table 3.10) are only filled by the team and rows above are automatically filled by inverse of corresponding entity. Table 3.11 reports the values of consistency index and consistency ratio. Table 3.12 provides the criteria weightages.

The comparisons ranked for criterions gave CR of 0.07, i.e. CR < 0.1 hence, pairwise ranking is acceptable.

Secondly, the challenges/issues are compared pairwise for each criterion separately and criteria wise weightage for each criterion is derived in Table 3.13.

The CR is less than 0.1 for all the matrices hence the pairwise comparisons is valid. The weightages of all the challenges/issues are derived in Table 3.14 using the pair-wise comparisons made in Table 3.13. The step-by-step AHP procedure explained in Sect. 3.4 is applied for computing the weightages in Table 3.14.

The weightage for criteria is already calculated in Table 3.12. For the convinience, this is reproduced in Table 3.15.

Now the above two matrices are multiplied to get final weightages of challenges/issues of humanitarian supply chain as presented in Table 3.16.

The result obtain here indicates that "distribution planning (DP)" is of topmost priority followed by "Information communication system (ICS)" which is marginally below DP.

Application 2: AHP and Principal Component Analysis (PCA) for prioritization of sensors in an electronic nose for Tea Industry

Tea is considered as the beverage having some medicinal properties for preventing obesity, reducing the risk of cancer, being anti-allergic and antibacterial antiquaries. There are various types of tea and each variety of tea exhibits its own taste and flavour. India is one of the largest producers and a major exporter of tea in the world. Various classifications of commonly available tea such as White tea, Green tea, Oolong, Black tea are based on various production and processing techniques involved.

Table 3.8 Saaty's table for CR calculation

Order of Matrix (n)	1	2	3	4	5	6	7	8	9	10	11	12	13	14	15
Random Index (RI)	0.00	0.00	0.58	0.90	1.12	1.24	1.32	1.41	1.45	1.49	1.51	1.48	1.56	1.57	1.59

Table 3.9 Criteria definition

Issues	Criteria							
	Reliability	Effectiveness	Ease to monitoring	Leanness	Controllability	Flexibility	Agility	Risk controllability
Distribution planning	To distribute the material to each and every affected person	To be able to reach to all affected person	To be easily and centrally monitored	To facilitate all affected people at minimum cost	To be easily controllable, as many organization are involved	To face any unplanned demand	To be able to establish and act quickly	To be Minimum to avoid any fraud, losses, wastage, etc.
Information and communication system	To collect, segregate; communicate correct information from every point in network	To be very effective for success of HSC	To be easily and centrally monitored	To become effectively workable at minimum cost	To be easily controllable, as many organization are involved	To be able to easily modify as per the requirement	To be able to establish and act quickly	To be Minimum to avoid any fraud, losses, wastage, etc.
Sourcing and supplier management	To be able to provide good quality material at lowest possible price	To be able to supply at desired place at desired time in desired quantity	To be easily monitored, as many supplier and customers involved	To be able to provide required quantity & quality at lowest price	To be easily controllable, as many supplier and customers involved	To be able meet the unpredictable demands	To be able to meet every requirement in lowest time	To be Minimum to avoid any fraud, losses, wastage, etc.
Supply chain coordination and integration	To deliver right material at right time at right place in right quantity to right person	To make supplies at desired place at desired time	To be easily controllable, as network contains large no. of sources and destinations	To be able to deliver high quantity at desired place at lowest price	To be easily controllable, as network contains large no. of sources and destinations	To be able meet the unpredictable demands from unpredictable place	To be able to make delivery in lowest possible time	To be Minimum to avoid any fraud, losses, wastage, etc.

(continued)

Table 3.9 (continued)

Issues	Criteria							
	Reliability	Effectiveness	Ease to monitoring	Leanness	Controllability	Flexibility	Agility	Risk controllability
Transportation, model choice, routing	To deliver right material at right time at right place in right quantity to right person	To be able to fulfil the requirement of all affected people	Easy to monitor as lot of agencies are involved	To be able to run the transportation network in lowest cost	To be able to easily operate and control as lot of agencies are involved	To be able meet the unpredictable demands from unpredictable place	To be able to respond quickly	Risk should be Minimum for fraud, losses, wastage, etc.

Table 3.10 Criteria selection

	R	E	EoM	L	C	F	A	RC
R	1.0000	2.0000	3.0000	4.0000	5.0000	6.0000	7.0000	8.0000
E	0.5000	1.0000	2.0000	3.0000	4.0000	5.0000	6.0000	8.0000
EoM	0.3333	0.5000	1.0000	3.0000	4.0000	5.0000	7.0000	8.0000
L	0.2500	0.3333	0.3333	1.0000	3.0000	4.0000	5.0000	6.0000
C	0.2000	0.2500	0.2500	0.3333	1.0000	3.0000	4.0000	5.0000
F	0.1667	0.2000	0.2000	0.2500	0.3333	1.0000	3.0000	4.0000
A	0.1429	0.1667	0.1429	0.2000	0.2500	0.3333	1.0000	4.0000
RC	0.1250	0.1250	0.1250	0.1667	0.2000	0.2500	0.2500	1.0000

Table 3.11 Consistency index and consistency ratio

λ_{max}	2.7513
CI	0.1073
RI	1.4100
CR	0.0761

Table 3.12 Criteria weightages

R	0.3104
E	0.2160
EoM	0.1807
L	0.1142
C	0.0755
F	0.0494
A	0.0343
RC	0.0196

The main issue in classifying different varieties of tea is identifying the main chemicals present in them, which decide the flavour of tea.

An electronic nose system can measure quality parameters of different tea varieties based on their chemical composition. An electronic nose system consists of a sensor array for detecting the chemicals present in the vaporized tea, micro-pump for supplying tea vapours to the sensor array, solenoid valves for passing/stopping the flow of tea vapours to the sensor array, illumination heating halogen bulbs for heating and vaporizing tea sample, software for PC-based data acquisition and rating the tea samples as per the sensor array output data.

Centre for Development of Advanced Computing (C-DAC) has used PCA technique, in the Olfaction software for E-nose developed by them, for analysing the chemical composition data of different tea varieties. They have used this technique for identifying eight principal chemicals which decide the flavour of different tea varieties and finally for rating of different tea varieties. In this paper, we have presented

Table 3.13 AHP calculations

Effectiveness

	DP	ICS	SSM	SSCI	TMR
DP	1.0000	3.0000	0.2500	0.5000	2.0000
ICS	0.3333	1.0000	0.1250	0.2000	0.5000
SSM	4.0000	8.0000	1.0000	2.0000	6.0000
SCCI	2.0000	5.0000	0.5000	1.0000	4.0000
TMR	0.5000	2.0000	0.1667	0.2500	1.0000

λmax	2.3901
CI	0.0088
RI	1.1200
CR	0.0079

Reliability

	DP	ICS	SSM	SSCI	TMR
DP	1.0000	0.3333	6.0000	3.0000	6.0000
ICS	3.0000	1.0000	7.0000	4.0000	6.0000
SSM	0.1667	0.1429	1.0000	0.3333	0.5000
SCCI	0.3333	0.2500	3.0000	1.0000	2.0000
TMR	0.1667	0.1667	2.0000	0.5000	1.0000

λmax	2.5023
CI	0.0436
RI	1.1200
CR	0.0389

Ease of monitoring

	DP	ICS	SSM	SSCI	TMR
DP	1.0000	2.0000	7.0000	4.0000	6.0000
ICS	0.5000	1.0000	5.0000	2.0000	4.0000
SSM	0.1429	0.2000	1.0000	0.3333	0.5000
SCCI	0.2500	0.5000	3.0000	1.0000	2.0000
TMR	0.1667	0.2500	2.0000	0.5000	1.0000

λmax	2.3614
CI	0.0122
RI	1.1200
CR	0.0109

Leanness

	DP	ICS	SSM	SSCI	TMR
DP	1.0000	2.0000	0.2500	0.1667	0.5000
ICS	0.5000	1.0000	0.2000	0.1429	0.3333
SSM	4.0000	5.0000	1.0000	0.5000	2.0000
SCCI	6.0000	7.0000	2.0000	1.0000	4.0000
TMR	2.0000	3.0000	0.5000	0.2500	1.0000

λmax	2.3614
CI	0.0122
RI	1.1200
CR	0.0109

Controllability

	DP	ICS	SSM	SSCI	TMR
DP	1.0000	0.5000	0.2000	0.3333	0.1429
ICS	2.0000	1.0000	0.2500	0.5000	0.1667
SSM	5.0000	4.0000	1.0000	2.0000	0.5000
SCCI	3.0000	2.0000	0.5000	1.0000	0.2500
TMR	7.0000	6.0000	2.0000	4.0000	1.0000

λmax	2.6898
CI	0.0371
RI	1.1200
CR	0.0331

(continued)

Table 3.13 (continued)

Agility

	DP	ICS	SSM	SCCI	TMR
DP	1.0000	2.0000	3.0000	0.2000	0.5000
ICS	0.5000	1.0000	2.0000	0.1667	0.2500
SSM	0.3333	0.5000	1.0000	0.1429	0.2000
SCCI	5.0000	6.0000	7.0000	1.0000	2.0000
TMR	2.0000	4.0000	5.0000	0.5000	1.0000

λmax	2.3614
CI	0.0122
RI	1.1200
CR	0.0109

Risk controllability

	DP	ICS	SSM	SSCI	TMR
DP	1.0000	4.0000	7.0000	6.0000	2.0000
ICS	0.2500	1.0000	3.0000	2.0000	0.5000
SSM	0.1429	0.3333	1.0000	0.5000	0.2000
SCCI	0.1667	0.5000	2.0000	1.0000	0.2500
TMR	0.5000	2.0000	5.0000	4.0000	1.0000

λmax	2.4473
CI	0.0175
RI	1.1200
CR	0.0156

Flexibility

	DP	ICS	SSM	SSCI	TMR
DP	1.0000	0.5000	2.0000	0.2500	0.1667
ICS	2.0000	1.0000	3.0000	0.5000	0.2500
SSM	0.5000	0.3333	1.0000	0.2000	0.1429
SCCI	4.0000	2.0000	5.0000	1.0000	0.2500
TMR	6.0000	4.0000	7.0000	4.0000	1.0000

λmax	2.3614
CI	0.0122
RI	1.1200
CR	0.0109

Table 3.14 HSCM criteria versus challenges pairwise comparison

	R	E	EoM	L	C	F	A	RC
DP	0.2866	0.1365	0.4672	0.0786	0.0510	0.0766	0.1308	0.4672
ICS	0.4759	0.0488	0.2651	0.0510	0.0786	0.1329	0.0776	0.1381
SSM	0.0468	0.4745	0.0510	0.2651	0.2651	0.0495	0.0504	0.0510
SCCI	0.1207	0.2626	0.1381	0.4672	0.1381	0.2271	0.4807	0.0786
TMR	0.0700	0.0776	0.0786	0.1381	0.4672	0.5138	0.2605	0.2651

Table 3.15 Criteria weightages

R	0.3104
E	0.2160
EoM	0.1807
L	0.1142
C	0.0755
F	0.0494
A	0.0343
RC	0.0196

Table 3.16 HSCM challenges priorities

Result	
DP	0.2331
ICS	0.2298
SSM	0.1817
SCCI	0.2122
TMR	0.1432

the application of this technique. The data (chemical composition for different tea varieties) used is self-generated for explaining the techniques.

We have then presented the application of AHP technique, a multi-criteria decision-making management technique, for analysis of the same data and compared the results from both the techniques highlighting the flaw in PCA, covered by the merits of AHP over PCA.

Requirement for data analysis: Problem statement

Before understanding the PCA and AHP techniques and their application for data analysis, let us first acknowledge the requirement for such data analysis for our case that is prioritization of sensors in an E-nose for tea industry. We define our problem as to develop an E-nose for rating the quality of say 50 different tea varieties based on the percentage of different chemicals present in them. The sensor array of E-nose should be capable of finding out the chemical composition of these tea varieties from the vapours produced by their solutions (with water). Let us say different tea varieties contain 70 different chemical compounds on an average. It will require 70 different sensors to be installed in a small space (considering E-nose is portable and of the size of an oven). The large number of sensors will increase the cost of E-nose and affect its affordability. The data matrix will be of 3500 elements (50×70). Working with such a large data matrix is practically impossible because they require lot of data storage space, processing complexity and time. Lastly, large number of sensors will affect the reliability due to increase in instances of sensor failure and increase in the maintenance cost as sensors will require replacement after completing their useful life.

Now suppose the percentages of only 10 chemical compounds (out of total 70) show significant variations for different tea varieties, whereas the remaining 60 chemicals show very less variation. Hence, we can have a good classification of the different tea varieties based on these 10 chemicals and ignore the rest. These 10 chemicals then become our principal components on decision criteria. Hence, we can have only 10 different sensors in our E-nose, meeting the size and cost requirements. The data matrix will now consist of only 600 elements thus reducing the requirement of data storage space, processing complexity and time.

Principal component analysis

Principal component analysis is a method used to identify patterns present within a data and identify the similarities and differences in the data. Identifying such patterns in a data consisting of higher dimensions is a difficult task, especially given the fact that graphical representation of such higher dimension data is not available. PCA provides a powerful technique for analyzing such large dimensional data. PCA helps to compress the data in table once the patterns in the data are identified which thereby leads to the reduction in the number of dimensions while retaining maximum amount of information, i.e. without losing significant information in the data.

The sample data used for PCA analysis is shown in Fig. 3.2 and Table 3.17. Table 3.18 reports the adjusted data.

The steps involved in PCA analysis are exaplained as below.

Step 1: Subtract the mean

First, calculate the mean value of each dimension and subtract the calculated mean value from each value for the corresponding dimension. For example, if all the X values have an average X_{av}, which is subtracted from all values of X, then similarly, a value of Y_{av} is subtracted from all the Y values. The result is a data set which has a combined mean of zero.

Step 2: Calculate the covariance matrix

Now calculate the covariance matrix for the given data. In the above example, we have a dimensional data, therefore, our resultant covariance matrix will have to be 2 X 2.

Fig. 3.2 PCA analysis data

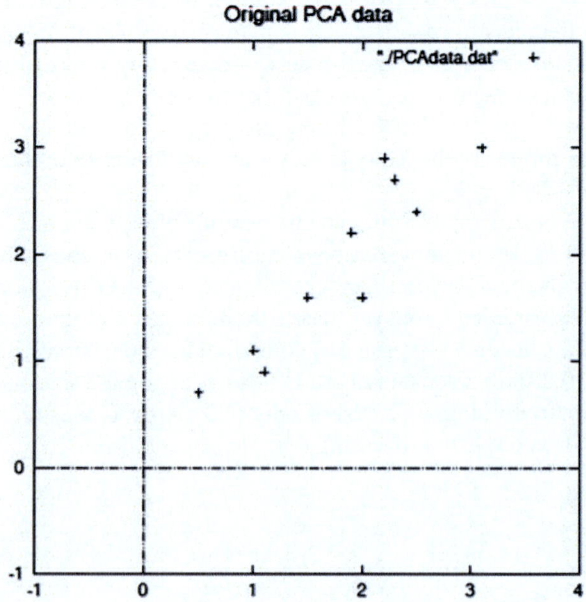

Table 3.17 Sample data

X	Y
2.5	2.4
0.5	0.7
2.2	2.9
1.9	2.2
3.1	3.0
2.3	2.7
2.0	1.6
1.0	1.1
1.5	1.6
1.1	0.9
Mean X_{av} 1.8	Y_{av} 1.9

Table 3.18 Data adjustment

$X_{new} = X - X_{av}$	$Y_{new} = Y - Y_{av}$
0.69	0.49
−1.31	−1.21
0.39	0.99
0.09	0.29
1.29	1.09
0.49	0.79
0.19	−0.31
−0.81	−0.81
−0.31	−0.31
−0.71	−1.01

$$\mathrm{Cov}(X, Y) = \{\sum_{i=1 \text{ to n}} (X_i - X_{av})(Y_i - Y_{av})\}/n - 1$$

$$\mathrm{Cov} = \begin{pmatrix} \mathrm{cov}(x, x) & \mathrm{cov}(x, y) \\ \mathrm{cov}(y, x) & \mathrm{cov}(y, y) \end{pmatrix}$$

$$\mathrm{Cov} = \begin{pmatrix} 0.616555556 & 0.615444444 \\ 0.615444444 & 0.716555556 \end{pmatrix}$$

From the above covariance matrix, we observe that the non-diagonal elements are positive, which indicates that both the X and Y variables have a positive correlation, i.e. increase in one of the value (X) will lead to increase in the other value (Y).

Step 3: Calculate the eigenvectors and eigenvalues of the covariance Matrix

Since the covariance matrix is square, we can calculate the eigenvectors and eigenvalues for this matrix.

$$\text{Eigen Values} = \begin{pmatrix} 0.0490833989 \\ 1.2840277100 \end{pmatrix}$$

$$\text{Eigen Vectors} = \begin{pmatrix} -0.735178656 & -0.677873399 \\ 0.677873399 & -0.735178656 \end{pmatrix}$$

Both the eigenvectors calculated above are unit eigenvectors, i.e. their lengths are 1 which is an important requirement in PCA. While using various math packages, this needs to be ensured and fortunately, many of the available packages provide unit eigenvectors while calculating eigenvectors.

From Fig. 3.3, we observe that there is a strong pattern in the considered sample data. Also, as noted based on the covariance matrix, there is a positive relationship between the variables, i.e. they both increase together. The eigenvector values are also plotted above the data, which are represented as diagonal dotted lines and are perpendicular to each other. The eigenvector lines provide information about the inherent pattern existing in the data. One of the eigenvectors, which passes through the middle of the data, represents how the data sets are related. The second eigenvector provides information about the other pattern in the data, that while the points considered all follow the main eigenvector line, they are not exactly on the line but

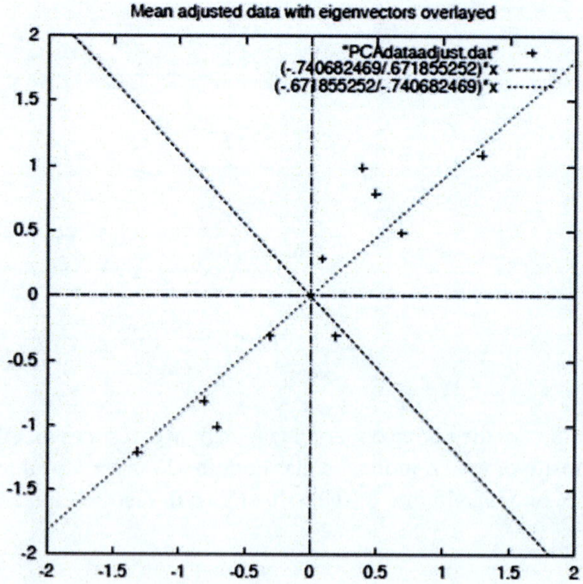

Fig. 3.3 A plot of the normalized data (mean subtracted) with the eigenvectors of the covariance matrix overlaid on top

off the main line by a small amount. Thus through the use of these two eigenvector lines we are able to characterize the available data. Further steps in PCA involve transformation of the data in order to express the data in terms of these lines.

Step 4: Choosing components and forming a feature vector

The next step involves choosing a set of principal components which results in a compression of data and also in a reduced dimensionality of the overall data set. When we observe the calculated eigenvectors and eigenvalues, we notice that the values of eigenvector and eigenvalues are significantly different from each other. The eigenvector with the highest numerical value is the principal component of the data set. In the considered example, the eigenvector with the larger value was the one which actually passed through the middle of the data thus signifying the most important relationship between various dimensions of the data set.

As we observe that eigenvectors with the highest values are the most significant component of the data, the eigenvalues are arranged in the descending order, i.e. from highest value to the lowest. This orders the order of significance of each component of the data. Based on this ordering we can choose the first few components while discarding the remaining components. Even though, this results in loss of some information from the initial data, the overall loss of information is generally low. This leads to the final data set which has comparatively lesser dimension than the actual data while ensuring minimal loss of information. If initially there were n dimensions in the data, and we choose to select only "p" principal components based on the calculated eigenvectors and eigenvalues, then the final data will have only p dimensions.

Based on the above chosen final data, we form a feature vector. Once, the set of eigenvectors that are to be retained from the initial list of eigenvectors is decided; we construct a matrix with these selected eigenvectors in the columns.

$$\text{Feature Vector} = \left(\text{eig}_1\ \text{eig}_2\text{eig}_3 \ldots\ \text{eig}_p\right)$$

For our example data set, from the 2 eigenvectors we choose only one, which has the highest eigenvector value while ignoring the less significant component and we are left with a single column:

$$\begin{pmatrix} -0.677873399 \\ -0.735178656 \end{pmatrix}$$

Step 5: Deriving the new data set

After identifying the principal components based on the eigenvectors and forming a feature vector, we multiply the transpose of the feature vector by the transpose of the original data set to obtain the final data.

$$\text{Final Data} = \text{Row Feature Vector} \times \text{Row Data Adjust}$$

Table 3.19 Transformed data (Single Eigenvector)

X
−0.827970186
1.77758033
−0.992197494
−0.274210416
−1.67580142
−0.912949103
0.0991094375
1.14457216
0.438046137
1.22382056

where Row Feature Vector is the transpose of the feature vector such that the eigenvectors are now in the rows instead of columns with the highest value at the first row. Row Data Adjust represents the transpose of the mean-adjusted data, where each row represents a different dimension. Table 3.19 reports the transformed data for the example.

The example row data adjust or the transformed data only has a single dimension. While our initial data set had two dimensions, we have reduced the dimensions of our data set by 1. We have effectively thrown away one of the dimensions and thus our plot of the adjusted data will be points on a single line where the other axis (i.e. the eigenvector) has been completely discarded. Thus, we have transformed our data successfully in terms of the patterns within our data which were identified by the eigenvector lines that closely describe the patterns.

Application of Principal Component Analysis.

See Table 3.20.

The above data is a factual data of chemical composition of five different tea varieties. The data gives percentages of 10 prominently found chemicals in these tea varieties, though in reality a variety may comprise of hundreds of different chemicals.

Table 3.20 Data set of chemical composition of five different tea varieties

Tea Varieties	Chemical Present (in %)									
	C1	C2	C3	C4	C5	C6	C7	C8	C9	C10
T1	12.4	32.1	12.5	8.3	4.5	10.2	2.1	8.1	3.1	1.7
T2	9.3	48.5	3.5	2.4	8.4	3.1	5.2	8	3	1.6
T3	17.3	39.7	2.1	6.3	2.1	6.4	8.4	8.2	3.1	1.3
T4	25.2	26.2	4.6	5.3	10.3	1.3	9.2	7.9	3.2	1.1
T5	20.6	44.5	1.1	1.3	6.1	8.2	1.1	8.3	3	0.9
C_{AVG}	16.96	38.2	4.76	4.72	6.28	5.84	5.2	8.1	3.08	1.32

As only prominent chemicals are shown, the percentages may not add to 100% in each row.

PCA analysis will be used to find out if these tea varieties can be classified on the basis of fewer than 10 chemical compounds thus compressing the data and reducing the number of sensors required.

Step 1: Subtract the mean

The new values (after subtracting mean value of column from each individual value) of the ten prominently found chemicals in Tea are summarized in Table 3.21.

Step 2: Calculate the covariance matrix

This is reported in Table 3.22.

Step 3: Calculate the eigenvectors and eigenvalues of the covariance matrix

See Table 3.23.

Table 3.21 Data adjustment

C1new	C2new	C3new	C4new	C5new	C6new	C7new	C8new	C9new	C10new
−4.56	−6.1	7.74	3.58	−1.78	4.36	−3.1	0	0.02	0.38
−7.66	10.3	−1.26	−2.32	2.12	−2.74	0	−0.1	−0.08	0.28
0.34	1.5	−2.66	1.58	−4.18	0.56	3.2	0.1	0.02	−0.02
8.24	−12	−0.16	0.58	4.02	−4.54	4	−0.2	0.12	−0.22
3.64	6.3	−3.66	−3.42	−0.18	2.36	−4.1	0.2	−0.08	−0.42

Table 3.22 Covariance matrix

−4.56	−6.1	7.74	3.58	−1.78	4.36	−3.1	0	0.02	0.38
−7.66	10.3	−1.26	−2.32	2.12	−2.74	0	−0.1	−0.08	0.28
0.34	1.5	−2.66	1.58	−4.18	0.56	3.2	0.1	0.02	−0.02
8.24	−12	−0.16	0.58	4.02	−4.54	4	−0.2	0.12	−0.22
3.64	6.3	−3.66	−3.42	−0.18	2.36	−4.1	0.2	−0.08	−0.42

Table 3.23 Eigenvalues

Eigen Values =	2.77E+00
	−1.11E+01
	5.03E+00
	3.43E+00
	−1.37E+00
	2.91E+00
	−8.05E−01
	7.99E−03
	7.74E−02
	4.30E−02

It can be seen that last three eigenvalues are very small compared to first seven. This means that variations in the percentages of last three chemicals (C8–C10) in our original data are very less for all five tea varieties. This implies that these chemicals will not be very much effective in identifying the tea varieties from each other, and hence, can be ignored. We can now transform the data with Row Feature Vector comprising of eigenvectors corresponding to only first seven eigenvalues.

This means that we need to sense only seven chemicals and hence need only seven sensors in our E-nose thus meeting our size and cost requirements. Our data matrix reduces from a 50-element matrix (5 rows × 10 columns) to 35 elements (5 rows × 7 columns) thus reducing the requirement of data storage space, processing complexity and time.

Need for AHP analysis for electronic nose problem

The primary flaw with PCA is that it does not take into consideration the importance of different chemicals in deciding the quality of tea. Let us again consider the above example where we finally reduced the number of sensors required in our E-nose from 10 to 7 using PCA. Suppose we can use only 5 sensors in our E-nose, and hence, we need to drop 2 more chemicals (read as sensors) out of remaining 7. A straightforward decision, based on PCA technique, will be to drop C5 and C7, the one's with lowest eigenvalues. But what if C5 and C7 are the chemicals, which prominently enhance the flavour of any tea variety? Hence, dropping these two chemicals will no doubt reduce the number of sensors and our E-nose will be able to distinguish between tea varieties with now five sensors but it will no longer be able to identify a good quality tea from a bad quality. For example, say C5 is Linalool Oxide, which is responsible for sweetness of tea and C7 is Geraniol, which is responsible for floral aroma in tea. These chemicals are obviously important contributors in enhancing the quality of tea. This demands a careful investigation.

AHP provides a reasonable solution to this problem by considering the prominence of different chemicals in enhancing the flavour of tea while prioritizing the sensors in an E-nose.

Application of AHP

This section reports an application of AHP to the same problem considered for PCA. The purpose is to prioritize the 7 sensors shortlisted by PCA. Let us assume that composition of the best-quality tea grown in India is available to us as presented in Table 3.24.

The eigenvector of the relative importance or value of C1, C2, C3, C4, C5, C6, C7 is (0.19, 0.29, 0.07, 0.10, 0.14, 0.08, 0.13). Arranging the chemical in their order of relative importance, from highest to lowest, we get (C2, C1, C5, C7, C4, C6, C3). Hence, it is quite evident that, if only five sensors are to be chosen, they should be the ones which are able to sense C2, C1, C5, C7 and C4. This will ensure that our E-nose not distinguishes between the five varieties but also tells us that which variety has a good flavour and which one is of poor quality. Whereas using only PCA analysis, we would have omitted the sensors corresponding to C5 and C7 chemicals limiting the functionality of our E-nose. Further, using these 5 chemicals (i.e. corresponding

Table 3.24 Percentage of chemicals in tea

(a)

C1	C2	C3	C4	C5	C6	C7
17.3	26.7	6.1	9.4	12.5	7.3	11.4

(b)

	C1 (17.3%)	C2 (26.7%)	C3 (6.1%)	C4 (9.4%)	C5 12.5%)	C6 (7.3%)	C7 (11.4%)	7th root of product of values	Eigenvector
C1 (17.3%)	1.00	0.65	2.84	1.84	1.38	2.37	1.52	1.50	0.19
C2 (26.7%)	1.54	1.00	4.38	2.84	2.14	3.66	2.34	2.31	0.29
C3 (6.1%)	0.35	0.23	1.00	0.65	0.49	0.84	0.54	0.53	0.07
C4 (9.4%)	0.54	0.35	1.54	1.00	0.75	1.29	0.82	0.81	0.10
C5 12.5%)	0.72	0.47	2.05	1.33	1.00	1.71	1.10	1.08	0.14
C6 (7.3%)	0.42	0.27	1.20	0.78	0.58	1.00	0.64	0.63	0.08
C7 (11.4%)	0.66	0.43	1.87	1.21	0.91	1.56	1.00	0.99	0.13
TOTAL								7.85	1.00

sensors) we try to rate the 5 tea varieties with composition of best quality tea as reference. Taking the data for only chemicals C2, C1, C5, C7 and C4, the pairwise comparison matrix is developed in Table 3.25.

Now, writing pairwise comparison matrices of 5 varieties of tea with respect to C1, C2, C4, C5 and C7.

For Chemical (Sensor) C1.

See Table 3.26.

For Chemical (Sensor) C2.

See Table 3.27.

For Chemical (Sensor) C4:

See Table 3.28.

Table 3.25 Pairwise comparison of tea chemical composition

	C1 (17.3%)	C2 (26.7%)	C4 (9.4%)	C5 (12.5%)	C7 (11.4%)	5th root of product of values	Eigenvector
C1 (17.3%)	1.00	0.65	1.84	1.38	1.52	1.20	0.22
C2 (26.7%)	1.54	1.00	2.84	2.14	2.34	1.85	0.35
C3 (6.1%)	0.35	0.23	0.65	0.49	0.54	0.65	0.12
C4 (9.4%)	0.54	0.35	1.00	0.75	0.82	0.87	0.16
C5 12.5%)	0.72	0.47	1.33	1.00	1.10	0.79	0.15
TOTAL						5.37	1.00

Table 3.26 Pairwise comparison of variety of tea for criteria—chemical (C1)

	T1 (12.4%)	T2 (9.3%)	T3 (17.3%)	T4 (25.2%)	T5 (20.6%)	5th root of product of values	Eigenvector
T1 (12.4%)	1.00	1.33	0.72	0.49	0.60	0.78	0.15
T2 (9.3%)	0.75	1.00	0.54	0.37	0.45	0.58	0.11
T3 (17.3%)	1.40	1.86	1.00	0.69	0.84	1.08	0.20
T4 25.2%)	2.03	2.71	1.46	1.00	1.22	1.58	0.30
T5 (20.6%)	1.66	2.22	1.19	0.82	1.00	1.29	0.24
Total						5.31	1.00

Table 3.27 Pairwise comparison of variety of tea for criteria—chemical (C2)

	T1 (32.1%)	T2 (48.5%)	T3 (39.7%)	T4 (26.2%)	T5 (44.5%)	5th root of product of values	Eigenvector
T1 (32.1%)	1.00	0.66	0.81	1.23	0.72	0.86	0.17
T2 (48.5%)	1.51	1.00	1.22	1.85	1.09	1.30	0.25
T3 (39.7%)	1.24	0.82	1.00	1.52	0.89	1.06	0.21
T4 26.2%)	0.82	0.54	0.66	1.00	0.59	0.70	0.14
T5 (44.5%)	1.39	0.92	1.12	1.70	1.00	1.19	0.23
Total						5.12	1.00

Table 3.28 Pairwise comparison of variety of tea for criteria—chemical (C4)

	T1 (8.3%)	T2 (2.4%)	T3 (6.3%)	T4 (5.3%)	T5 (1.3%)	5th root of product of values	Eigenvector
T1 (8.3%)	1.00	3.46	1.32	1.57	6.38	2.15	0.35
T2 (2.4%)	0.29	1.00	0.38	0.45	1.85	0.62	0.10
T3 (6.3%)	0.76	2.63	1.00	1.19	4.85	1.63	0.27
T4 5.3%)	0.64	2.21	0.84	1.00	4.08	1.37	0.22
T5 (1.3%)	0.16	0.54	0.21	0.25	1.00	0.34	0.06
Total						6.10	1.00

For Chemical (Sensor) C5:
See Table 3.29.
For Chemical (Sensor) C7:
See Table 3.30.

Tea Chemical Composition Matrix (TCCM).

See Table 3.31.

Relative Value Vector (RVV) of the five chemicals in enhancing the flavour of tea.

See Table 3.32.

Table 3.29 Pairwise comparison of variety of tea for criteria—chemical (C5)

	T1 (4.5%)	T2 (8.4%)	T3 (2.1%)	T4 (10.3%)	T5 (6.1%)	5th root of product of values	Eigenvector
T1 (4.5%)	1.00	0.54	2.14	0.44	0.74	0.82	0.14
T2 (8.4%)	1.87	1.00	4.00	0.82	1.38	1.53	0.27
T3 (2.1%)	0.47	0.25	1.00	0.20	0.34	0.38	0.07
T4 (10.3%)	2.29	1.23	4.90	1.00	1.69	1.88	0.33
T5 (6.1%)	1.36	0.73	2.90	0.59	1.00	1.11	0.19
Total						5.72	1.00

Table 3.30 Pairwise comparison of variety of tea for criteria—chemical (C7)

	T1 (2.1%)	T2 (5.2%)	T3 (8.4%)	T4 (9.2%)	T5 (1.1%)	5th root of product of values	Eigenvector
T1 (2.1%)	1.00	0.40	0.25	0.23	1.91	0.54	0.08
T2 (5.2%)	2.48	1.00	0.62	0.57	4.73	1.33	0.20
T3 (8.4%)	4.00	1.62	1.00	0.91	7.64	2.14	0.32
T4 (9.2%)	4.38	1.77	1.10	1.00	8.36	2.35	0.35
T5 (1.1%)	0.52	0.21	0.13	0.12	1.00	0.28	0.04
Total						6.63	1.00

Table 3.31 Tea chemical composition matrix

	C1	C2	C4	C5	C7
T1	0.15	0.17	0.35	0.14	0.08
T2	0.11	0.25	0.10	0.27	0.20
T3	0.20	0.21	0.27	0.07	0.32
T4	0.30	0.14	0.22	0.33	0.35
T5	0.24	0.23	0.06	0.19	0.04

Table 3.32 Relative value vector

C1	0.22
C2	0.35
C4	0.12
C5	0.16
C7	0.15

Table 3.33 Total quality rating for variety of tea

T1	0.105
T2	0.196
T3	0.209
T4	0.247
T5	0.177

Tea Quality Rating = TCCM ×RVV.

See Table 3.33.

Hence, using AHP we have evaluated that the 5 tea varieties in order of their quality are as follows: T4, T3, T2, T5, T1.

3.6 Advantages of AHP

It has a simple structure which is capable enough to address adequate complexity of the real-life problem.

It can easily be integrated with other qualitative and quantitative approaches. This includes its integration with linear programming, goal programming, data envelopment analysis, game theory, conjoint analysis and SWOT analysis.

It enables decision makers to accommodate their experience and intuitions in terms of relative importance of one attribute on another and develop better confidence in the results derived through the technique.

Its strength lies in executing the decision-making process based on the consensus on various assessment factors and their significance on the overall prioritization decisions. This helps the decision makers to converge their decision with a shared and common understanding very fast.

A hierarchical structure of AHP provides an intuitive appeal and a way to better understand the complexity of the problem. This enables decision maker to think about various consequences associated with decision making.

3.7 Limitations of AHP

It is difficult at times, to obtain consensus on experts' opinions as it involves large number of pairwise comparisons.

It needs a reliable data based on experience, knowledge and judgement which are quite subjective for each decision maker.

Does not take into account the uncertainty.

Time consuming with large number of attributes and alternatives.

AHP is successfully employed in many of the multi-criteria decision-making problems. The use of technique is explored in majority of the fields such as engineering, humanities, psychology, management and others. This technique is extensively useful when the decision problem has linear hierarchies. The technique has an inherent limitation that it is not possible to accommodate nonlinear structure of the problem. The next chapter will provide insights into the MCDM technique—Analytic Network Process (ANP) which accommodates nonlinear relationships among criteria and alternatives.

Selected Bibliography and Further Reading

1. Saaty, R.W. (1987). The analytic hierarchy processd what it is and how it is used. *Mathematical Modelling, 9*(3), 161–176.
2. Saaty, T. L. (1980). *The analytical hierarchy process*. New York, NY: McGraw-Hill.
3. Thanki, S. J., Kannan, G., & Thakkar, J. J. (2016, November 1). An investigation on lean-green implementation practices in Indian SMEs using analytical hierarchy process (AHP) approach. *Journal of Cleaner Production,135*, 284–298.
4. Thakkar, J. J., Deshmukh, S. G., & Kanda, A. (2006). Implementing six sigma in service sector using AHP and Alderfer's motivational model: A case of educational services. *International Journal of Six Sigma and Competitive Advantage, 2*(4), 353–376.
5. Vinodh, S., Arvind, K. R., & Somanaathan, M. (2011). Tools and techniques for enabling sustainability through lean initiatives. *Clean Technologies and Environmental Policy, 13*(3), 469–479.
6. Saaty, T. L. (1977). A scaling method for priorities in hierarchical structures. *Journal of Mathematical Psychology, 15*(3), 234–281.
7. Saaty, T. L. (1990). How to make a decision: The analytic hierarchy process. *European Journal of Operational Research, 48*(1), 9–26.
8. Saaty, T. L. (2006). Rank from comparisons and from ratings in the analytic hierarchy/network processes. *European Journal of Operational Research, 168*(2), 557–570.
9. Zahedi, F. (1986). The analytic hierarchy process—A survey of the method and its applications. *Interfaces, 16*(4), 96–108.
10. Vaidya, O. S., & Kumar, S. (2006). Analytic hierarchy process: An overview of applications. *European Journal of Operational Research, 169*(1), 1–29.
11. Vargas, L. G. (1990). An overview of the analytic hierarchy process and its applications. *European Journal of Operational Research, 48*(1), 2–8.
12. Vinodh, S., Shivraman, K. R., & Viswesh, S. (2012). AHP-based lean concept selection in a manufacturing organization. *Journal of Manufacturing Technology Management, 23*(1), 124e136.

Chapter 4
Analytic Network Process (ANP)

4.1 Background

With the advancement in the fields of project engineering and management, there has been an advanced shift from linear weighing to multi-criteria decision-making models (MCDM). MCDM is a technique that helps in choosing the best option among the available ones. A set of criteria are used to determine the dominant option. In this chapter, we would appreciate the features and application of Analytic Network Process (ANP). The basic concepts of AHP serve as an initial point for ANP. The ANP, as the general form of AHP, deals with complex decisions where there is interdependence in decision model. In analytical network process, a network is designed to deal with the dependence and feedbacks within the same group (intradependencies) or among the different clusters (interdependencies). The resulting network does not rely on the hierarchical approach but gives the solution more accurately and with greater flexibility considering the mutual relationship of the elements.

For example, a company wants to select an appropriate method to evaluate the project performance. ANP can be used as an important tool to determine the initial viability by conducting a detailed feasibility study within the selected projects. The selection of projects may depend upon various other factors such as preferences, profit margin and risk taking behaviour. The final score would help in the decision-making process.

The analytic network process (ANP), a more general form of the analytic hierarchy process (AHP) is used in multi-criteria decision analysis and is widely applicable in various disciplines of Engineering, Science and Technology.

AHP structures a decision problem into a hierarchy with a goal, decision criteria and alternatives, whereas the ANP structures it as a network. Both the techniques then use a system of pairwise comparisons to measure the weights of the components of the structure and to finally rank the alternatives in the decision. The AHP is a general theory of measurement. It is used to derive relative priorities on absolute

J. J. Thakkar, *Multi-Criteria Decision Making*, Studies in Systems, Decision and Control 336, https://doi.org/10.1007/978-981-33-4745-8_4

scales. In its general form, the AHP is a nonlinear framework and carries out both deductive and inductive thinking without use of the syllogism. This is made possible by taking several factors into consideration simultaneously, allowing for dependence and feedback, and making numerical trade-offs to arrive at a synthesis or conclusion. Paired comparison judgements in the AHP/ANP are applied to pairs of homogeneous elements.

The basic structure does not have a sequential hierarchy but resembles a network with different cycles connecting the various interdependent components while the loops connect the intradependent ones. It also consists of sources and sinks.

4.2 Qualitative Versus Quantitative Approach to Decision Making

The ANP approach involves both qualitative and quantitative forms.

Qualitative Approach To Decision Making

There are four major steps involved in the qualitative approach:

- The first step is to identify the decision problem. It is the primary objective function for which the decision needs to be made.
- The second step is to make sure that the decision problem can be solved with ANP as network problems. The problems that are based on hierarchy can be solved simply by AHP.
- The third step is to break the problem into manageable and measurable forms. The topmost part comprises of the decision problem while the lowermost step comprises of the alternatives and the overall scenario.
- The final step is to determine the decision-making body. This varies from organization to organization. In some cases, a group of top officials decide while in other cases the work is evenly divided among the top and middle level managers.

Quantitative Approach To Decision Making

After deciding the qualitative variables, the quantitative approach needs to be adopted for solving the problem further. The steps involved in this are as follows:

- A nine-point priority scale is given below. The decision makers must make a pairwise comparison of the quantitative questionnaire and assign the points accordingly.

Intensity of importance	Definition	Explanation
1	Equal importance	Two activities contribute equally to the objective
2	Weak or slight moderate	
3	Moderate importance	Experience and judgement slightly favour one activity over another
4	Moderate plus	
5	Strong importance	Experience and judgement strongly favour one activity over another
6	Strong plus	
7	Very strong	An activity is favoured very strongly over another; its dominance demonstrated in practice
8	Very, very strong	
9	Extreme importance	The evidence favouring one activity over another is of the highest possible order of affirmation

Source The Analytic Hierarchy and Analytic Network Measurement Processes: Applications to Decisions under Risk *by Thomas L. Saaty*

- Once the relative importance is calculated from the above table, the eigenvector (also referred to as "weights") is then calculated for the developed matrices.
- The third step is to measure the inconsistency of every matrix. This is done by calculating the consistency ratio (CR). Three acceptable levels of CR is given.

3 × 3 matrix	0.05
4 × 4 matrix	0.08
Others	0.1

- Matrices that do not satisfy the above criteria should to be rated again by the decision makers.
- The next step is to place the eigenvector calculated in the previous step to form a super matrix. There are four elements incorporated in the formation of a super matrix: relationships to the final objective; comparisons among factors and clusters; comparisons of alternative relationships with respect to the factors; and an identity matrix for all alternatives.
- The final step is to calculate the stable weights from the super matrix. The stable weights are determined by raising the super matrix to a high power until the weights are converged and stabilized.

4.3 Step-by-Step Procedure of ANP

I. Depict the decision problem at the point of interest including its objectives, criteria and sub-criteria, actors and their targets and the conceivable results of that choice. Give points of interest of impacts that decide how that decision may turn out.

II. Decide the control criteria and sub-criteria in the four control hierarchies one each for the advantages, opportunities, expenses and dangers of that choice and get their needs from matched correlations grids. In a case where a control standard or sub-criterion has a worldwide need of 3% or less, you may consider disposing it for further analysis. The product naturally bargains as it were with those criteria or sub-criteria that have subnets under them. For advantages and opportunities, ask what gives the most advantages or shows the best chance, to impact satisfaction of that control rule. For costs and risks, ask what acquires the most cost or faces the most serious risk. Here and there (once in a while), the correlations are made essentially with respect to advantages, opportunities, costs and risks in total without utilizing control criteria and sub-criteria. Figure 4.1 represents an interdependence network structure of the hierarchy typically evaluated using ANP method. In the structural relationships of nonlinear hierarchy, there are two-way relationships between criteria 1 and 2, and criteria 3 and 4. There are one-way (outer) dependence relationships between criteria 1 and 3, criteria 2 and 3 and criteria 2 and 4. There is also a self-loop of criteria 3.

III. Decide the most broad system of groups (or parts) and their components that apply to all the control criteria, to better sort out the advancement of the model. Additionally you can number and orchestrate the groups and their components advantageously (maybe in a section). Utilize the indistinguishable name to speak to the same group and the same components for all the control criteria.

IV. For every control model or sub-criterion, decide the groups of the general criticism framework with their components and interface them as indicated by their external and internal reliance impacts.

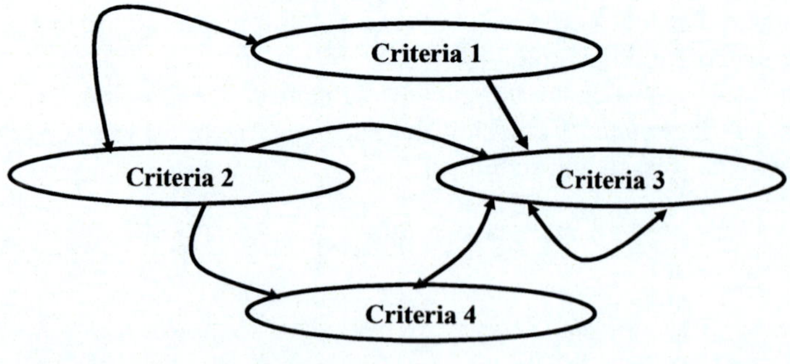

Fig. 4.1 Interdependent network hierarchy for ANP analysis

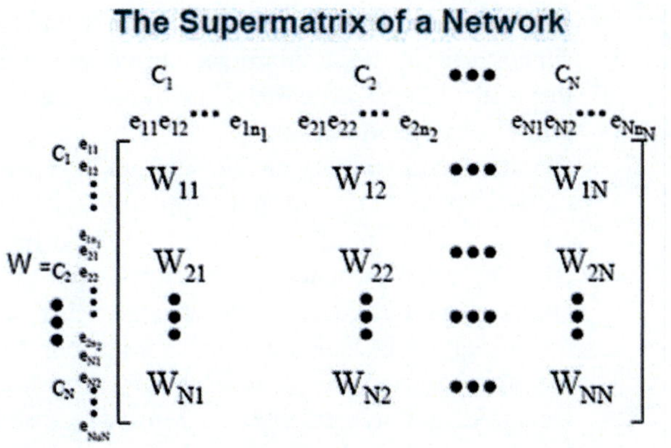

Fig. 4.2 Formation of supermatrix for ANP

V. Decide the methodology you need to follow in the examination of every group or component, impacting (the favoured methodology) different groups and components with appreciation to a standard. The sense (being affected or impacting) must apply to every one of the criteria for the four control progressive systems.

VI. For every control rule, build the supermatrix by laying out the bunches in the request they are numbered along with every one of the components in every bunch both vertically on the left and on a level plane at the top. Enter in the proper position the needs got, from the matched correlations as subcolumns of the relating segment of the supermatrix (Fig. 4.2).

VII. Perform combined examinations on the components inside the clusters itself as per their impact on every component in another cluster they are associated to (external reliance) or on components in their own particular group (inward reliance). In making correlations, you should have a criterion in mind. Correlations of elements based on which component emphatically impacts another component increasingly and how it is contrasted with a control criterion or sub-criterion of the control pecking order as a main priority.

VIII. Perform paired comparisons on the clusters as they impact every group to which they are associated concerning the given control rule. The determined weights are utilized to weigh the components of the comparing segment squares of the supermatrix. Dole out a zero when there is no impact and acquire the weighted section stochastic supermatrix.

IX. Compute the breaking point needs of the stochastic supermatrix as indicated by whether it is irreducible (primitive or imprimitive [cyclic]) or it is reducible with one being a basic or a various root and whether the framework is cyclic or not. Two kinds of results are conceivable. In the first, every section of the network is indistinguishable and each gives the relative needs of the components from which the needs of the components in every bunch are standardized to one. In the second, the point of confinement cycles in squares as far as possible are summed up and arrived at the midpoint, and again standardized to one for every bunch. In spite of the fact that the need vectors are entered in the supermatrix in the standardized structure, the farthest point needs are placed in a romanticized structure in the light of the fact that the control criteria does not rely on the choices.

X. Blend the constraining needs by weighing each romanticized limit vector by the weight of its control paradigm and including the subsequent vectors for each of the four merits: Benefits (B), Opportunities (O), Costs (C) and Risks (R). There are presently four vectors, one for each of the four benefits. An answer containing minor estimations of the benefits is acquired by shaping the proportion BO/CR for every option from the four vectors. The option with the biggest proportion is decided for a few choices.

XI. Decide key criteria and their needs to rate each of the four benefits. Standardize the four evaluations and utilize them to ascertain the general blend of the four vectors. For every option, subtract the expenses and dangers from the entirety of the advantages and opportunities. At different times one may subtract the expenses from one option and dangers from another and after that add them to the weighted formal and opportunities.

XII. Perform affectability examination on the last result and decipher the after effects of affectability showing how extensive or little these proportions are.

The priorities derived from pairwise comparison matrices are entered as parts of the columns of a supermatrix. The supermatrix speaks to the impact need of a component on the left of the network on a component at the highest point of the framework with deference to a specific control foundation. A supermatrix alongside a case of one of its general section networks is shown in Fig. 4.3. The part C1 in the supermatrix incorporates all the need vectors inferred for hubs that are "parent" nodes in the C1 bunch. Figure 4.4 gives the supermatrix of a progression and Fig. 4.5 demonstrates the kth power of that supermatrix which is the same as hierarchic piece in the $(k + 1)$ position.

Fig. 4.3 Supermatrix of a network and detail of a component in it

W_{ij} Component of Supermatrix

$$W_{ij} = \begin{bmatrix} W_{i1}^{(j_1)} & W_{i1}^{(j_2)} & \cdots & W_{i1}^{(jn_j)} \\ W_{i2}^{(j_1)} & W_{i2}^{(j_2)} & \cdots & W_{i2}^{(jn_j)} \\ \vdots & \vdots & & \vdots \\ W_{in_i}^{(j_1)} & W_{in_i}^{(j_2)} & \cdots & W_{in_i}^{(jn_j)} \end{bmatrix}$$

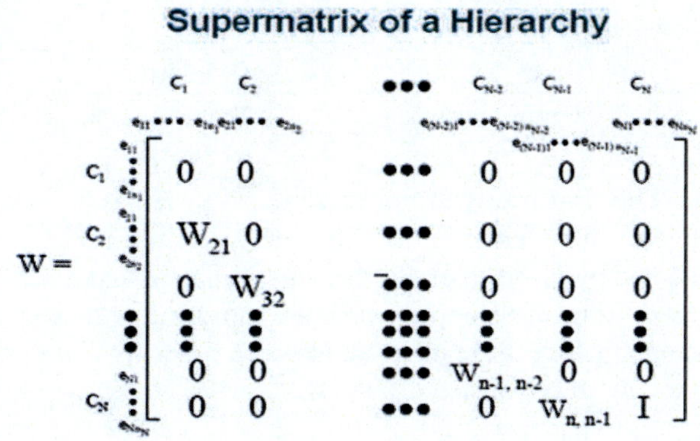

Fig. 4.4 Supermatrix of a hierarchy

$$W^k = \begin{bmatrix} 0 & 0 & \cdots & 0 & 0 & 0 \\ 0 & 0 & \cdots & 0 & 0 & 0 \\ \vdots & \vdots & \vdots & \vdots & \vdots & \vdots \\ 0 & 0 & \cdots & 0 & 0 & 0 \\ W_{n,n-1}W_{n-1,n-2}\cdots W_{32}W_{21} & W_{n,n-1}W_{n-1,n-2}\cdots W_{32} & \cdots W_{n,n-1}W_{n-1,n-2} & W_{n,n-1} & I \end{bmatrix}$$

Fig. 4.5 Limit supermatrix of a hierarchy

4.4 Illustrative Examples on ANP

Application 1: Risk Prioritization for Construction Company

Sunny Construction is a reputed Delhi-based firm that deals with housing and commercial projects. The overall turnover of the firm is around 600 crores. Its major customers include real estate developers and government agencies.

The company holds strength in terms of its management, quality and cost efficiencies, while it has lost a lot in terms of environment and safety concerns. The firm wants to hunt for bigger projects in other infrastructure fields and thus wants to overview the kind of risks that it faced in its previous projects, as it had lost a lot of investment in terms of money, time, quality, etc.

Sl No.	Risks	Definition
1	Project delay	The delay from the actual project time
2	Inflation	Rise in price of resources like labour, materials, etc
3	Poor management	Lack of communication inter-firm and intra-firm
4	Clearance and bureaucracy	Project approval and environmental clearances and other factors
5	Pollution	Pollution due to construction and material preparation, transport, etc

(continued)

(continued)

Sl No.	Risks	Definition
6	Unsafe operations	The safety of humans, buildings, etc

The firm wants to prioritize these risks based on the following criteria—costs, quality, time, environment and safety as major criteria. Now these factors are inter-related by all risks under the same problem and hence hierarchy process cannot be used. Thus the company decided to prioritize these risks using analytical network model, where these interrelations can be easily understood and a deeper study of the model can be formed (Table 4.1).

Step 1: The formulated problem is converted into a hierarchical structure where the sub-factors are changed to alternatives to measure using ANP technique. The formulation is based on 5 criteria: Cost, Time, Safety, Quality and Environment. The sub-factors are the prioritizations of risks given in terms of alternatives to define the hierarchy of the risks involved. The hierarchical structure is formulated in (Fig. 4.6).

Step 2: Assuming interdependencies among various factors involved, comparisons of these factors are done using the scale 1–9 made with respect to the goal (Table 4.2).

Step 3: After doing the pairwise comparisons, the interdependencies of the factors are attained and the normalized values of each of these tables are obtained. The interdependency tables of criteria are given below. More realistic and appropriate results are obtained with the scaling of alternatives using the ANP technique. The resultant eigenvectors of each of these tables are obtained and represented in the last column. The following tables are obtained using the software: Super decisions (Table 4.3).

Step 4: The interdependencies of the sub-criteria factors which are considered in terms of the prioritizations of the risks are given in the following tables (Tables 4.4, 4.5, 4.6, 4.7, 4.8, 4.9, 4.10, 4.11, 4.12, 4.13 and4.14).

Step 5: In this step, interdependent priorities of factors are calculated. The overall unweighted supermatrix is formed from the interdependent factors with goal "prioritization of risks" as the objective (Fig. 4.7).

Step 6: The weighted supermatrix of the same is given below (Fig. 4.8).

Table 4.1 Interrelations of risks and criteria

Sl. no.		Cost	Time	Quality	Safety	Environment
1	Project delay	✓	✓	✓	✓	
2	Inflation	✓	✓	✓	✓	
3	Poor management		✓	✓	✓	✓
4	Clearance and bureaucracy	✓	✓			✓
5	Pollution	✓			✓	✓
6	Unsafe operations	✓	✓		✓	

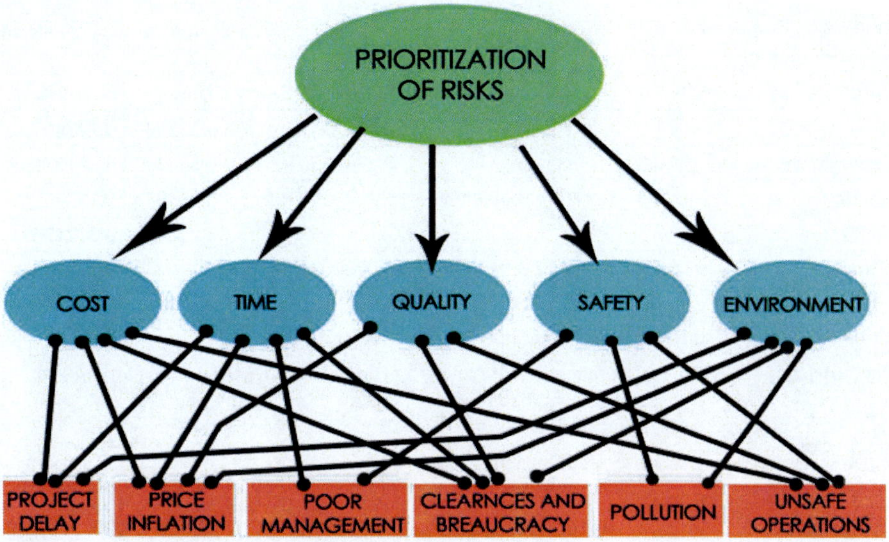

Fig. 4.6 Interrelation hierarchy for risk prioritization

Table 4.2 Saaty's fundamental comparison scale

Intensity of importance	Verbal definition	Explanation
1	Equally important	Two decision elements have equal influence on the superior decision element
3	Moderately more important	One decision element has moderately more influence than the other
5	Strongly or essentially more important	One decision element has strongly more influence than the other
7	Very strong or demonstrated importance	One decision element has very strongly more influence than the other
9	Extremely more important	One decision element has extremely has extremely more influence than the other
2, 4, 6, 8	Intermediate values of judgement	

Step 7: The limit matrix is calculated by powering the weighted super matrix to the power of $k+1$, where k is an arbitrary constant. The prioritization of the risks is shown with the goals in the limit matrix (Fig. 4.9).

Step 8: The prioritizing of risks is given in Table 4.15.

Table 4.3 Comparisons wrt "prioritization of risks involved in infrastructure project" node in "criteria" cluster

Inconsistency—0.0909	Cost	Environment	Quality	Safety	Time	Normalized
Cost	1	3	2	0.5	0.333333	0.17380704
Environment	0.333333333	1	0.333333	0.5	0.333333	0.08128535
Quality	0.5	3.000003	1	0.5	0.333333	0.1331184
Safety	2	2	2	1	2	0.31213681
Time	3.000003	3.000003	3.000003	0.5	1	0.29965239

Table 4.4 Comparisons wrt "cost" node in "risks involved" cluster

Inconsistency—0.0908	Clearances and bureaucracy	Price inflation	Project delay	Unsafe operations	Normalized
Clearances and bureaucracy	1	0.333333	2	0.333333	0.136471812
Price inflation	3.000003	1	4	0.333333	0.290150052
Project delay	0.5	0.25	1	0.333333	0.092813262
Unsafe operations	3.000003	3.000003	3.000003	1	0.480564875

Table 4.5 Comparisons wrt "environment" node in "risks involved" cluster

Inconsistency—0.0806	Clearances and bureaucracy	Pollution	Price inflation	Project delay	Normalized
Clearances and bureaucracy	1	0.333333	3	2	0.257417718
Pollution	3.000003	1	3	2	0.450860045
Price inflation	0.333333333	0.333333333	1	0.333333	0.093933913
Project delay	0.5	0.5	3.000003	1	0.197788323

Table 4.6 Comparisons wrt "quality" node in "risks involved" cluster

Inconsistency—0.0515	Poor management	Price inflation	Unsafe operations	Normalized
Poor management	1	3	0.5	0.33251588
Price inflation	0.333333333	1	0.333333	0.13964787
Unsafe operations	2	3.000003	1	0.52783624

Table 4.7 Comparisons wrt "safety" node in "risks involved" cluster

Inconsistency—0.0000	Poor management	Unsafe operations	Normalized
Poor management	1	0.333333	0.24999981
Unsafe operations	3.000003	1	0.75000019

Table 4.8 Comparisons wrt "time" node in "risks involved" cluster

Inconsistency—0.0327	Clearances and bureaucracy	Poor management	Price inflation	Project delay	Normalized
Clearances and bureaucracy	1	3	2	0.333333	0.244892437
Poor management	0.333333333	1	0.5	0.25	0.092593902
Price inflation	0.5	2	1	0.333333	0.15448751
Project delay	3.000003	4	3.000003	1	0.50802615

Table 4.9 Comparisons wrt "clearances and bureaucracy" node in "criteria" cluster

Inconsistency—0.0515	Cost	Environment	Time	Normalized
Cost	1	0.333333	0.333333	0.13964782
Environment	3.000003	1	2	0.52783621
Time	3.000003	0.5	1	0.33251597

Table 4.10 Comparisons wrt "pollution" node in "criteria" cluster

Inconsistency—0.0515	Cost	Environment	Safety	Normalized
Cost	1	0.333333	0.5	0.15705571
Environment	3.000003	1	3	0.5936338
Safety	2	0.333333333	1	0.24931049

Table 4.11 Comparisons wrt "poor management" node in "criteria" cluster

Inconsistency—0.0441	Environment	Quality	Safety	Time	Normalized
Environment	1	0.333333	0.333333	0.5	0.105715701
Quality	3.000003	1	0.333333	1	0.214797604
Safety	3.000003	3.000003	1	3	0.490361101

After applying the ANP method, we find that the unsafe operations should be given the most priority if interrelations between the risks and the criteria are involved. The

Table 4.12 Comparisons wrt "price inflation" node in "criteria" cluster

Inconsistency—0.0386	Cost	Quality	Safety	Time	Normalized
Cost	1	3	2	1	0.358115753
Quality	0.333333333	1	2	0.5	0.173313464
Safety	0.5	0.5	1	0.333333	0.122772555
Time	1	2	3.000003	1	0.345798228

Table 4.13 Comparisons wrt "project delay" node in "criteria" cluster

Inconsistency—0.0806	Cost	Quality	Safety	Time	Normalized
Cost	1	3	3	0.5	0.316824511
Quality	0.333333333	1	2	0.333333	0.150466797
Safety	0.333333333	0.5	1	0.5	0.120367337
Time	2	3.000003	2	1	0.412341355

Table 4.14 Comparisons wrt "unsafe operations" node in "criteria" cluster

Inconsistency—0.0370	Cost	Safety	Time	Normalized
Cost	1	0.333333	3	0.25828488
Safety	3.000003	1	5	0.6369857
Time	0.333333333	0.2	1	0.10472942

results would have been different if AHP model was used. ANP model has provided the priorities of risks involved in the project of Sunny Constructions ltd. as follows.

Unsafe operations	1
Project delay	2
Poor management	3
Price inflation	4
Clearances and bureaucracy	5
Pollution	6

Application 2: Application of ANP for prioritization of risks in defence organization

- We have taken a case of a typical defence organization where the aim is to develop quantitative risk assessment model for estimation and prioritization of risks involved in the activities.
- Ammunition and explosives brings with it the potential for an undesired explosion resulting in
  - personnel injuries or death
  - damage or destruction to stores, equipment and infrastructure
- To reduce the risk of an undesired event
  - identifying the risk
  - analysing it in terms of probability and consequence
  - mitigating the dangerous activities
  - having the proper authority accept and approve the risk
  - ensuring that the risk is properly monitored

Cluster Node Labels		criteria					goal	risks involved					
		cost	environment	quality	safety	time	priortization of risks involved in infra structure project	clearances and Bureaucracy	pollution	poor management	price inflation	project delay	unsafe operations
criteria	cost	0.000000	0.000000	0.000000	0.000000	0.000000	0.173807	0.139648	0.157056	0.000000	0.358116	0.316825	0.258285
	environment	0.000000	0.000000	0.000000	0.000000	0.000000	0.081285	0.527836	0.593634	0.105716	0.000000	0.000000	0.000000
	quality	0.000000	0.000000	0.000000	0.000000	0.000000	0.133118	0.000000	0.000000	0.214798	0.173313	0.150467	0.000000
	safety	0.000000	0.000000	0.000000	0.000000	0.000000	0.312137	0.000000	0.249311	0.490361	0.122773	0.120367	0.636986
	time	0.000000	0.000000	0.000000	0.000000	0.000000	0.299652	0.332516	0.000000	0.189126	0.345798	0.412341	0.104729
goal	priortization of risks involved in infra structure project	0.000000	0.000000	0.000000	0.000000	0.000000	0.000000	0.000000	0.000000	0.000000	0.000000	0.000000	0.000000
risks involved	clearances and Bureaucracy	0.136472	0.257418	0.000000	0.000000	0.244893	0.000000	0.000000	0.000000	0.000000	0.000000	0.000000	0.000000
	pollution	0.000000	0.450860	0.000000	0.000000	0.000000	0.000000	0.000000	0.000000	0.000000	0.000000	0.000000	0.000000
	poor management	0.000000	0.000000	0.332516	0.250000	0.092594	0.000000	0.000000	0.000000	0.000000	0.000000	0.000000	0.000000
	price inflation	0.290150	0.093934	0.139648	0.000000	0.154488	0.000000	0.000000	0.000000	0.000000	0.000000	0.000000	0.000000
	project delay	0.092813	0.197788	0.000000	0.000000	0.508026	0.000000	0.000000	0.000000	0.000000	0.000000	0.000000	0.000000
	unsafe operations	0.480565	0.000000	0.527836	0.750000	0.000000	0.000000	0.000000	0.000000	0.000000	0.000000	0.000000	0.000000

Fig. 4.7 Unweighted supermatrix

	Cluster Node Labels	criteria					goal	risks involved					
		cost	environment	quality	safety	time	priortization of risks involved in infra structure project	clearances and Bureaucracy	pollution	poor management	price inflation	project delay	unsafe operations
criteria	cost	0.000000	0.000000	0.000000	0.000000	0.000000	0.173807	0.139648	0.157056	0.000000	0.358116	0.316825	0.258285
	environment	0.000000	0.000000	0.000000	0.000000	0.000000	0.081285	0.527836	0.593634	0.105716	0.000000	0.000000	0.000000
	quality	0.000000	0.000000	0.000000	0.000000	0.000000	0.133118	0.000000	0.000000	0.214798	0.173313	0.150467	0.000000
	safety	0.000000	0.000000	0.000000	0.000000	0.000000	0.312137	0.000000	0.249311	0.490361	0.122773	0.120367	0.636986
	time	0.000000	0.000000	0.000000	0.000000	0.000000	0.299652	0.332516	0.000000	0.189126	0.345798	0.412341	0.104729
goal	priortization of risks involved in infra structure project	0.000000	0.000000	0.000000	0.000000	0.000000	0.000000	0.000000	0.000000	0.000000	0.000000	0.000000	0.000000
risks involved	clearances and Bureaucracy	0.136472	0.257418	0.000000	0.000000	0.244893	0.000000	0.000000	0.000000	0.000000	0.000000	0.000000	0.000000
	pollution	0.000000	0.450860	0.000000	0.000000	0.000000	0.000000	0.000000	0.000000	0.000000	0.000000	0.000000	0.000000
	poor management	0.000000	0.000000	0.332516	0.250000	0.092594	0.000000	0.000000	0.000000	0.000000	0.000000	0.000000	0.000000
	price inflation	0.290150	0.093934	0.139648	0.000000	0.154488	0.000000	0.000000	0.000000	0.000000	0.000000	0.000000	0.000000
	project delay	0.092813	0.197788	0.000000	0.000000	0.508026	0.000000	0.000000	0.000000	0.000000	0.000000	0.000000	0.000000
	unsafe operations	0.480565	0.000000	0.527836	0.750000	0.000000	0.000000	0.000000	0.000000	0.000000	0.000000	0.000000	0.000000

Fig. 4.8 Weighted supermatrix

Cluster	Node Labels	criteria: cost	criteria: environmen t	criteria: quality	criteria: safety	criteria: time	goal: priortization of risks involved in infra structure project	risks involved: clearances and Bureaucracy	risks involved: pollution	risks involved: poor management	risks involved: price inflation	risks involved: project delay	risks involved: unsafe operations
criteria	cost	0.112300	0.112300	0.112300	0.112300	0.112300	0.112300	0.112300	0.112300	0.112300	0.112300	0.112300	0.112300
	environmen t	0.048924	0.048924	0.048924	0.048924	0.048924	0.048924	0.048924	0.048924	0.048924	0.048924	0.048924	0.048924
	quality	0.036474	0.036474	0.036474	0.036474	0.036474	0.036474	0.036474	0.036474	0.036474	0.036474	0.036474	0.036474
	safety	0.197339	0.197339	0.197339	0.197339	0.197339	0.197339	0.197339	0.197339	0.197339	0.197339	0.197339	0.197339
	time	0.104963	0.104963	0.104963	0.104963	0.104963	0.104963	0.104963	0.104963	0.104963	0.104963	0.104963	0.104963
goal	prioritization of risks involved in infra structure project	0.000000	0.000000	0.000000	0.000000	0.000000	0.000000	0.000000	0.000000	0.000000	0.000000	0.000000	0.000000
risks involved	clearances and Bureaucracy	0.053624	0.053624	0.053624	0.053624	0.053624	0.053624	0.053624	0.053624	0.053624	0.053624	0.053624	0.053624
	pollution	0.022058	0.022058	0.022058	0.022058	0.022058	0.022058	0.022058	0.022058	0.022058	0.022058	0.022058	0.022058
	poor management	0.071182	0.071182	0.071182	0.071182	0.071182	0.071182	0.071182	0.071182	0.071182	0.071182	0.071182	0.071182
	price inflation	0.058488	0.058488	0.058488	0.058488	0.058488	0.058488	0.058488	0.058488	0.058488	0.058488	0.058488	0.058488
	project delay	0.073423	0.073423	0.073423	0.073423	0.073423	0.073423	0.073423	0.073423	0.073423	0.073423	0.073423	0.073423
	unsafe operations	0.221224	0.221224	0.221224	0.221224	0.221224	0.221224	0.221224	0.221224	0.221224	0.221224	0.221224	0.221224

Fig. 4.9 Limit matrix

Table 4.15 Priority and ranking of risks affecting construction company

Risks involved	Ideals	Normal	Raw	Ranking
Clearances and bureaucracy	0.242398	**0.107248**	0.053624	5
Pollution	0.099709	**0.044116**	0.022058	6
Poor management	0.321764	**0.142364**	0.071182	3
Price inflation	0.264385	**0.116977**	0.058488	4
Project delay	0.331895	**0.146846**	0.073423	2
Unsafe operations	1	**0.442449**	0.221224	1

Different risk factors like storage, transportation, ammunition assembly and preparation and temperature conditioning are taken into consideration while applying the MCDM technique (Fig. 4.10).

Step-by-step application

Human Error (H)

See Table 4.16.

 # Step 1.

First sum (add up) all the values in each column.

 # Step 2.

Next the values in each column are divided by the corresponding column sums.

 # Step 3.

Next convert fractions to decimals and find the average of each row.

Machine Failure (MF)

See Table 4.17.

Improper Handling (IH)

See Table 4.18.

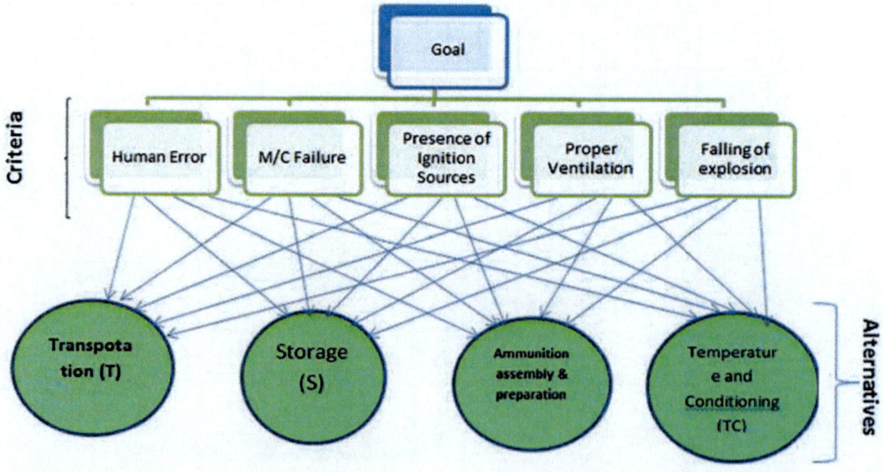

Fig. 4.10 Interdependence hierarchy for risk prioritization in defence industry

Table 4.16 Criteria weightages

	T	S	AAP	TC	Row average
T	1	5	1/6	1/4	0.146
S	1/5	1	1/7	1/2	0.074
AAP	6	7	1	2	0.522
TC	4	2	1/2	1	0.258
Sum of each column	11.2	15	1.81	3.25	
CR = 4.24%					

Table 4.17 Criteria row averages for MF

	T	S	AAP	TC	Row average
T	1	3	3	1/5	0.302
S	1/3	1	1/4	3	0.138
AAP	1/3	4	1	5	0.325
TC	5	1/3	1/5	1	0.236
CR = 15%					

Table 4.18 Criteria row averages for IH

	T	S	AAP	TC	Row Average
T	1	2	1/3	2	0.238
S	1/2	1	1/3	1/3	0.107
AAP	3	3	1	2	0.442
TC	1/2	3	1/2	1	0.213
CR = 7.6%					

Presence of Ignition Sources (PI)

See Table 4.19.

Proper Ventilation (PV)

See Table 4.20.

Falling of Explosive (FE)

Table 4.19 Criteria row averages for PI

	T	S	AAP	TC	Row average
T	1	1/4	1/5	1/4	0.266
S	4	1	7	2	0.345
AAP	5	1/7	1	3	0.218
TC	4	1/2	1/3	1	0.171
CR = 9.32%					

Table 4.20 Criteria row averages for PV

	T	S	AAP	TC	Row average
T	1	1/6	1/5	1/4	0.066
S	6	1	1/5	2	0.240
AAP	5	5	1	3	0.529
TC	4	1/2	1/3	1	0.163
CR = 12.4%					

Table 4.21 Criteria row averages for PI

	T	S	AAP	TC	Row average
T	1	1/5	4	3	0.248
S	5	1	4	4	0.432
AAP	1/4	1/4	1	2	0.143
TC	1/3	1/4	1/2	1	0.167
CR = 16.28%					

See Tables 4.21, 4.22, 4.23, 4.24.

Table 4.22 Criteria weightages

	H	M	IH	PI	PV	FE	Row average
H	1	4	2	5	2	1/3	0.245
M	1/4	1	3	1/3	2	1/4	0.088
IH	1/2	1/3	1	1/4	2	1/5	0.075
PI	1/5	3	4	1	6	4	0.275
PV	1/2	1/2	1/2	1/6	1	3	0.099
FE	3	4	5	1/4	1/3	1	0.217

Table 4.23 Alternative weightages

	HE (0.245)	ME (0.088)	IH (0.075)	PI (0.275)	PV (0.0995)	FE (0.217)	Row average
Transportation	0.147	0.302	0.238	0.266	0.066	0.248	0.213 (Rank - 3)
Storage	0.074	0.138	0.107	0.345	0.240	0.432	0.251 (Rank - 2)
AAP	0.522	0.325	0.442	0.218	0.529	0.143	**0.333 (Rank - 1)**
Temperature and conditioning	0.258	0.236	0.213	0.171	0.163	0.177	0.201 (Rank - 4)

Table 4.24 Super decision matrix

		Criteria						Alternatives			
	GOAL	HE	ME	IH	PI	PV	FE	T	S	AAP	TC
GOAL	0	0	0	0	0	0	0	0	0	0	0
HE	0	0	0	0	0	0	0	0	0	0	0
ME	0	0	0	0	0	0	0	0	0	0	0
IH	0	0	0	0	0	0	0	0	0	0	0
PI	0	0	0	0	0	0	0	0	0	0	0
PV	0	0	0	0	0	0	0	0	0	0	0
FE	0	0	0	0	0	0	0	0	0	0	0
T	0.213	0.146	0.302	0.238	0.266	0.066	0.248	1	0	0	0
S	0.251	0.074	0.138	0.107	0.345	0.24	0.432	0	1	0	0
AAP	0.333	0.522	0.325	0.442	0.218	0.529	0.143	0	0	1	0
TC	0.201	0.258	0.236	0.213	0.171	0.163	0.177	0	0	0	1

4.5 Advantages of ANP

- AHP solely uses the hierarchy relations for decision making, fails to consider the interaction among various components and does not accept rank reversal whereas ANP addresses all these issues.
- ANP helps in understanding the interrelationships, has a clear outlook and is a simple technique.
- Both quantitative and qualitative factors are accounted in ANP and hence facilitates effective decision making.
- In case of multiple stakeholders, it helps in reaching a common solution due to its structure and can also be useful as a consensus building tool.
- ANP is the ideal tool to gain deeper understanding of a specific problem and its association with related factors.

4.6 Limitation of ANP

- The main challenge faced in this method of decision making is that it becomes difficult to prioritize the elements of the network and the alternatives of the decision.
- The process involves cycles and as cycling is an infinite process, the operations needed to compute the priorities are complex.
- The presence of feedback loops makes the process of achieving desired consensus difficult.
- Verification of result due to feedback loops and interrelations, is difficult.

This chapter discussed the salient features and application of analytic network process (ANP). The next chapter will provide the details on MCDM technique, "Technique for Order Preference and Similarity to Ideal Solution" (TOPSIS). This technique identifies the closeness to ideal solution in ranking the alternatives.

Selected Bibliography and Further Reading

1. Akintola S Akintoye, M. J. (1997). Risk analysis and management in construction. *International Journal of Project Management, 15,* 31–38.
2. Dagdeviren, I. Y. (2007). Using the analytic network process(ANP) in a SWOT analysis-a case study for a textile firm. *Information Sciences, 177,* 3364–3382.
3. Lesmes, D., Castillo, M., & Zarama, R. (2009). Application of the analytic network process (Anp) to establish weights in order o re-accredit a program of a university. In *Proceedings of the International Symposium on the Analytic Hierarchy Process.*
4. Longo, G., Padoano, E., & Rosato, P. (2009). Considerations on the application of Ahp/Anp methodologies to decisions concerning a railway infrastructure. In *Proceedings of the International Symposium on the Analytic Hierarchy Process.*
5. Saaty, T. L. (2006). The Analytic Network Process.
6. Saaty, T. L. (2008). The analytic hierarchy and analytic network measurement processes: applications to decisions under risk. *European Journal of Pure and Applied Mathematics,* 122-196.
7. Zhu, Q., & Dou, Y. (2010). A portfolio-based analysis for green supplier management using the analytical network process. *Supply Chain Management: An International Journal,* 306-319.

Chapter 5
Technique for Order Preference and Similarity to Ideal Solution (TOPSIS)

5.1 Background

TOPSIS is a multi-criteria decision-making tool developed by Yoon and Hwang and has been appraised by various levels of decision makers. It takes a compensatory aggregation approach for identifying the best alternative among the identified set of alternatives. The method is based on the concept that the best alternative should have the least geometric distance from a positive-ideal solution and similarly farthest from any negative-ideal solution. The alternatives are initially graded based on their similarity with an ideal solution, which is a solution that is best in all aspects and therefore practically may not exist. The alternative that has a higher similarity to the best solution is rated higher than alternatives that have a lower similarity value. Basically, the TOPSIS approach involves identifying the distance of each alternative from the ideal solution and choosing the best alternative based on this distance. The overall methodology involves comparing a set of alternatives based on weights assigned to each criterion for evaluation, then calculating normalized values of these scores, which are then used to calculate the geometric distance from an ideal solution. The complete steps in a typical TOPSIS application are shown in Fig. 5.1.

5.2 Step-by-Step Procedure of TOPSIS

Step 1

Since different criteria will have different scales for measurement, it becomes essential to standardize the scale for each factor (Eastman et al. 1995). The most commonly used approach for balancing the scales of different criteria is the linear scale transformation approach which is a deterministic method to transform input data

J. J. Thakkar, *Multi-Criteria Decision Making*, Studies in Systems, Decision and Control 336, https://doi.org/10.1007/978-981-33-4745-8_5

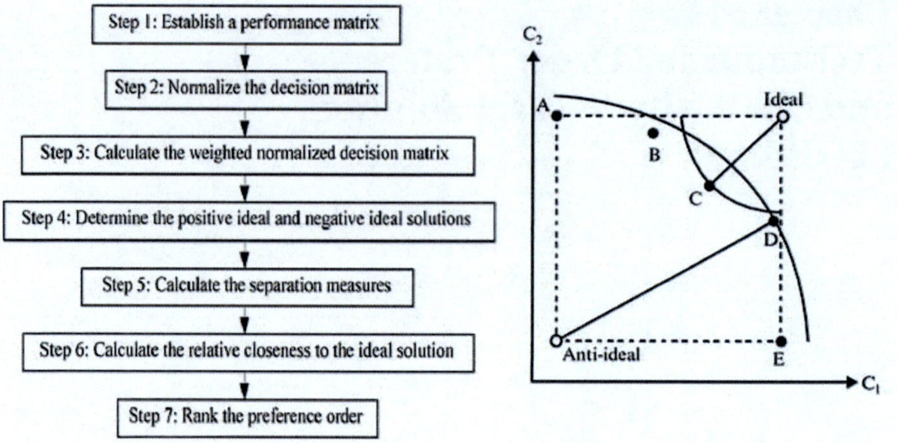

Fig. 5.1 TOPSIS procedure

(Malczewski 1999; Chakhar and Mousseau 2008). Various linear scale transformation approaches exist. The normalization method used is shown by the equation below.

$$r_{ij} = x_{ij}^2 / \sqrt{\sum_{1}^{n} x_{ij}^2}$$

Step 2

The weights are assigned to the various criteria according to the importance of them.

$$w_{ij} \quad j = 1, 2 \ldots n$$

Step 3

The weighted normalized decision matrix is obtained by calculating the product of normalized decision matrix and the associated weights. The weighted normalized value v_{ij} can be computed as below:

$$v_{ij} = w_{ij} \times r_{ij}$$

Step 4

Determine the positive-ideal solution and negative-ideal solution.

$$A^* = \{(\max\ v_{ij} | j \in J), (\min\ v_{ij} | j \in J')\}$$

$$A^- = \{(\min\ v_{ij} | j \in J), (\max\ v_{ij} | j \in J')\}$$

$J = 1, 2, 3\ldots, n$ where J is associated with the benefit criteria.
$J' = 1, 2, 3\ldots, n$ where J' is associated with the cost criteria.

Step 5

Calculate the separation measure. The separation of each alternative from the positive-ideal one is given by:

$$S_i^* = \sqrt{\sum_1^n \left(v_{ij} - v_j^*\right)^2 j = 1 \text{ where } i = 1, 2\ldots, m}$$

Similarly, the separation of each alternative from the negative-ideal one is given by:

$$S_i^- = \sqrt{\sum \left(v_{ij} - v_j^-\right)^2 j = 1 \text{ where } i = 1, 2\ldots, m}$$

Step 6

Calculate the relative closeness to the ideal solution. The relative closeness of Ai with respect to $A*$ is defined as:

$$C_i^* = S_i^- / (S_i^* + S_i^-), 0 \le Ci^* \le 1 \text{ where } i = 1, 2, \ldots, m$$

The larger the C_i^* value, the better is the performance of the alternatives.

Step 7

Arrange the options in the decreasing order of the C_i^* values to rank the alternatives.

5.3 Illustrative Application of TOPSIS for Humanitarian Disaster

Modern businesses are forced to make large number of critical business decisions within a short span of time due to their evolving challenges and problems. This short turnover in decision making requires managers to use tools and techniques that enable quick decision making while minimizing risks and maximizing the quality of these decisions. TOPSIS is one of such tools developed for decision making and applied by various practitioners. In this section, we present a numerical application of TOPSIS methodology for a real-life case situation. The case example under consideration is an organization that is involved in humanitarian operations. The objective is to identify the key issues and thus aid in improving the efficiency of relief programmes. Humanitarian organizations are generally involved in delivering aid to people in need, during disasters. Such humanitarian organizations need their supply chains to be both agile and efficient so that they are able to tackle sudden occurrences especially in rural regions or far-off places in developing countries.

We consider the example of World Food Programme that undertakes Seasonal Targeted Assistance programme that aims to provide aid in the country of Zimbabwe. Zimbabwe as a country has been affected by various calamities such as natural disasters, political instability and also economic turmoil. The country faces droughts, and thus has poor agricultural productivity. It also has one of the highest prevalence of HIV/AIDS in the world.

World Food Programme is one of the organizations that provide humanitarian services to the country through a programme called the Seasonal Targeted Assistance programme to ensure food security for the needy. It is one of the world's largest humanitarian aid organizations and the United Nations' largest frontline relief agency. WFP logistics consists of about 2000 employees all over the world and its supply chain begins at the procuring stage which includes both from private organizations or government. The cargo thus obtained is then shipped using various means to the destination locations. Its operations also include infrastructure development in the region to improve the overall transport facilities in the region. WFP faces continuous challenges in the relief delivery as it has to manage the lag between the occurrence of disasters and the availability of relief materials. The supply chain not only faces delays due to unavailability but also due to external factors such as checks and controls in the neighbouring countries and inadequate transportation. The information flow is impaired inadequate information and communication technology which results in lack of coordination across supply chain. An adequate transport infrastructure caused unprecedented delays in transportation of goods leading to increased costs and wastages.

Due to various issues and the long timeline of the programme, WFP has decided to improve its operation efficiency by focusing on the key issues. An application of TOPSIS is demonstrated for prioritizing the key issues faced by WFP.

The vital attributes that are taken into consideration are:

- Responsiveness
- Flexibility
- Cost
- Stock out
- Proper assessment and planning
- Coordination
- Sustainability
- Quality management
- Security
- Procurement
- Programme reach

The key issues faced by WFP are as follows:

1. Quantity procured
2. Procurement process
3. Transportation means
4. Skillset of volunteers
5. Donation processing time
6. Situation assessment
7. Coordination and flow of information
8. Facility location for distribution centres

The first step in TOPSIS is to obtain decision matrix. Based on the consensus of the experts, it is obtained in Table 5.1.

Step 1: The normalization is done and the following matrix is obtained (Table 5.2).

Step 2: The weights for each success factor are assigned based on a group consensus as given in Table 5.3.

Step 3: The next step is to calculate the weighted normalized matrix by multiplying with corresponding column weight. Table 5.4 reports this.

Step 4: The ideal and non-ideal solution are obtained by taking the maximum and minimum of each column, respectively (Tables 5.5, 5.6).

Step 5: S^* = The Euclidean distance of each issue for ideal solution (Table 5.7, 5.8).

Step 6: C_i^* The relative closeness for each of criteria is calculated in Table 5.9.

Step 7: Therefore, based upon the value of the C_i^* the criteria are arranged in the order of the most important to least important.

$$3 > 2 > 7 > 5 > 1 > 4 > 8 > 6$$

The results of TOPSIS algorithm indicate that the transportation issues should be handled first followed by the procurement process related issues. The result of TOPSIS algorithm is synchronous with the major problems faced by WFP in day-to-day operations, i.e. the transportation of goods and the associated time delay. For an

Table 5.1 Decision matrix

	Responsiveness	Flexibility	Cost	Stock out	Proper assessment and planning	coordination	Sustainability	Quality management	Security	Procurement	Programme reach
Quantity procured	4	3	6	9	7	6	7.5	6	6	4	8.5
Procurement process	5	6	3	6	7.5	8	5.5	8.5	5	9	6.5
Transportation means	7.5	7.5	7	6.5	6	7	6	7.5	6	7.5	9
Skillset of volunteers	6	5	1	7.5	8.5	8	7	7	7.5	5.5	7.5
Donation processing time	8	7	7	5	3	3	5	3	7	6	8.5
Situation assessment	3	4	3.5	8	9	7	8	6	5	6	7
Coordination and flow of information	8	7	5	6.5	6	8.5	4.5	4	5	5	6.5
Facility location for distribution centres	6.5	7	5	6	3	4	5.5	3	4	6.5	6

Table 5.2 Normalized matrix

0.23	0.18	0.42	0.46	0.37	0.32	0.43	0.36	0.37	0.22	0.40
0.28	0.35	0.21	0.31	0.40	0.42	0.31	0.50	0.31	0.50	0.31
0.42	0.44	0.49	0.33	0.32	0.37	0.34	0.45	0.37	0.42	0.42
0.34	0.29	0.07	0.38	0.46	0.42	0.40	0.42	0.46	0.31	0.35
0.45	0.41	0.49	0.26	0.16	0.16	0.28	0.18	0.43	0.33	0.40
0.17	0.24	0.24	0.41	0.48	0.37	0.45	0.36	0.31	0.33	0.33
0.45	0.41	0.35	0.33	0.32	0.45	0.26	0.24	0.31	0.28	0.31
0.37	0.41	0.35	0.31	0.16	0.21	0.31	0.18	0.24	0.36	0.28

Table 5.3 Weightages assigned to success factors

0.1	0.1	0.1	0.1	0.05	0.1	0.05	0.1	0.1	0.1	0.1

Table 5.4 Weighted normalized matrix

0.023	0.018	0.042	0.046	0.019	0.032	0.021	0.036	0.037	0.022	0.040
0.028	0.035	0.021	0.031	0.020	0.042	0.016	0.050	0.031	0.050	0.031
0.042	0.044	0.049	0.033	0.016	0.037	0.017	0.045	0.037	0.042	0.042
0.034	0.029	0.007	0.038	0.023	0.042	0.020	0.042	0.046	0.031	0.035
0.045	0.041	0.049	0.026	0.008	0.016	0.014	0.018	0.043	0.033	0.040
0.017	0.024	0.024	0.041	0.024	0.037	0.023	0.036	0.031	0.033	0.033
0.045	0.041	0.035	0.033	0.016	0.045	0.013	0.024	0.031	0.028	0.031
0.037	0.041	0.035	0.031	0.008	0.021	0.016	0.018	0.024	0.036	0.028

Table 5.5 Ideal solution

A* =										
0.05	0.04	0.05	0.05	0.02	0.04	0.02	0.05	0.05	0.05	0.04

Table 5.6 Non-ideal solution

A^- =										
0.017	0.018	0.007	0.026	0.008	0.016	0.013	0.018	0.024	0.022	0.028

Table 5.7 Euclidean distance for ideal solution

0.050
0.043
0.023
0.052
0.054
0.052
0.046
0.057

S^- = distance from non-ideal solution.

Table 5.8 Euclidean distance for non-ideal solution

0.052
0.058
0.072
0.052
0.061
0.043
0.057
0.044

Table 5.9 Relative closeness for each criteria

0.51
0.58
0.76
0.50
0.53
0.45
0.55
0.44

efficient humanitarian supply chain, a sustainable transportation system is essential to minimize delays in goods movement leading to food shortage. WFP should devise new transportation models which includes identifying better road network, reviving the rail network or the usage of air transport in worst cases to improve the movement of the goods and reduce the delays. WFP is also facing issues with the procurement of food materials. It is due to the various checks involved with the import of donations. WFP should evolve a mechanism to improve warehouse capacity utilization and hence supply chain responsiveness.

5.4 Advantages of TOPSIS

- One of the major advantages of TOPSIS compared with other techniques is the ability to rapidly identify the best among the alternatives.
- The calculations involved are simple and easy to understand for practicing managers.
- TOPSIS exhibits efficiency in the overall computation and presents an ability to measure the relative performance for each alternative in a simple mathematical form.

5.5 Limitations of TOPSIS

- TOPSIS is based on the assumption of monotonically increasing or decreasing the criteria.
- Since Euclidean distances are used, correlation among the attributes is not taken into consideration.
- It is difficult to determine the weights of criteria and also to ensure consistency of judgement.
- The solution is restricted to a better efficiency within the selected attributes. This is an issue in case of non-identification of key attributes.

This chapter discussed the application of TOPIS. The technique identifies the closeness of the alternatives to the ideal solution in ranking the alternatives. The next chapter will discuss the salient features and use of ELECTRE technique which works on the principle of outranking of the alternatives.

Selected Bibliography and Further Reading

1. Behzadian, M., Otaghsara, S. K., Yazdani, M., & Ignatius, J. (2012). A state-of the-art survey of TOPSIS applications. *Expert Systems with Applications, 39*(17), 13051–13069.
2. Boran, F. E., Genç, S., Kurt, M., & Akay, D. (2009). A multi-criteria intuitionistic fuzzy group decision making for supplier selection with TOPSIS method. *Expert Systems with Applications, 36*(8), 11363–11368.
3. Lai, Y. J., Liu, T. Y., & Hwang, C. L. (1994). Topsis for MODM. *European Journal of Operational Research, 76*(3), 486–500.
4. Oloruntoba, R., & Gray, R. (2006). Humanitarian aid: An agile supply chain? *Supply Chain Management: An International Journal, 11*(2), 115–120.
5. Shih, H. S., Shyur, H. J., & Lee, E. S. (2007). An extension of TOPSIS for group decision aking. *Mathematical and Computer Modelling, 45*(7–8), 801–813.
6. Tripathy, S., & Tripathy, D. K. (2016). Multi-attribute optimization of machining process parameters in powder mixed electro-discharge machining using TOPSIS and grey relational analysis. *Engineering Science and Technology: An International Journal, 19,* 62–70.
7. Van Wassenhove, L. N. (2006). Humanitarian aid logistics: Supply chain management in high gear. *Journal of the Operational Research Society, 57*(5), 475–489.

Chapter 6
Elimination Et Choice Translating Reality (ELECTRE)

6.1 Background

Elimination Et Choix Traduisant He realite (ELECTRE) is a decision-making method that was proposed by Bernard Roy and his team at SEMA consultancy company. This technique slowly evolved into ELECTRE, which has been constantly evolving even today from ELECTRE I, ELECTRE II, ELECTRE III, ELECTRE IV, ELECTRE IS, ELECTRE TRI, etc. The ELECTRE approach has been successfully applied to various applications ranging from food, health care, infrastructure management, finance, water resources management, etc. ELECTRE method is classified as the "outranking method" of decision making.

The ELECTRE approach consists of two major steps: first, outranking relations which provide the basis for pairwise comparison of various alternative course of actions available followed by the second exploitation phase that elaborates the various recommendations that are made based on the first step. ELECTRE procedure can be used for making various types of decisions such as selection, prioritization or sorting which also decides the type of recommendation made at the first stage. Generally, ELECTRE methods are applied to identify unacceptable solutions from a set of solutions.

The ELECTRE method employs the concept of ranking various alternatives based on a set of attributes or criteria. A solution or alternative is considered to exhibit dominance over other alternatives if it performs better than other alternatives in some set of criteria and at least equal to other alternatives in the remaining set of criteria. The ranking is done based on pairwise comparisons and they are ranked relative to each other. Outranking relationship between two solutions suggests that even when one of the alternatives does not dominate the other quantitatively, the decisionmaker is likely to consider one alternative as definitely better than the other.

J. J. Thakkar, *Multi-Criteria Decision Making*, Studies in Systems, Decision and Control 336, https://doi.org/10.1007/978-981-33-4745-8_6

Application of the ELECTRE method commences with comparison between each alternatives pairwise, i.e. each alternative is compared with other alternatives two at a time. The decisionmaker then expresses his/her preference between alternatives that are being considered. The decisionmaker may have a weak preference or a strong preference for one or some of the alternatives. This may be presented through either physical/monetary values or a threshold value as difference between these alternatives. There may be scenario where the decisionmaker may not be able to present any preference relations. This binary relation between various alternatives is called outranking relations. If the outranking relation between all alternatives is available, it is said to be complete. As some of these outranking relations may not be available, it can be incomplete.

Next step involves assigning weights for different factors or criteria against which all the solutions are evaluated. These weights signify the order of importance within the set of criteria. The procedure results in a set of binary outranking relations relative to each set of alternatives based on the various criteria considered. As the outranking relations may not be complete, ELECTRE method does not always identify the best alternative. Usually, the process terminates with a set of prioritized alternatives. Application of some other MCDM approach may be possible to identify the best alternative among these. Thus, ELECTRE provides an approach to get a reduced list of alternatives by elimination of the least important alternatives. A typical procedure of ELECTRE is illustrated in Fig. 6.1.

6.2 Step-by-Step Procedure of ELECTRE

Step 1: Normalizing the Decision Matrix

The first step usually involves transformation of various units of different factors in the decision criteria into a dimensionless quantity that is readily comparable using the Formula given below:

$$xij = \frac{aij}{\sqrt{\sum_{i=1}^{M} aij^2}}$$

The normalized matrix X of the dimensionless values is defined as follows:

$$X = \begin{Bmatrix} x11 & x12 \cdots & x1N \\ x21 & x22 \cdots & x2N \\ xM1 & xM2 \cdots & xMN \end{Bmatrix}$$

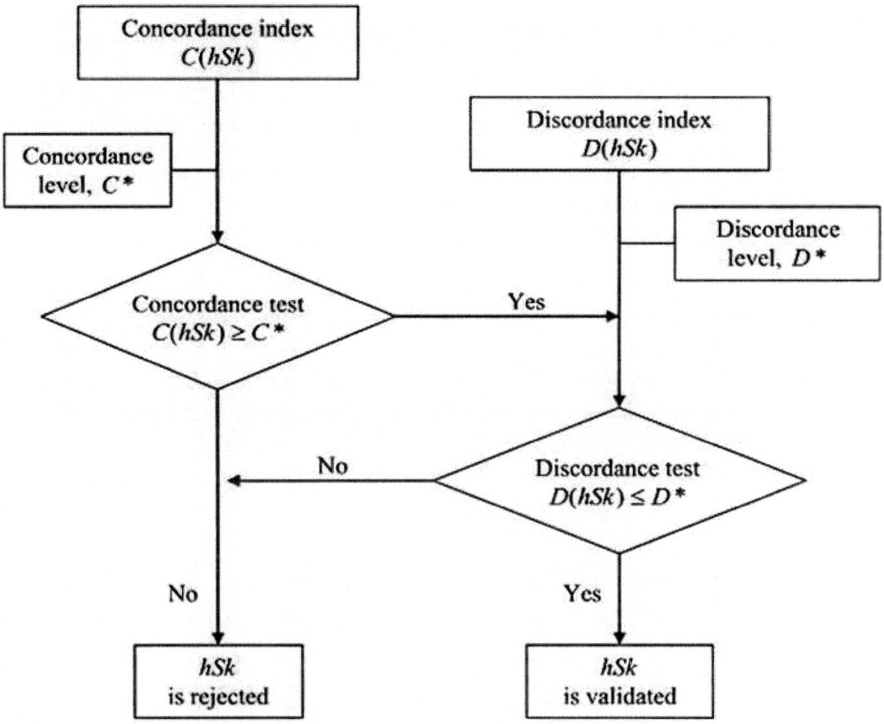

Fig. 6.1 ELECTRE process

where M is the number of alternatives, N is the number of criteria and xij is the new and dimensionless preference measure of the i-th alternative in terms of the j-th criterion.

Step 2: Weighting the Normalized Decision Matrix

The values from the previous step are weighted by multiplying the matrix X with the weights associated with each factor/criteria.

The weighted matrix Y is given by $Y = XW$, where

$$Y = \begin{Bmatrix} y11 & y12 \cdots & y1N \\ y21 & y22 \cdots & y2N \\ yM1 & yM2 \cdots & yMN \end{Bmatrix} = \begin{Bmatrix} w1y11 & w2y12 \cdots & wNy1N \\ w1y21 & w2y22 \cdots & wNy2N \\ w1yM1 & w2yM2 \cdots & wNyMN \end{Bmatrix}$$

and $W = \begin{Bmatrix} y11 & y12 \cdots & y1N \\ y21 & y22 \cdots & y2N \\ yM1 & yM2 \cdots & yMN \end{Bmatrix}$ Also, $\sum_{i=1}^{M} wi = 1.$

Step 3: Determine the Concordance and Discordance Sets

The concordance set *Ckl* of two alternatives *Ak* and *Al*, where *M.k*, $l \geq 1$, is defined as the set of all criteria for which *Ak* is preferred to *Al*. That is, the following is true.

Ckl = {*j, such that: ykj* ≥ *ylj*}, for $j = 1, 2, 3, \ldots, N$. The complementary subset is called the discordance set and it is described as follows:
Dkl = {*j, such that: ykj* < *ylj*}, for $j = 1, 2, 3\ldots., N$.

Step 4: Construct the Concordance and Discordance Matrices

Concordance index represents the relative value of all the elements in concordance matrix *C*. The concordance index *ckl* is the sum of the weights associated with the criteria contained in the concordance. That is, the following is true:

$$Ckl = \sum\nolimits_{j \in Ckl} Wj \text{ for } j'1, 2, 3, \ldots, N.$$

The concordance index indicates the relative importance of alternative *Ak* with respect to alternative *Al*. Apparently, 0 < *ckl* > 1. Therefore, the concordance matrix *C* is defined as follows:

$$\mathrm{C} = \begin{Bmatrix} - & C12 \cdots & C1M \\ C21 & - \cdots & C2M \\ CM1 & CM2 \cdots & - \end{Bmatrix}$$

It should be noted here that the entries of matrix *C* are not defined when $k = l$.

The discordance matrix *D* expresses the degree that a certain alternative *Ak* is worse than a competing alternative *Al*. The elements *Dkl* of the discordance matrix are defined as follows:

$$Dkl = \frac{\text{Max}|ykj - ylj|j \in Dkl}{\text{Max}|ykj - ylj|}$$

The discordance matrix is defined as follows:

$$D = \begin{Bmatrix} - & d12 \cdots & d1M \\ d21 & - \cdots & d2M \\ dM1 & dM2 \cdots & - \end{Bmatrix}$$

As before, the entries of matrix *D* are not defined when $k = l$.

It should also be noted here that the previous two $M \times M$ matrices are not symmetric.

Step 5: Determine the Concordance and Discordance Dominance Matrices

The concordance dominance matrix is constructed by means of a threshold value for the concordance index. For example, *Ak* will only have a chance to dominate *Al* if its corresponding concordance index *ckl* exceeds at least ascertain threshold value *c*. That is, the following is true:

$$ckl > c$$

The threshold value *c* can be determined as the average concordance index. That is, the following relation is true:

$$\underline{c} = \frac{1}{M(M-1)} \sum^{M}_{\substack{k=1 \\ \text{and } k \neq l}} \sum^{M}_{\substack{l=1 \\ \text{and } l \neq k}} ckl$$

Based on the threshold value, the concordance dominance matrix *F* is determined as follows:

$$fkl = 1, \text{if } ckl \geq c,$$
$$fkl = 0, \text{if } ckl < c.$$

Similarly, the discordance dominance matrix *G* is defined by using a threshold value *d*, where *d* is defined as follows:

$$\underline{\mathrm{d}} = \frac{1}{M(M-1)} \sum^{M}_{\substack{k=1 \\ \text{and} k \neq l}} \sum^{M}_{\substack{l=1 \\ \text{and} l \neq k}} \mathrm{d}kl$$

6.3 Steps of Fuzzy ELECTRE

Step 1 First step begins with a formulation of panel of decision makers (DMs) with expertise in the respective domain. Suppose there are k decisionmakers (i.e., $D_1, D_2, \ldots, D_k$) who are entrusted with the task of ranking (y_{jk}) of each criterion (i.e. $C_1, C_2, \ldots, C_n$) in an increasing order. The weights for each criterion is defined as fuzzy triangular numbers $w_j = (l_j, m_j, u_j)$ for $K = 1, 2, \ldots, k$ and $j = 1, 2, \ldots, n$. The aggregated fuzzy weight of the data set can be determined as follows

$$l_j = \min_k\{y_{jk}\} \quad m_j = \frac{1}{k}\sum_{k=1}^{k} y_{jk}, \quad u_j = \max_k\{y_{jk}\}$$

Then, the aggregated fuzzy importance weight for each criterion is normalized as:

$$\tilde{w}_j = \left(w_{j1}, w_{j2}, w_{j3}\right),$$

where

$$w_{j1} = \frac{1/l_j}{\sum_{j=1}^{n} 1/l_j} \quad w_{j2} = \frac{1/m_j}{\sum_{j=1}^{n} 1/m_j}$$

$$w_{j3} = \frac{1/u_j}{\sum_{j=1}^{n} 1/u_j}$$

The normalized aggregated fuzzy weight matrix which is represented by W is calculated as shown above $W = [w_1, w_2, w_n]$.

Step 2 A decision matrix is formed as:

$$X = \begin{bmatrix} X_{11} & X_{12} & \cdots & X_{1n} \\ X_{21} & X_{22} & \cdots & X_{21} \\ \cdots & \cdots & \cdots & \cdots \\ X_{m1} & X_{m2} & \cdots & X_{mn} \end{bmatrix}$$

Step 3 The decision matrix thus obtained is normalized. The normalized decision matrix after forming the decision matrix, normalization is applied. Then, the normalized decision matrix is obtained as below:

$$X = \begin{bmatrix} r_{11} & r_{12} & \cdots & r_{1n} \\ r_{21} & r_{22} & \cdots & r_{21} \\ \cdots & \cdots & \cdots & \cdots \\ r_{m1} & r_{m2} & \cdots & r_{mn} \end{bmatrix}$$

Step 4 The weighted normalized decision matrix is then obtained by taking the product of the weights of each factor and the values of normalized decision matrix. The weighted normalized decision matrix V for each criterion is defined as:

$$V = \left[v_{ij}\right]_{mxn} \text{for } i = 1, 2, \ldots, m \text{ and}$$
$$\text{j} = 1, 2, \ldots, n, \text{ where } v_{ii} = r_i \times w_i.$$

$$V^1 = \begin{bmatrix} v_{11}^1 & v_{12}^1 & \cdots & v_{1n}^1 \\ v_{21}^1 & v_{22}^1 & \cdots & v_{2n}^1 \\ \cdots & \cdots & \cdots & \cdots \\ v_{m1}^1 & v_{m2}^1 & \cdots & v_{mn}^1 \end{bmatrix},$$

$$V^2 = \begin{bmatrix} v_{11}^2 & v_{12}^2 & \cdots & v_{1n}^2 \\ v_{21}^2 & v_{22}^2 & \cdots & v_{2n}^2 \\ \cdots & \cdots & \cdots & \cdots \\ v_{m1}^2 & v_{m2}^2 & \cdots & v_{mn}^2 \end{bmatrix}$$

$$V^3 = \begin{bmatrix} v_{11}^3 & v_{12}^3 & \cdots & v_{1n}^3 \\ v_{21}^3 & v_{22}^3 & \cdots & v_{2n}^3 \\ \cdots & \cdots & \cdots & \cdots \\ v_{m1}^3 & v_{m2}^3 & \cdots & v_{mn}^3 \end{bmatrix}.$$

Here, v_{ij} denotes normalized positive triangular fuzzy numbers.

Step 5 Next, the concordance and discordance indices are computed for varying weights of each criterion. The concordance index acts show how reliable the pairwise judgements are which is calculated using the Formula:

$$C_{pq}^1 = \sum_{j^*} w_{j1},\ C_{pq}^2 = \sum_{j^*} w_{j2,},\ C_{pq}^2 = \sum_{j^*} w_{j3},$$

where $J*$ are the attributes contained in the concordance set $C(p, q)$.

Step 6 The discordance index, on the other hand, represents the degree of disagreement in $(Ap \rightarrow Aq)$ and is defined as

$$D_{pq}^1 = \frac{\sum_{j+} \left| v_{pj+}^1 - v_{qj+}^1 \right|}{\sum_j \left| v_{pj}^1 - v_{qj}^1 \right|}$$

$$D_{pq}^2 = \frac{\sum_{j+} \left| v_{pj+}^2 - v_{qj+}^2 \right|}{\sum_j \left| v_{pj}^2 - v_{qj}^2 \right|}$$

$$D_{pq}^3 = \frac{\sum_{j+} \left| v_{pj+}^3 - v_{qj+}^3 \right|}{\sum_j \left| v_{pj}^3 - v_{qj}^3 \right|}$$

where J^+ is the attribute contained in the discordance set $D(p, q)$, and v_{ij} is the weighted normalized evaluation of the alternative i on the criterion j.

Step 7 The final concordance and discordance indexes are computed using the following Formula:

$$C^*_{pq} = \sqrt[z]{\prod_{z=1}^{Z} C^z_{pq}}$$

$$D^*_{pq} = \sqrt[2]{\prod_{z=1}^{Z} D^z_{pq}} \text{ where,} \quad Z = 3.$$

Above formulas are part of the defuzzification procedure in the fuzzy ELECTRE method. Higher concordance index C_{pq} represents dominance relationship of alternative A_p over the alternative A_q while a smaller final discordance index D_pq. The outranking relation is obtained by applying the following equation procedure to gain the kernel as the subset of the best alternatives:

$$\text{If } C(p, q) \geq C \text{ and } D(p, q) \geq \text{D}$$

where C and D are the averages of C_{pq} and D_{pq}, respectively.

6.4 Illustrative Applications of ELECTRE

Application 1: Selection of the best location for construction of a new hospital

M&T Ltd. decided to establish a hospital in Osaka, Japan. The company formulated a panel of five experts consisting the people with diverse background in engineering, project management, social and political science to decide about the most suitable location for constructing the hospital. The experts conducted surveys on five different locations (L1 to L5) in Osaka where construction possibilities exist. The committee identified nine most important criteria (C1 to C9) as listed in Table 6.1 for prioritizing the locations. The panel decided to use ELECTRE for the purpose.

Criteria For Project Selection (Worst–Best)

See Table 6.1.

Decision matrix after using a numerical scale for intangibles

See Table 6.2.

Normalized Decision Matrix

Obtain the normalized decision matrix, R, using the relationship

$$rij = xij/\text{sq root(sum, } i = 1 \ldots m \text{ of } xij2)$$

See Table 6.3.

Table 6.1 Project selection criteria and weightages

Criteria	Weightage
C1 = General land use conformity (political restrictions, by laws, people, etc.)	10–27
C2 = Site development potential (parcel shape and geometry, parking potential, expansion scenarios, etc.)	30–40
C3 = Community relationship (service catchment area, provisions for allied services long-term cares, pharmacy, office, site amenities trails, park, restaurants, shopping, etc.)	20–32
C4 = Educational and research or supportive institutes	10–30
C5 = Accessibility (visibility, proximity to EMS/police/patient transfer sites/disaster preparedness, roadways)	15–35
C6 = Site conditions (topography, drainage, heritage and environment features—rivers/lakes/streams, vegetation, protected wetlands, etc.)	10–30
C7 = Servicing (established or potential, redundant services for electrical and water requirement)	10–30
C8 = Microclimate (wind, noise, air quality)	5–24
C9 = Utilities and other costs (rents, parking cost, electricity, water, etc.) (in crore of rupees)	15–37

(# Point given out of 100 by a panel of five experts).

Table 6.2 Decision matrix

	C1	C2	C3	C4	C5	C6	C7	C8	C9
L1	25	33	24	13	22	17	20	16	21
L2	20	27	29	21	20	15	24	19	30
L3	26	19	25	24	18	11	19	9	22
L4	17	37	22	23	21	14	19	17	23
L5	24	25	19	16	20	16	21	15	27

Table 6.3 Normalized matrix

	C1	C2	C3	C4	C5	C6	C7	C8	C9
L1	0.493	0.510	0.446	0.292	0.486	0.515	0.432	0.464	0.378
L2	0.394	0.411	0.539	0.473	0.441	0.454	0.518	0.551	0.540
L3	0.513	0.289	0.465	0.540	0.397	0.333	0.410	0.261	0.396
L4	0.335	0.564	0.409	0.518	0.463	0.424	0.410	0.493	0.414
L5	0.473	0.381	0.353	0.360	0.441	0.485	0.454	0.435	0.486

Weighted Decision Matrix

Obtain the weighted decision matrix V (Table 6.4) by multiplying each column of R by the corresponding weight.

Table 6.4 Weighted normalized matrix

	C1	C2	C3	C4	C5	C6	C7	C8	C9
L1	0.0986	0.0510	0.044	0.029	0.0486	0.051	0.043	0.046	0.378
L2	0.0788	0.0411	0.053	0.047	0.0441	0.045	0.051	0.0551	0.540
L3	0.1026	0.0289	0.046	0.054	0.0397	0.033	0.0410	0.026	0.396
L4	0.067	0.0564	0.040	0.051	0.0463	0.042	0.041	0.0493	0.414
L5	0.0946	0.0381	0.035	0.036	0.0441	0.048	0.045	0.043	0.486

(LI, L2, L3, L4 and L5 are locations).

Table 6.5 Concordance and discordance sets

	1	2	3	4	5	6	7	8	9	C	D
C12	1	1			1	1				(1, 2, 5, 6)	(3, 4, 7, 8, 9)
C13		1				1	1	1		(2, 6, 7, 8)	(1, 3, 4, 5, 9)
C14					1	1	1			(5, 6, 7)	(1, 2, 3, 4, 8, 9)
C15	1	1	1		1	1		1		(1, 2, 3, 5, 6, 8)	(4, 7, 9)
C23	1	1	1		1	1	1	1	1	(1, 2, 3, 5, 6, 7, 8, 9)	(4)
C24	1		1		1		1	1	1	(1, 3, 5, 7, 8, 9)	(2, 4, 6)
C25		1	1	1	1					(2, 3, 4, 5)	(1, 6, 7, 8, 9)
C34			1	1			1			(3, 4, 7)	(1, 2, 5, 6, 8, 9)
C35			1	1						(3, 4, 7)	(1, 2, 5, 6, 8, 9)
C45		1	1	1				1		(2, 3, 4, 8)	(1, 5, 6, 7)

$$W = \begin{pmatrix} 0.2 & 0.1 & 0.1 & 0.1 & 0.1 & 0.1 & 0.1 & 0.1 & 0.1 \end{pmatrix}$$

Concordance and Discordance Sets

See Table 6.5.

Calculate the *concordance matrix, C*

–	0.5	0.3	0.3	0.7
0.7	–	1.0	0.8	0.4
0.7	0.1	–	0.3	0.3
0.9	0.3	0.9	–	0.4
0.3	0.7	0.8	0.5	–

e.g $C12 = \Sigma Wj$

$j \; \varepsilon \; C12 = 1 + 2 + 5 + 6 = 0.5$

Calculate the discordance matrix, *D*

$$\max I v1j - \text{v}2I$$

$$\text{d}12 = j\varepsilon D12$$

$$\max\ v1j - v2j$$

–	0.32	0.48	0.105	0.542
1	–	1	1	1
1	0.512	–	0.421	0.34
1	0.461	0.522	–	1
1	0.571	0.362	0.42	–

Application 2: Prioritization of Banking websites

State Bank of India (SBI) is an Indian multinational, public sector banking and financial services organization. It is one of the largest financial institution in the country with headquarters in Mumbai, Maharashtra. The company is government owned and had net worth of more than INR 20,000 billion. SBI operates branches all over the country and has in total more than 15,000 branches, with some of them (more than 200 branches) in overseas locations across 36 different countries. SBI is the 43rd largest bank in the world and ranked 221st in the Fortune Global 500 list of the world's biggest corporations of 2020.

Different Types of Services the bank provides include:

Individual banking:

- Account enquiries
- Savings accounts/current accounts
- Provision of cards, i.e. debit and credit cards
- Insurance*
- Wealth management

Business banking: Different types of business banking services include:

- Business loans
- Checking accounts
- Savings accounts
- Debit and credit cards
- Merchant services (credit card processing, reconciliation, check collection)
- Cash management (payroll services, deposit services, etc.)

Digital banking: Banks will typically offer digital banking services that include:

- Mobile banking
- Deposit of mobile cheques
- Text alerts for important messages or transactions
- Obtaining e-statements
- Bill and due payments

Loans: Common types of loans that banks provide include:

- Personal loans
- Home equity loans
- Home equity lines of credit
- Home loans
- Business loans

Customer Segment

Geographic—Tier-1, Tier-2, Tier-3, Tier-4, rural and semi-urban branches.

Age—10–59, 60 and above.

Income—Marginal, low, middle, high.

Occupation—Unskilled workers, skilled workers, shop owners, businessmen.

SWOT analysis of SBI

SBI has its roots since 1806 which was later transformed under various names; finally, SBI was established after the act in parliament on May 1955.

Strengths

- SBI is the largest bank in India in terms of market share, revenue and assets.
- It has more than 13,000 outlets and 25,000 ATM centre.
- The bank has its presence in 32 countries engaging currency trade all over the world
- SBI being a government organization and also being one of the oldest, it possesses a first mover advantage in the banking sector.
- SBI has constantly tried to keep up with the modern technology and has recently revised its vision and mission statements prioritizing development in the information and communication technology.

Weakness

- SBI lacks existing infrastructure in terms of technology-driven services especially in comparison with private banks.
- Being a government organization, employees show aversion to improvement initatives.
- The bank spends large amount of money on rented buildings and compensation to its employees.

Opportunities

- SBI has been merged with five banks, namely: State Bank of Hyderabad, State bank of Patiala, State bank of Bikaner and Jaipur, State of bank of Travancore and State bank of Mysore which provides new business avenues.
- There exists plan to expand its operations in overseas markets due to increased inflow from outside India, especially Asian markets.

Threats

- Declining net profit.
- Reducing market share.
- Competition from private banks like HDFC, AXIS bank, etc.

Case example

This section demonstrates an application of ELECTRE for "PRIORITIZATION OF BANKING WEBSITES"

The six bank alteratives (B1 to B6) and nine criteria (C1 to C9) are considered for demonstrating an application of ELECTRE.

1. AXIS (B1)
2. SBI (B2)
3. HDFC (B3)
4. PNB (B4)
5. ICICI (B5)
6. BOI (B6)

The following nine criteria (C1 to C9) are considered with their respective weightages (W1 to W9).

1. Security (C1, W1 = 10%)
2. Accounts Services (Deposit and Retail) (C2, W2 = 15%)
3. Statement (Cards and ATMs) (C3, W3 = 10%)
4. Payment Transfer (C4, W4 = 10%)
5. Personal and Corporate Finance (MSME) (C5, W5 = 15%)
6. Investment and Business Solution (Tax) (C6, W6 = 10%)
7. Capital Market Service (Trade and Forex) (C7, W7 = 10%)
8. Collection Solutions (C8, W8 = 10%)
9. Customer Care (C9, W9 = 10%)

Application of MCDM Technique

SL. NO.	CODE	CRITICAL CRITERIA	DEFINITION
1	C1	SECURITY	ABILITY TO PROVIDE FOR SECURE AND SAFE TRANSACTION
2	C2	ACCOUNTS SERVICES (DEPOSIT & RETAIL)	ABILITY TO PROVIDE FOR FAST AND RELIABLE SERVICES FOR A/C ACTIVITIES
3	C3	STATEMENT (CARDS & ATMs)	ABILITY TO PROVIDE FOR USER FRIENDLY & HIGHLY ACCESSIBLE CARD SERVICES
4	C4	PAYMENT TRANSFER	ABILITY TO PROVIDE FOR EASY & RELIABLE VALUE TRANSFER TO SELF & OTHERS
5	C5	PERSONAL & CORPORATE FINANCE (MSME)	ABILITY TO PROVIDE FOR FINANCE OF DIFFERENT STRUCTURE TO DIFFERENT CUSTOMER
6	C6	INVESTMENT & BUSINESS SOLUTION(TAX)	ABILITY TO PROVIDE FOR SECURE & RELIABLE ENVIRON FOR INVESTMENT & BUSINESS
7	C7	CAPITAL MARKET SERVICE(TRADE & FOREX)	ABILITY TO FUND DIFFERENT ECONOMIC ACTIVITIES
8	C8	COLLECTION SOLUTIONS	ABILITY TO PROVIDE FOR SECURE & RELIABLE ENVIRON FOR COLLECTION SOLUTIONS
9	C9	CUSTOMER CARE	ADAPTABILITY TO PROVIDE BACK UP SUPPORT STRUCTURE FOR CUSTOMERS

Fig. 6.2 Critical criteria and their code (Formulation step 1)

Fig. 6.3 Bank names and their respective code (Formulation step 2)

SL.NO.	CODE	BANK NAME
1	B1	AXIS
2	B2	SBI
3	B3	HDFC
4	B4	PNB
5	B5	ICICI
6	B6	BOI

Application of ELECTRE Technique

The procedural steps of ELECTRE as discussed in the previous section are applied for prioritization of banking websites (see Figs. 6.2, 6.3, 6.4, 6.5, 6.6, 6.7, 6.8, 6.9, 6.10, 6.11, 6.12, 6.13, 6.14, 6.15, 6.16 and 6.17) .

SL. NO.	CODE	CRITICAL CRITERIA	WEIGHTAGE(%)	WEIGHTAGE(DECIMAL)
1	C1	SECURITY	10%	0.10
2	C2	ACCOUNTS SERVICES (DEPOSIT & RETAIL)	15%	0.15
3	C3	STATEMENT (CARDS & ATMs)	10%	0.10
4	C4	PAYMENT TRANSFER	10%	0.10
5	C5	PERSONAL & CORPORATE FINANCE (MSME)	15%	0.15
6	C6	INVESTMENT & BUSINESS SOLUTION(TAX)	10%	0.10
7	C7	CAPITAL MARKET SERVICE(TRADE & FOREX)	10%	0.10
8	C8	COLLECTION SOLUTIONS	10%	0.10
9	C9	CUSTOMER CARE	10%	0.10
			100%	1.00

Fig. 6.4 Critical criteria and their respective weightage (Formulation step 3)

Fig. 6.5 Scales for intangibles for cost and benefit attributes (Formulation step 4)

SCALE FOR INTANGIBLES			
COST ATTRIBUTES			BENEFIT ATTRIBUTES
VERY HIGH	1	1	VERY LOW
	2	2	
HIGH	3	3	LOW
	4	4	
AVERAGE	5	5	AVERAGE
	6	6	
LOW	7	7	HIGH
	8	8	
VERY LOW	9	9	VERY HIGH

Option B5 Most dominating (Analysis step 12)

If we discuss the result obtained from the point of view of a project manager, then it reveals the immense importance of the ELECTRE method of multi-criteria decision making. We have obtained option B5 or ICICI Bank as the most favourable option or rather we should say that we have eliminated other five options as their dominance is not as strong as option B5 or ICICI Bank.

Through this ELECTRE method it could have been the case that we may not be able to get a single most favourable option but we are lucky to obtain a single case as a most favourable option.

	C1	C2	C3	C4	C5	C6	C7	C8	C9
B1	HIGH	AVERAGE	HIGH	VERY HIGH	LOW	AVERAGE	HIGH	VERY HIGH	AVERAGE
B2	AVERAGE	HIGH	LOW	AVERAGE	HIGH	HIGH	HIGH	HIGH	HIGH
B3	VERY HIGH	VERY LOW	HIGH	VERY HIGH	LOW	AVERAGE	HIGH	HIGH	HIGH
B4	VERY LOW	AVERAGE	VERY HIGH	HIGH	AVERAGE	AVERAGE	VERY HIGH	VERY HIGH	VERY HIGH
B5	HIGH	LOW	AVERAGE	HIGH	HIGH	HIGH	AVERAGE	AVERAGE	VERY HIGH
B6	AVERAGE	AVERAGE	VERY LOW	LOW	AVERAGE	LOW	AVERAGE	AVERAGE	AVERAGE

Fig. 6.6 Basic decision matrix (Analysis step 1)

	C1	C2	C3	C4	C5	C6	C7	C8	C9
B1	7	5	7	8.5	3	5.5	7	9	5
B2	5.5	7	2	4	3.5	7	3	7.5	7
B3	8	3	6.5	9	8	5	3.5	7	8
B4	1	6	9	6.5	5	6	8	8.5	8.5
B5	8.5	3.5	5	7	2.5	8	5.5	6	9
B6	5	5.5	1	3	6	9	5	5	4.5

Fig. 6.7 Scaled decision matrix (Analysis step 2)

	C1	C2	C3	C4	C5	C6	C7	C8	C9
B1	0.4504	0.3934	0.4922	0.5178	0.2429	0.3256	0.5085	0.5035	0.2833
B2	0.3539	0.5508	0.1406	0.2437	0.2834	0.4145	0.2179	0.4196	0.3966
B3	0.5148	0.2361	0.4571	0.5482	0.6478	0.2960	0.2543	0.3916	0.4533
B4	0.0643	0.4721	0.6328	0.3959	0.4049	0.3553	0.5811	0.4755	0.4816
B5	0.5470	0.2754	0.3516	0.4264	0.2024	0.4737	0.3995	0.3357	0.5099
B6	0.3217	0.4328	0.0703	0.1827	0.4859	0.5329	0.3632	0.2797	0.2550

Fig. 6.8 Normalized decision matrix (Analysis step 3)

WEIGHT	0.10	0.15	0.10	0.10	0.15	0.10	0.10	0.10	0.10
	C1	C2	C3	C4	C5	C6	C7	C8	C9
B1	0.0450	0.0590	0.0492	0.0518	0.0364	0.0326	0.0509	0.0504	0.0283
B2	0.0354	0.0826	0.0141	0.0244	0.0425	0.0414	0.0218	0.0420	0.0397
B3	0.0515	0.0354	0.0457	0.0548	0.0972	0.0296	0.0254	0.0392	0.0453
B4	0.0064	0.0708	0.0633	0.0396	0.0607	0.0355	0.0581	0.0476	0.0482
B5	0.0547	0.0413	0.0352	0.0426	0.0304	0.0474	0.0400	0.0336	0.0510
B6	0.0322	0.0649	0.0070	0.0183	0.0729	0.0533	0.0363	0.0280	0.0255

	HIGHEST
	LOWEST

Fig. 6.9 Weighted decision matrix (Analysis step 4)

C12	{1,3,4,5,8}	D12	{2,6,7,9}
C13	{2,3,5,6,8}	D13	{1,4,7,9}
C14	{1,4,5,7,8}	D14	{2,3,6,9}
C15	{2,3,4,8}	D15	{1,5,6,7,9}
C16	{1,3,4,5,8,9}	D16	{2,6,7}
C21	{2,6,7,9}	D21	{1,3,4,5,8}
C23	{2,5,6,7,8}	D23	{1,3,4,9}
C24	{1,2,5,6,7}	D24	{3,4,8,9}
C25	{2,7,8}	D25	{1,3,4,5,6,9}
C26	{1,2,3,4,7,8,9}	D26	{5,6}
C31	{1,4,7,9}	D31	{2,3,5,6,8}
C32	{1,3,4,9}	D32	{2,5,6,7,8}
C34	{1,4,7}	D34	{2,3,5,6,8,9}
C35	{3,4,7,8}	D35	{1,2,5,6,9}
C36	{1,3,4,7,8,9}	D36	{2,5,6}
C41	{2,3,6,9}	D41	{1,4,5,7,8}
C42	{3,4,8,9}	D42	{1,2,5,6,7}
C43	{2,3,5,6,8,9}	D43	{1,4,7}
C45	{2,3,8}	D45	{1,4,5,6,7,9}
C46	{2,3,4,5,8,9}	D46	{1,6,7}
C51	{1,5,6,7,9}	D51	{2,3,4,8}
C52	{1,3,4,5,6,9}	D52	{2,7,8}
C53	{1,2,5,6,9}	D53	{3,4,7,8}
C54	{1,4,5,6,7,9}	D54	{2,3,8}
C56	{1,3,4,5,8,9}	D56	{2,6,7}
C61	{2,6,7}	D61	{1,3,4,5,8,9}
C62	{5,6}	D62	{1,2,3,4,7,8,9}
C63	{2,5,6}	D63	{1,3,4,7,8,9}
C64	{1,6,7}	D64	{2,3,4,5,8,9}
C65	{2,6,7}	D65	{1,3,4,5,8,9}

Fig. 6.10 Determination of concordance and discordance sets (Analysis step 5)

	1	2	3	4	5	6
1		0.55	0.60	0.55	0.45	0.65
2	0.45		0.60	0.60	0.35	0.75
3	0.40	0.40		0.30	0.40	0.60
4	0.45	0.40	0.70		0.35	0.70
5	0.55	0.55	0.60	0.65		0.65
6	0.35	0.25	0.40	0.30	0.35	

Fig. 6.11 Determination of concordance matrix c (Analysis step 6)

	1	2	3	4	5	6	7	8	9
B2-B1	0.0097	0.0236	0.0352	0.0274	0.0061	0.0089	0.0291	0.0084	0.0113
B3-B1	0.0064	0.0236	0.0035	0.0030	0.0607	0.0030	0.0254	0.0112	0.0170
B4-B1	0.0386	0.0118	0.0141	0.0122	0.0243	0.0030	0.0073	0.0028	0.0198
B5-B1	0.0097	0.0177	0.0141	0.0091	0.0061	0.0148	0.0109	0.0168	0.0227
B6-B1	0.0129	0.0059	0.0422	0.0335	0.0364	0.0207	0.0145	0.0224	0.0028
B3-B2	0.0161	0.0472	0.0316	0.0305	0.0547	0.0118	0.0036	0.0028	0.0057
B4-B2	0.0290	0.0118	0.0492	0.0152	0.0182	0.0059	0.0363	0.0056	0.0085
B5-B2	0.0193	0.0413	0.0211	0.0183	0.0121	0.0059	0.0182	0.0084	0.0113
B6-B2	0.0032	0.0177	0.0070	0.0061	0.0304	0.0118	0.0145	0.0140	0.0142
B4-B3	0.0450	0.0354	0.0176	0.0152	0.0364	0.0059	0.0327	0.0084	0.0028
B5-B3	0.0032	0.0059	0.0105	0.0122	0.0668	0.0178	0.0145	0.0056	0.0057
B6-B3	0.0193	0.0295	0.0387	0.0365	0.0243	0.0237	0.0109	0.0112	0.0198
B5-B4	0.0483	0.0295	0.0281	0.0030	0.0304	0.0118	0.0182	0.0140	0.0028
B6-B4	0.0257	0.0059	0.0563	0.0213	0.0121	0.0178	0.0218	0.0196	0.0227
B6-B5	0.0225	0.0236	0.0281	0.0244	0.0425	0.0059	0.0036	0.0056	0.0255

Fig. 6.12 Calculation for discordance matrix d (Analysis step 7)

	1	2	3	4	5	6
1		0.8265	0.4186	0.5136	1.0000	0.4912
2	1.0000		0.5789	1.0000	0.5106	1.0000
3	1.0000	1.0000		0.8090	1.0000	0.7630
4	0.6292	0.7379	1.0000		1.0000	0.4576
5	0.7812	1.0000	0.2175	0.6114		0.5553
6	1.0000	0.5830	1.0000	1.0000	1.0000	

Fig. 6.13 Determination of discordance matrix d (Analysis step 8)

c= ΣΣckl / 30 = 14.9/30= 0.4966
ENTRIES > 0.4966 TREATED AS 1
ENTRIES < 0.4966 TREATED AS 0

	1	2	3	4	5	6
1		1	1	1	0	1
2	0		1	1	0	1
3	0	0		0	0	1
4	0	0	1		0	1
5	1	1	1	1		1
6	0	0	0	0	0	

Fig. 6.14 Determination of concordance dominance matrix (Analysis step 9)

d= ΣΣdkl / 30 = 23.48/30= 0.4966
ENTRIES < 0.7826 TREATED AS 1
ENTRIES > 0.7826 TREATED AS 0

	1	2	3	4	5	6
1		0	1	1	0	1
2	0		1	0	1	0
3	0	0		0	0	1
4	1	1	0		0	1
5	1	0	1	1		1
6	0	1	0	0	0	

Fig. 6.15 Determination of discordance dominance matrix (Analysis step 10)

	1	2	3	4	5	6	SUM
1		0	1	1	0	1	3
2	0		1	0	0	0	1
3	0	0		0	0	1	1
4	0	0	0		0	1	1
5	1	0	1	1		1	4
6	0	0	0	0	0		0

Fig. 6.16 Determination of aggregate dominance matrix (Analysis step 11)

AXIS		PRIORITY 2	
SBI		PRIORITY 3	
HDFC		PRIORITY 3	
PNB		PRIORITY 3	
ICICI		PRIORITY 1	MOST FAVOURABLE OPTION
BOI		PRIORITY 4	

Fig. 6.17 Final result—elimination of less favourable alternatives

Beneficiary

Mainly through this process of decision making one who is a willing customer for the banks can take his or her decision in more effective manner whereas in the previous situation many factors of decision were there which results into indecision for him or her.

Possible Consequences

In this particular case, there is only one most favourable option, rest options are eliminated as those were less favourable compared to it.

Prediction of Future Events

If we do an analytical comparison between the intangible scale of the criteria of ICICI bank with other banks then we can find that the dominance and as well as aggregate matrix has followed a particular pattern as in other case also.

From it we can draw the inference that if any other bank which has certain criterions in the same sector as ICICI bank and it is more efficient in banking website sector depending on the chosen criteria then it will be the most favourable option or vice versa, though at the same time there is an option also that if certain alternate criteria are reversed then the result may be reversed or we can find both ICICI and any other bank as dominating ones from the result of ELECTRE method.

6.5 Advantages of ELECTRE

- ELECTRE methods are relevant when decision criteria are more than three in the model.
- This process accommodates the qualitative nature of the criteria.
- This outranking method has the ability to take into account the ordinal scales without converting the original scale into abstract one with imposed arbitrary range.
- This method could be applied when a strong heterogeneity related to the nature of evaluation exists in criteria.

- When compensation is not possible for a given criteria of loss by a gain on another, a non-compensatory aggregation procedure requires, ELECTRE method works well in this situation.
- When small differences in evaluation are not significant but accumulation of several small differences are considerable, then discrimination thresholds are required, and that needs to form a comprehensive intransitive indifference binary relation. ELECTRE method is very much helpful for this type of decision-making problem.

6.6 Limitations of ELECTRE

- ELECTRE is a complex decision-making method compared with some of the MCDM techniques and also requires significant amount of primary data.
- ELECTRE method has a difficulty that it needs precise measurements of performance ratings and criteria weights.
- Scoring or assigning a number to an action is very fragile, so it may mislead to get the real result.
- The property of independence with respect to irrelevant alternatives can be violated at the time of outranking.
- If there is a preference for transitivity, ELECTRE method does not always fulfil the transitivity requirements.
- In ELECTRE method, thresholds may not be well defined in terms of its physical or psychological interpretation.

This chapter has discussed the use of ELECTRE method which works on the principle of outranking the criteria in evaluating the alternatives. The method works on the key principle of identifying concordance and discordance sets for elimination of the select alternatives. The next chapter will discuss another outranking method called Preference Ranking Organization Method for Enrichment Evaluations (PROMETHEE).

Selected Bibliography and Further Reading

1. Saracoglu, B. O. (2015). An experimental research of small hydropower plant investments selection in turkey by carrot2, DEXi. *DEXiTree. Journal of Investment and Management, 4*(1), 47–60.
2. Mojahed, M., Marjani, E.M., Afsari, A. (2013). Using ELECTRE-AHP as a mixed method for personnel selection. *Proceedings of the International Symposium on the Analytic Hierarchy Process.*
3. Capaldo, G., & Zollo, G. (2001). Applying fuzzy logic to staff assessment: A case study. *Omega; The International Journal of Management Science, 29,* 585–597.
4. Botti, L., Peypoch, N. (2013). Multi-criteria electre method and destination competitiveness, *Elsevier(6), 108–113*

5. Velasquez, M., & Hester, P. (2013). an analysis of multi-criteria decision making methods, *International Journal of Operations Research*. (*10*) *2*, 56–66.
6. Marbini, A., Tavan, M. (2011). An extension of the Electre I method for group decision-making under a fuzzy environment. *Elsevier (39), 373–386*
7. Ros, J. C. (2011). Introduction to decision deck-diviz: Examples and user guide. Retrieved from www.decision-deck.org
8. Papadopoulos, M. (2011). Overview and selection of multi-criteria evaluation methods for mitigation/adaptation policy instruments. Retrieved from www.promitheasnet.kepa.uoa.gr.

Chapter 7
Preference Ranking Organization Method for Enrichment Evaluations (PROMETHEE)

7.1 Background

Multi-criteria decision making (MCDM) has been one of the highly fertile ground for research in the domain of operational research (OR). Especially, in the last few decades, there has been rapid growth with various new methods for decision making and also improvements to existing methods being proposed continuously. MCDM problems usually deal with selection, sorting or prioritizing alternatives among a set of possible solutions.

The major goal in these approaches is to identify the best solution that is better than the other solutions and has the least amount of compromise. A good decision-making tool must not only consider various external criteria but also give due importance to the preferences of the decision maker.

The PROMETHEE method (Preference Ranking Organization Method for Enrichment Evaluations) proposed by Brans in 1982 has been further extended by Vincke and Brans (1985). It is one of the recent among MCDA methods. PROMETHEE is based on an outranking approach that ranks and selects one or a set of alternative actions among all possible alternative solutions while considering various set of criteria which are often conflicting in nature.

The method is used to evaluate the alternatives based on the pairwise comparison of the preference function values for a set of predefined criteria. This method tries to identify the best alternative based on ranking of alternatives using preference functions. The major features that enable application of the method are simplicity, clarity and balance. The PROMETHEE method can be applied for both partial ranking and complete ranking of the set of alternatives. The procedure of PROMETHEE is explained in Fig. 7.1.

J. J. Thakkar, *Multi-Criteria Decision Making*, Studies in Systems, Decision and Control 336, https://doi.org/10.1007/978-981-33-4745-8_7

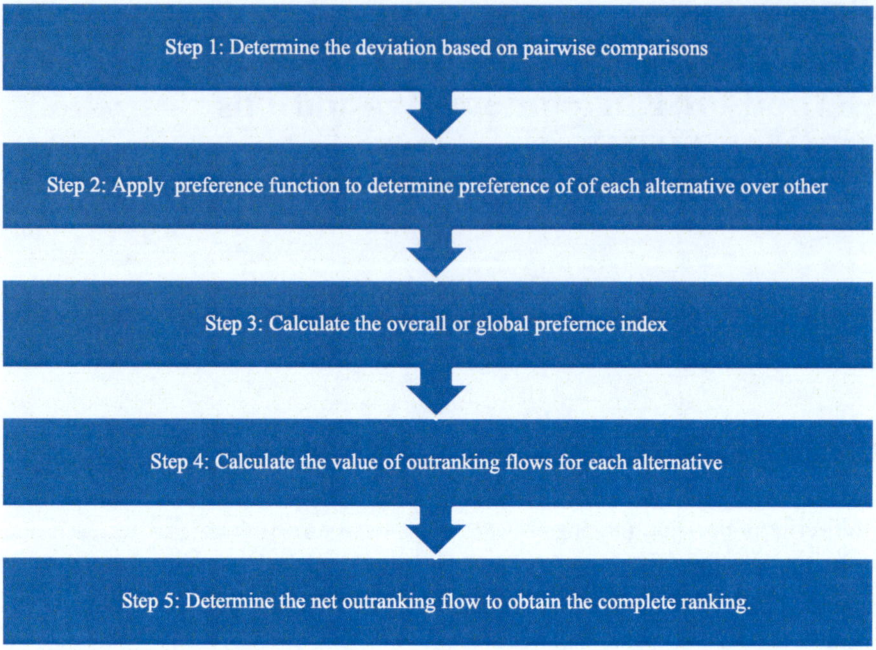

Fig. 7.1 Procedure of PROMETHEE

7.2 Steps in PROMETHEE

The PROMETHEE method is based on pairwise comparison of each alternative with other alternatives for the predefined set of criteria. There are two variants of PROMETHEE generally used. PROMETHEE I is used for partial ordering while complete ordering is done through PROMETHEE II, i.e. it provides ranking for all the alternatives. The procedure discussed in this section is PROMETHEE II which is used to obtain complete ranking of alternatives.

The procedural steps involved in PROMETHEE II method are enlisted as below:

Step 1: Decision matrix is normalized using the following equation:

$$R_{ij} = \left[x_{ij} - \min\ x_{ij}\right] / \left[\max\ x_{ij} - \min\ x_{ij}\right] \quad (i = 1, 2, \ldots, n,\ j = 1, 2, \ldots, m) \tag{7.1}$$

where X_{ij} is the performance measure of ith alternative with respect to jth criterion.

Equation (7.1) in case of non-beneficial criteria can be modified as below:

$$R_{ij} = \left[\max\ x_{ij} - x_{ij}\right] / \left[\max\ x_{ij} - \min\ x_{ij}\right] \tag{7.2}$$

Step 2: Pairwise calculation of differences is done for all alternatives based on their criteria values to obtain the evaluative differences for every alternative in comparison with other alternatives.

Step 3: Calculate the preference function, $P_j(i, i')$.

While as per Brans [3] there exist six types of preference functions, it is usually difficult for the decision maker to ascertain which preference function is suitable for each factor. A simplified preference function is given as below:

$$P_j\left(i, i'\right) = 0 \quad \text{if } R_{ij} R_{i'j} \tag{7.3}$$

$$P_j\left(i, i'\right) = \left(R_{ij} - R_{i'j}\right) \quad \text{if } R_{ij} R_{i'j} \tag{7.4}$$

Step 4: Calculate the aggregated preference function considering the weights assigned to each criteria.

$$\text{Aggregated preference function,} \quad \pi(i, i') = \left[\sum_{(j=1)}^{m} w_j \times P_j(i, i')\right] / \sum_{(j=1)}^{m} w_j \tag{7.5}$$

where w_j is the relative importance (weight) of *j*th criterion.

Step 5: Calculation of outranking flows/PROMETHEE I partial ranking

$$\varphi^{+}(i) = \frac{1}{n-1} \sum_{i'=1}^{n} \pi\left(i', i\right) \quad (i \neq i') \tag{7.6}$$

$$\varphi^{-}(i) = \frac{1}{n-1} \sum_{i'=1}^{n} \pi\left(i', i\right) \quad (i \neq i') \tag{7.7}$$

where "*n*" is the number of alternatives.

Step 6: Calculate the net outranking flow/PROMETHEE II complete ranking

$$\varphi^{+}(i) = \varphi^{+}(i) - \varphi^{-}(i) \tag{7.8}$$

Step 7: Values of $\varphi(i)$ are then used to rank all the alternatives. Higher value of $\varphi(i)$ indicates that the alternative is better than an alternative with lower $\varphi(i)$ value. The best alternative is thus the one which has the highest value of $\varphi(i)$.

7.3 Illustrative Application of PROMETHEE for the Selection of e-Retail Alternatives

In everyday life, a consumer encounters a type of confusion when he makes online purchase using a computer or a mobile phone. Customer generally gets in to a dilemma for choosing which websites will fulfil their requirements for buying a specific product. In this section, an application of PROMETHEE is demonstrated to select the best online retail website from **Flipkart**, **Amazon**, **Shopclues**, **Paytm** and **eBay**.

The situation considered here includes 10 online retail selection criteria and 5 alternative online retail websites. Table 7.1 examines five retail websites (alternatives (A)) for 10 criteria (C). The objective values for these criteria are assigned on 11-point scale, as given in Table 7.2. The 10 selection criteria considered here are variety (V), quality of product (Q), offers (O), search (S—how efficiently user able to search what he wants), delivery time (DT), service area (SA), product information (PI), payment option (PO), advertisement (AD), scalability (SC—perform efficiently through traffic peaks and valleys). Among the considered criteria, DT is a non-beneficial attribute while the rest of the attributes are beneficial.

The subjective data represented in Table 7.1 are converted into crisp scores using 11-point scale (as given in Table 7.2). Using Eqs. (7.1) or (7.2), the crisp values represented in Table 7.3 are then normalized (see Table 7.4).

We assumed the criteria weights for the considered criteria as given in Table 7.5.

Now, the preference functions are calculated for all the pairs of alternatives, using Eqs. (7.3) and (7.4), and are given in Table 7.5. Table 7.6 presents the aggregated preference function values for all the paired alternatives, as calculated using Eq. (7.5). The outranking flows for different location alternatives are now computed using Eqs. (7.6) and (7.7), respectively, and are shown in Table 7.7.

The net outranking flow values for different alternative websites and their relative rankings are given in Table 7.8. The best choice of online retail website for a customer is Amazon followed by Flipkart then eBay Table 7.9.

The result can be further improved if we combine PROMETHEE process with other MCDM techniques. In this case, we have considered the weights given to criteria by assumption but for more accuracy we can first use AHP process to find out the weights for different criteria and then solve the rest problem by PROMETHEE. Implementing GAIA method will further increase the accuracy of the result.

Above example shows how PROMETHEE II method can be used effectively in solving a real-life problem of online retail website selection. It also demonstrates how the results can help the decisionmaker to make better decisions while choosing the best alternative among all the available alternatives. The PROMETHEE method as can be seen is an easy and simple but effective method. The method as demonstrated above can be applied to any strategic problems where a decision related to choice of alternatives has to be made.

Table 7.1 Information for online retails alternatives

Alternatives (A)	Criteria (C)									
	Variety	Quality	Offer	Search	Delivery time	Service area	Product Info	Payment options	Advertisement	Scalability
Flipkart	H	EH	L	VH	VL	L	VH	L	H	VH
Amazon	AA	VH	AA	H	EL	A	EH	EH	VH	H
Shopclues	A	EL	H	BA	H	VH	L	H	L	VL
Paytm	A	AA	VH	A	A	A	AA	A	AA	H
eBay	VH	A	A	AA	VH	BA	H	VH	BA	A

Table 7.2 Point 11 fuzzy scale

Linguistic term	Crisp score
Exceptionally low	0.045
Extremely low	0.135
Very low	0.255
Low	0.335
Below average	0.410
Average	0.500
Above average	0.590
High	0.665
Very high	0.745
Extremely high	0.865
Exceptionally high	0.955

7.4 Advantages of PROMETHEE

- The interactive PROMETHEE method can be applied to multi-criteria decisions that involve both qualitative and quantitative data.
- Major advantage of PROMETHEE over methods like ELECTRE, AHP, etc., is the ability to evaluate alternatives which are actually difficult to compare due to some existing trade-off relation between evaluation standards as non-comparable alternatives.
- It also has an advantage over AHP that if new alternatives are added or removed, the pairwise comparison process need not be repeated again.

7.5 Limitations of PROMETHEE

- One of the major limitations with respect to PROMETHEE is that there is no formal guideline for selecting of criteria weights making an inherent assumption that the decision maker would be able to choose appropriate weights.
- PROMETHEE has issues with reversals in the ranks of alternatives when new alternatives are considered.

This chapter has discussed the use of outranking MCDM method—PROMETHEE. The next chapter will discuss the use of VIKOR which determines compromise solution with non-commensurable and conflicting criteria which can enable the decisionmaker to draw final conclusions about the problem.

Table 7.3 Objective data for online retail website selection problem

A	C									
	Variety	Quality	Offers	Search	Delivery time	Service area	Product Info	Payment options	Advertisement	Scalability
F	0.665	0.865	0.335	0.745	0.225	0.335	0.745	0.335	0.665	0.745
A	0.590	0.745	0.590	0.665	0.135	0.500	0.865	0.865	0.745	0.665
S	0.500	0.135	0.665	0.410	0.665	0.745	0.335	0.665	0.335	0.255
P	0.410	0.590	0.745	0.500	0.500	0.590	0.590	0.500	0.590	0.590
E	0.745	0.500	0.500	0.590	0.745	0.410	0.665	0.745	0.410	0.500

Table 7.4 Normalized decision matrix

C \ A	V	Q	O	S	DT	SA	PI	PO	AD	SC
F	0.761	1	0	1	0.852	0	0.773	0	0.804	1
A	0.537	0.835	0.621	0.761	1	0.402	1	1	1	0.836
S	0.268	0	0.804	0	0.131	1	0	0.622	0	0
P	0	0.623	1	0.268	0.401	0.621	0.481	0.311	0.621	0.683
E	1	0.5	0.402	0.537	0	0.182	0.622	0.773	0.182	0.5

Sample Calculation:
F-V: $R_{ij} = [x_{ij} - \min x_{ij}]/[\max x_{ij} - \min x_{ij}] = (0.665\text{-}0.410)/(0.745\text{-}0.410) = 0.761$
P-DT: $R_{ij} = [\max x_{ij} - x_{ij}]/[\max x_{ij} - \min x_{ij}] = (0.745\text{-}0.500)/(0.745\text{-}0.135) = 0.401$

Table 7.5 Criteria weights

Criteria	Q	DT	V	O	S	PI	PO	SA	AD	SC
Weights	0.2764	0.1668	0.1526	0.1267	0.0883	0.0706	0.0584	0.0258	0.0172	0.0172

Table 7.6 Preference functions for all the pairs of alternatives

C \ A	V	Q	O	S	DT	SA	PI	PO	AD	SC
F,A	0.224	0.165	0	0.239	0	0	0	0	0	0.164
F,S	0.493	1	0	1	0.721	0	0.773	0	0.804	1
F,P	0.761	0.337	0	0.732	0.451	0	0.292	0	0.183	0.317
F,E	0	0.5	0	0.463	0.852	0	0.151	0	0.618	0.5
A,F	0	0	0.621	0	0.148	0.402	0.227	1	0.196	0
A,S	0.269	0.835	0	0.761	0.869	0	1	0.378	1	0.836
A,P	0.537	0.212	0	0.493	0.599	0	0.519	0.689	0.379	0.153
A,E	0	0.335	0.219	0.224	1	0.220	0.378	0.227	0.818	0.336
S,F	0	0	0.804	0	0	1	0	0.622	0	0
S,A	0	0	0.183	0	0	0.598	0	0	0	0
S,P	0.268	0	0	0	0	0.379	0	0.311	0	0
S,E	0	0	0	0	0.131	0.818	0	0	0	0
P,F	0	0	1	0	0	0.621	0	0.311	0	0
P,A	0	0	0.379	0	0	0.219	0	0	0	0
P,S	0	0.623	0.196	0.268	0.270	0	0.481	0	0.621	0.683
P,E	0	0.123	0.598	0	0.401	0.439	0	0	0.439	0.183
E,F	0.239	0	0.402	0	0	0.182	0	0.773	0	0
E,A	0.463	0	0	0	0	0	0	0	0	0
E,S	0.732	0.5	0	0.537	0	0	0.622	0.150	0.182	0.5
E,P	1	0	0	0.269	0	0	0.141	0.462	0	0

Sample Calculation:
(F,A)-V: $P_j(i,i') = (R_{ij} - R_{i'j}) = (0.761\text{-}0.537) = 0.224$
(F,A)-AD: $P_j(i,i') = 0$ if $R_{ij} \leq R_{i'j}$

Table 7.7 Aggregated preference function

Value of Aggregate Preference Function for Pair "SP"
(0.268*0.1526)+(0.379*0.0258)+(0.311*0.0584) = 0.0688

Online retail	Flipkart	Amazon	Shopclues	Paytm	EBay
Flipkart	-	0.1037	0.6458	0.3783	0.3511
Amazon	0.1915	-	0.6082	0.3700	0.3723
Shopclues	0.1640	0.0386	-	0.0688	0.0429
Paytm	0.1608	0.0536	0.3221	-	0.1986
EBay	0.1372	0.0706	0.3617	0.2132	-

Table 7.8 Leaving and entering flows for different locations

Online retail	$\varphi^+(i)$	$\varphi^-(i)$
Flipkart	0.3697	0.1634
Amazon	0.3855	0.0666
Shopclues	0.0785	0.4844
Paytm	0.1838	0.2575
EBay	0.1956	0.2413

Sample Calculation:
$\varphi^+(i)$ for Flipkart:
(0.1037+0.6458+0.3783+0.3511)/4 = 0.3697
$\varphi^-(i)$ for Flipkart:
(0.1915+0.1640+0.1608+0.1372)/4 = 0.1634

Table 7.9 Net outranking flow values for different location alternatives

Online retail	Net outranking flow $[\varphi^+(i) - \varphi^-(i)]$	*Rank*
Flipkart	0.2063	2
Amazon	0.3189	1
Shopclues	-0.4059	5
Paytm	-0.0737	4
EBay	-0.0457	3

Selected Bibliography and Further Reading

1 Kabir, G., & Sumi, R. S. (2014). Integrating fuzzy analytic hierarchy process with PROMETHEE method for total quality management consultant selection.
2 Tomić, V., Marinković, Z., & Janošević, D. (2011). PROMETHEE method implementation with multi-criteria decisions.
3. BRANS, J. P., & Vincke, Ph. (1985). A preference ranking organisation method.
4. Behzadian, M., Kazemzadeh, R. B., Albadvi, A., Aghdasi, M. (2010). PROMETHEE: A comprehensive literature review on methodologies and applications.
5. Athawale, V. M., & Chakraborty, S. (2010). Facility location selection using PROMETHEE II method.

Chapter 8
VIseKriterijumska Optimizacija I Kompromisno Resenje (VIKOR)

8.1 Background

The VIKOR multi-criteria decision-making technique was initially proposed by S. Opricovic in 1979 which was followed by its first application in 1980. It was named VIKOR in 1990 from the Serbian: VIseKriterijumska Optimizacija I Kompromisno Resenje which stands for Multi-criteria Optimization and Compromise Solution. It is a powerful tool that can be used for various strategic decision-making problems in various environments such as social, economic or environmental. VIKOR approach consists of identifying various alternatives for a problem, establishing the priority among them and ranking them and selection of the best compromise solution based on the ranking.

VIKOR helps identify the best solution in terms of trade-off among alternatives based on criteria that are conflicting in nature and also difficult to quantify. Its major focus is on ranking of all the alternative solutions based on specified criteria so that the best solution with the least amount of compromise can be determined. For example, in any organization while undertaking new projects it is necessary to choose among competing projects for various conflicting criteria such as cost, time horizon, projected returns, etc. The procedural steps of VIKOR are presented in Fig. 8.1.

In many real life situations involving conflicting criteria like cost and quality, it is difficult to identify a solution that is better than all other alternatives. This leads to Pareto-optimal solutions. Pareto-optimal solutions are solutions where improving one attribute of the solution leads to degradation in other attributes. In such cases, it is necessary for the decision maker to perform a holistic analysis to identify the solution that dominates others and offers the best overall trade-off.

J. J. Thakkar, *Multi-Criteria Decision Making*, Studies in Systems, Decision and Control 336, https://doi.org/10.1007/978-981-33-4745-8_8

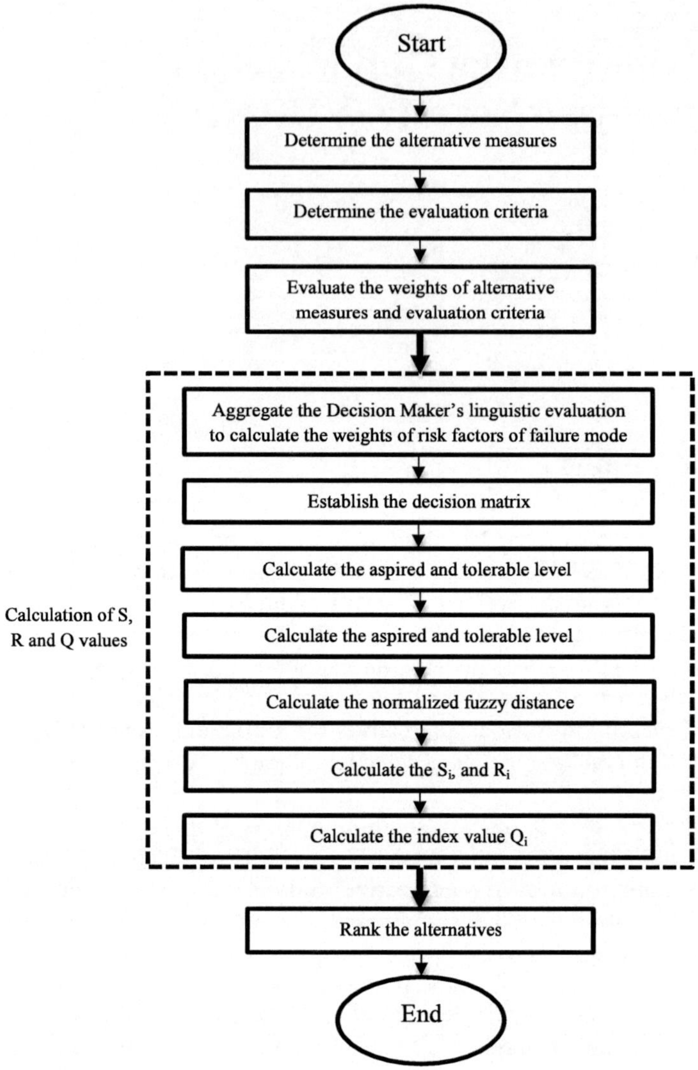

Fig. 8.1 VIKOR procedure

8.2 Step-by-Step Procedure of VIKOR

The procedure for VIKOR involves evaluating each alternative based on each of the criterion. The ranking process is done based on the measure of regret which involves calculating how far away the alternative is from an ideal solution. Thus, the closeness to the best possible solution is used as the measure for ranking these alternatives. Let $a_1, a_2, \ldots a_J$ represent the set of j alternatives that are to be considered with the value of the ith criteria for the alternative a_j is denoted by f_{ji}. Ranking is done based on the measure L_p—norm as given below.

$$L_{p,j} = \left\{ \sum_{i=1}^{n} [w_i (f_i^* - f_{ij})/(f_i^* - f_i^-)]^p \right\}^{1/p}, \quad 1 \le p \le \infty; \; j = 1, 2, \ldots, J$$

The procedure of ranking is described as follows:

- Determine the best f_i^* and the worst f_i^- values of all criterion functions, where $i = 1, 2, \ldots n$.

$$f_i^* = \max_j f_{ji}, \quad f_i^- = \min_j f_{ji}$$

- Compute the S_j (the maximum group utility) and R_j (the minimum individual regret of the opponent) values as follows where $j = 1, 2, \ldots J$

$$S_j = L_{1,j} = \sum_{i=1}^{n} w_i (f_i^* - f_{ji})/(f_i^* - f_i^-)$$

$$R_j = L_{\infty,j} = \max_j \left[\sum_{i=1}^{n} w_i (f_i^* - f_{ji})/(f_i^* - f_i^-) \right]$$

where w_i is the weight of the ith criterion which expresses the relative importance of criteria.

- Compute the value of Q_j as follows where $j = 1, 2, \ldots., J$

$$Q_j = v(S_j - S^*)/(S^- - S^*) + (1 - v)(R_j - R^*)/(R^- - R^*)$$

where $S^* = \min_j S_j, \quad S^- = \max_j S_j, \quad R^* = \min_j R_j, \quad R^- = \max_j R_j$

and v is introduced as weight of the strategy of $S_j$and R_j

- Rank the alternatives, sorting it by the S, R and Q values, in decreasing order. The results are three ranking lists.
- Alternative (a') is proposed as a compromise solution which is ranked the best by the minimum Q if the following two conditions are satisfied:

C1. "Acceptable advantage":

$Q(a'') - Q(a') \geq DQ$, where a'' is the alternative with second position in the ranking list by Q, $Q = 1/(J - 1)$ and J is the number of alternatives.

C2. "Acceptable stability in decision making":

Alternative a' must also be the best ranked by S or/and R. This compromise solution is stable within a decision-making process, which could be: "voting by majority rule" (when $v > 0.5$ is needed), or "by consensus" $v \approx 0.5$, or "with vote" ($v < 0.5$). Here, v is the weight of the decision-making strategy "the majority of criteria" (or "the maximum group utility"). In this paper, $v = 0.5$ is used in numerical examples.

If one of the conditions is not satisfied, then a set of compromise solutions is proposed, which consists of.

- Alternative a' and a'' if only condition C2 is not satisfied, or
- Alternative a', a'', …, $a^{(M)}$ if condition C1 is not satisfied; and $a^{(M)}$ is determined by the relation $(Q(a^{(M)}) - Q(a')) < DQ$ for maximum M (the positions of these alternatives are in closeness).

VIKOR was developed to enhance decision making for complex systems. This methodology tries to obtain a compromise ranking of alternatives for a set of conflicting criteria. The process calculates an index which measures the closeness of each solution to an ideal solution to determine the ranking of the alternatives as presented in Fig. 8.2.

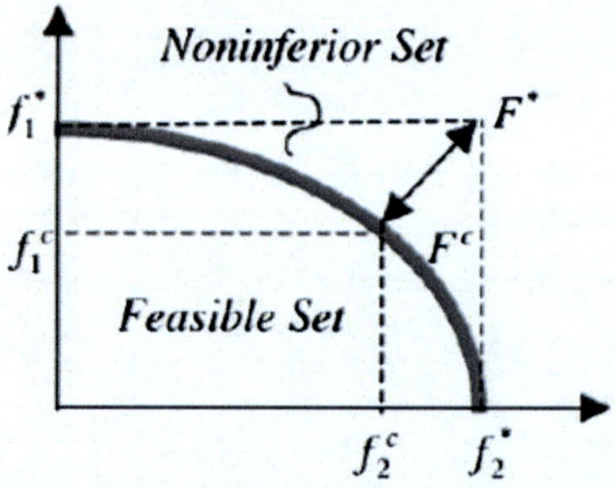

Fig. 8.2 Ideal and compromise solution

8.3 Illustrative Application of VIKOR

8.3.1 Application 1: Selection of Construction Project

Recent years have seen a rapid increase in the number of infrastructure construction projects. This demands evaluating the potential projects for possible returns. One of the major challenges in project management is to ensure an optimal utilization of various resources such as knowledge, expertise, tools and techniques to satisfy the stakeholders' needs. This demands consideration of following issues while evaluating the various project alternatives.

- Projected completion time
- Overall cost and quality
- Stakeholder expectations.

Construction project selection is a multi-dimension problem having conflicting criteria with varied level of significance in an organization.

The example considered here evaluates six construction projects for a set of criteria listed in Table 8.1.

The decision matrix in Table 8.2 is developed by a panel of senior executives involved in the evaluation of the projects. The weights for criteria (in Table 8.3) are chosen based on the opinions of the decision makers. The best fi* and the worst fi– values of all criterion functions are presented in Table 8.4. Subsequently, Si and Ri values for criteria are calculated as shown in Table 8.5.

Table 8.1 Criteria and their computation units

Criterion index	Criterion description	Computation units
C_1	Raw material supply	Score (1–10)
C_2	Proximity to customers	Score (1–10)
C_3	Proximity to supplier	Distance (km)
C_4	Proximity to schools, hospitals	Distance (km)
C_5	Transportation cost	Score (1–10)
C_6	Availability of skill labour	Score (1–10)
C_7	Labour cost	Currency (Thousand Lakhs)
C_8	Schedule	No. of years
C_9	Government facilities	Score (1–10)
C_{10}	Construction cost (investment cost)	Score (1–10)

Table 8.2 Decision matrix

	C_1	C_2	C_3	C_4	C_5	C_6	C_7	C_8	C_9	C_{10}
Project 1	5	3	19.7	8.7	1	3	1500	1	1	7
Project 2	8	5	101	96.4	5	9	800	6	10	1
Project 3	7	2	43.7	23.6	2	7	1200	2	4	6
Project 4	6	1	13.6	24.8	4	10	1200	3	2	5
Project 5	10	7	7.6	12.8	2	9	800	1	7	9

Table 8.3 Weights assigned to criteria

Criteria	C_1	C_2	C_3	C_4	C_5	C_6	C_7	C_8	C_9	C_{10}
Weights	0.112	0.028	0.019	0.024	0.090	0.266	0.155	0.063	0.051	0.193

Table 8.4 The best f_i^* and the worst f_i^- values of all criterion functions

Criteria	Effect	f_i^* (best value)	f_i^- (worst value)
C_1	+	10	5
C_2	–	1	7
C_3	–	7.6	101
C_4	–	8.7	96.4
C_5	–	1	5
C_6	+	10	3
C_7	–	800	1500
C_8	–	1	6
C_9	+	10	1
C_{10}	–	1	9

Table 8.5 Calculation of S_i and R_i for criteria

	C_1	C_2	C_3	C_4	C_5	C_6	C_7	C_8	C_9	C_{10}
Project 1	0.112	0.009	0.003	0.000	0.000	0.266	0.155	0.000	0.051	0.145
Project 2	0.045	0.019	0.019	0.024	0.090	0.038	0.000	0.063	0.000	0.000
Project 3	0.067	0.005	0.008	0.004	0.023	0.114	0.088	0.013	0.034	0.121
Project 4	0.089	0.000	0.001	0.004	0.068	0.000	0.088	0.025	0.045	0.097
Project 5	0.000	0.028	0.000	0.001	0.023	0.038	0.000	0.000	0.017	0.193

The ranking of alternative solutions is obtained using the VIKOR approach. We obtain $S^* = 0.297$, $S^- = 0.739$, $R^* = 0.090$, $R^- = 0.266$. The valuation results and final ranking of all the alternative locations are presented in Table 8.6.

Table 8.6 CALCULATION OF S_i AND R_i FOR CRITERIA

Alternatives	S_j	Rank	R_j	*Rank*	Q_j $(v = 0.5)$	Rank
Project 1	0.739	5	0.266	5	1.000	5
Project 2	0.297	1	0.090	1	0.000	1
Project 3	0.475	4	0.121	3	0.288	3
Project 4	0.418	3	0.097	2	0.154	2
Project 5	0.300	2	0.193	4	0.296	4

Table 8.7 VIKOR ranking

Rank	Q_j	Alternatives
1	0.000	*Project 2 (P2)*
2	0.154	*Project 4 (P4)*
3	0.288	*Project 3 (P3)*
4	0.296	*Project 5 (P5)*
5	1.000	*Project 1 (P1)*

C1. Acceptable Advantage: $DQ = 1/(5-1) = 0.25$.

$Q(F^{(2)} - F^{(1)}) \geq DQ \rightarrow 0.154 - 0 < 0.25$ but $Q(F^{(2)} - F^{(1)}) \geq DQ \rightarrow 0.288 - 0 \geq 0.25$ therefore the positions of project 2 and project 4 are "in closeness".

C2. Acceptable stability in decision making:

Project 2 is the alternative with the best ranking among the set of projects based on Q, S and R. This is the solution that offers the best compromise among all the projects and is stable within the overall decision-making process. Based on the RC_j values, we can rank the alternatives from best to worst as P2, P4, P3, P5 and P1 (Table 8.7). It is advisable for the decision-making team to re-evaluate the top two alternatives (P2, P4) as they are quite close to each other.

8.3.2 Application 2: Knowledge Management

Knowledge Management (KM) is about making the right knowledge available to the right people. It aims to create a learning environment throughout the organization and enables the use of knowledge as and when required for enhancing the overall efficiency and effectiveness of organizational processes. According to Drucker [3], it is "the coordination and exploitation of organizational knowledge resources, in order to create benefit and competitive advantage". KM plays a very important role in R&D operations as it requires taking on new challenges or providing solutions to age-old problems.

Table 8.8 List of knowledge management criteria

C1	Cost	**C6**	Time
C2	Profit	**C7**	Collaboration
C3	Productivity	**C8**	Communication
C4	Decision making	**C9**	Skill development
C5	Product quality	**C10**	Innovation

Problem Description

The following knowledge management strategies (S1 to S5) are considered for the purpose of evaluation.

1. Knowledge Acquisition (S1): It involves acquiring knowledge-base from 3rd party sources corresponding to the needs and demands of the project at hand.
2. Knowledge Creation (S2): It involves developing new systems from the available resources
3. Knowledge Sharing (S3): Adopting a culture that encourages the sharing of information, based on the concept that knowledge should be shared and updated to remain relevant
4. Knowledge Reuse (S4): Nurture and document the information generated by various ongoing and completed projects inside the organization to guide the future projects
5. Knowledge Assessment (S5): The evaluation or analysis of a person's or group's knowledge of a particular subject or subjects.

Following are the criteria used in the study (Table 8.8):

Calculation

The values of S, R and Q are calculated (as presented in Table 8.9 using the formulae explained in the previous section. An equal weightage is given to each of the criteria. The preference ranking in Table 8.9 indicates S4 (Knowledge Reuse) as the most preferred knowledge management strategy which is followed by S3 (Knowledge Sharing).

8.4 Advantages of VIKOR

- VIKOR allows the use of variables with different units of measurement and criteria of different types.
- VIKOR ranks the alternatives by considering degree of satisfaction of each criterion.

Table 8.9: f_{ji} values for all the pairs

Strategy	Criteria										VIKOR			Total	Preference ranking
	C1	C2	C3	C4	C5	C6	C7	C8	C9	C10	S value	R value	Q value		
S1	0.58	0.63	0.87	0.18	0.29	0.67	0.30	0.41	0.15	0.64	0.57866049	0.10000	0.5	1.2	4
S2	0.97	0.28	0.83	0.70	0.23	0.78	0.25	0.32	0.55	0.17	0.5384603	0.10000	0.4	1.0	3
S3	0.28	0.84	0.25	0.24	0.74	0.32	0.22	0.80	0.92	0.75	0.50237214	0.10000	0.3	0.9	2
S4	0.33	0.61	0.66	0.49	0.64	0.69	0.72	0.24	0.97	0.66	0.3838558	0.10000	0.0	0.5	1
S5	0.69	0.86	0.62	0.66	0.99	0.05	0.31	0.84	0.15	0.19	0.46895269	0.10000	0.7	1.3	5
Best	0.97	0.86	0.87	0.70	0.99	0.78	0.72	0.84	0.97	0.75	0.58	0.10			
Worst	0.28	0.28	0.25	0.18	0.23	0.05	0.22	0.24	0.15	0.17	0.38	0.10000			
Weight	0.10	0.10	0.10	0.10	0.10	0.10	0.10	0.10	0.10	0.10					

8.5 Limitations of VIKOR

- The original concept of VIKOR is to compromise two different kinds of regret, i.e., S and R values, to obtain the preference ranking. The normalization of regret gets influenced by both the best and worst values of the criteria and hence marginalizes the quality of the solution.

This chapter has illustrated the use of MCDM technique—VIKOR which identifies the compromise solution in prioritizing the alternatives. The next chapter will discuss the salient features and use of an important MCDM technique—Decision-Making Trial and Evaluation Laboratory (DEMATEL).

Selected Bibliography and Further Reading

1. Barda, O. H., Dupuis, J., & Lencioni, P. (1990). Multi-criteria location of thermal power plants. *European Journal of Operational Research, 45*(2–3), 332–346.
2. Opricovic, S., & Tzeng, G. H. (2004). Compromise solution by MADM methods: A comparative analysis of VIKOR and TOPSIS. *European Journal of Operational Research, 156*(2), 445–455.
3. Drucker, P. (1999). Knowledge-worker productivity: the biggest challenge. *California Management Review*, 79–94.
4. Frost, A. (2015, Feburary 9). *Knowledge management tools*. Retrieved from https://www.knowledge-management-tools.net/
5. Jih-Jeng Huang, G.-H. T.-H. (2009). A revised VIKOR Model for multiple criteria decision making—The perspective of regret theory. *Communications in Computer and Information Science*, 761–768.

Chapter 9
Decision-Making Trial and Evaluation Laboratory (DEMATEL)

9.1 Background

Decision-Making Trial and Evaluation Laboratory (DEMATEL) technique was first employed by Fontela and Gabus in 1976. It has managed to solve many global complex problems in scientific, political and economic domains by considering experts' judgements. BMI institute applied DEMATEL method to execute big and complicated project in GRC and "Science and Human Affairs Program of the Battelle Memorial Institute of Geneva". DEMATEL become more popular method in Japan because it is a widespread technique which is able to evaluate and formulate all intertwined cause and effect relationships in each structural model. It is useful for visualizing the structure of complicated causal relationships with matrices or digraphs.

DEMATEL is a type of structural modelling approach that is useful in analysing the cause and effect relationships among the constituents of a system. DEMATEL can be applied to confirm the existence of a relationship/interdependence among components or reflect the relative level of relationships within them. Thus, it can be used to tackle complicated problems that involve lot of interdependencies or relationships. The DEMATEL method not only transforms the relationships between factors into a cause and effect group but also finds the critical factors of a complex system with the help of impact relationship diagram. A procedure of DEMATEL is illustrated in Fig. 9.1.

DEMATEL is a comprehensive method for building and analysing a structural model involving causal relationships among complex factors. The basic steps of DEMATEL include the construction of the evaluation hierarchy, selection of expert team, calculation of the total relation matrix, determining the degree of influence, and $R + C$ and $R - C$ are calculated to draw the influence map of the total relationship.

J. J. Thakkar, *Multi-Criteria Decision Making*, Studies in Systems, Decision and Control 336, https://doi.org/10.1007/978-981-33-4745-8_9

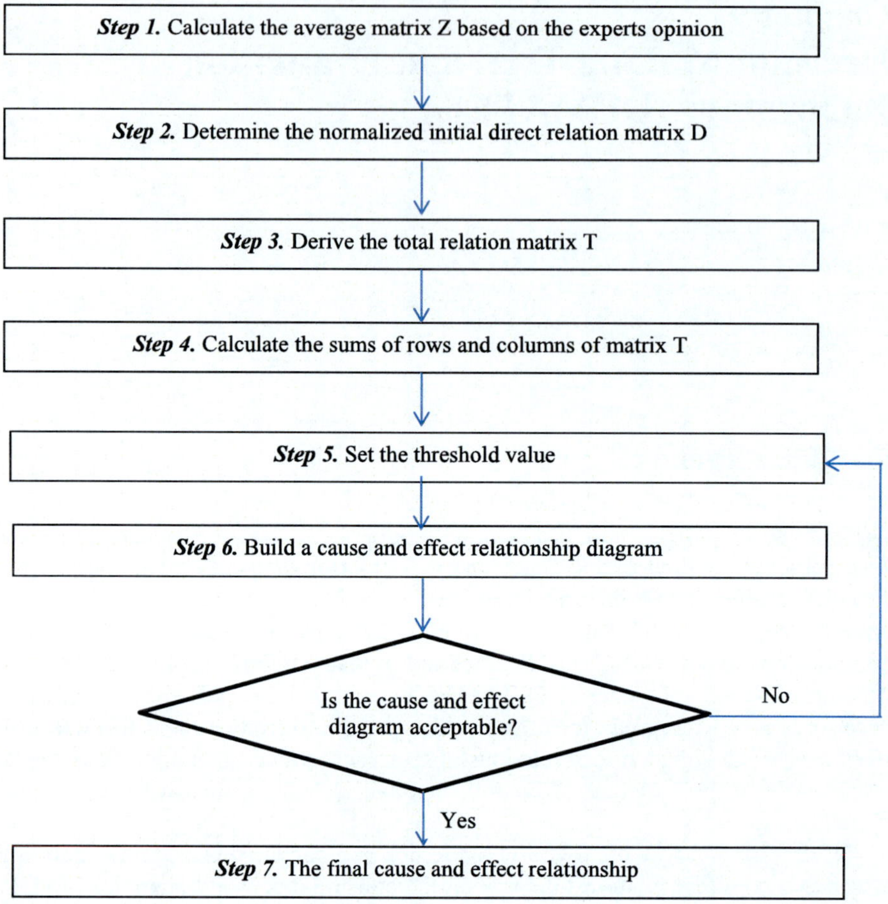

Fig. 9.1 DEMATEL Procedure

9.2 Step-by-Step Procedure of DEMATEL

Step 1: Expert panel and dimensions/criteria are determined.

In this step, a panel of experts is selected to gather subjective opinions. Based on the literature and expert opinion, challenges are determined and discussed.

Step 2: Initial direct-relation matrix is constructed.

This matrix is created by carrying out pairwise comparisons between the challenges/criteria. The following scale is used to rate the relationship between the criteria.

0: No influence.
1: Very low influence.
2: Low influence.
3: High influence.
4: Very high influence.
x_{ij}^k represents the extent of influence of criteria i on criteria j, as viewed by expert k. Let, there are M experts and n criteria, then, there will be M non-negative matrices of nXn dimension. For $i = j$, the x_{ij}^k value is 0, signifying no influence. The average matrix A is formed by the following formula:

$$a_{ij} = \frac{1}{M} \sum_{k=1}^{M} x_{ij}^k$$

Step 3: Normalized initial direct matrix D is constructed.

$$D = A \times S$$

where $S = \min\left(\frac{1}{\max \sum_{j=1}^{n} a_{ij}}, \frac{1}{\max \sum_{i=1}^{n} a_{ij}}\right)$
Every element in matrix D lies between 0 and 1.

Step 4: Total relation matrix T is generated.

$$T = (I - D)^{-1}$$

where I is the identity matrix.

Step 5: Prominence and Relation are calculated for each criterion.

$$R_i = \sum_{j=1}^{n} t_{ij}$$

$$C_j = \sum_{i=1}^{n} t_{ij}$$

$R_i + C_j$ is called the "Prominence" and $R_i - C_j$ is called the "Relation".

Step 6: Cause and effect diagram is created by mapping the dataset $(R_i + C_j : R_i - C_j)$.

9.3 Illustrative Application of DEMATEL

9.3.1 Application 1: Applying DEMATEL Method for Selection of Lean Tools (5S, Jidoka, Heijunka, JIT, Kaizen, Kanban, OEE, Poka-Yoke, SMED, TPM, Value Stream Mapping) for a Typical Small Size Manufacturing Company in India

Step 1:

Compute the average matrix. The direct influence between any two factors is evaluated by each respondent by providing an integer score of 0, 1, 2, ...,10 (represents "increasing influence"). The notation of x_{ij} indicates the degree to which the respondent believes factor i affects factor j. For $i = j$, the diagonal elements are set to zero, which indicates no influence. An $n * n$ non-negative matrix will be established for each respondent as $X^k = [x_{ij}{}^k]$, where k is the number of respondents with $1 \leq k \leq H$, and n is the number of elements in the system. If there are H respondents, $X_1, X_2, X_3, \ldots,$ and X_H will be established. To aggregate all opinions from H respondents, the average matrix $A = [a_{ij}]$ can be constructed as follows:

$$a_{ij} = \frac{1}{H}\sum_{k=1}^{H} x_{ij}^{k}$$

The causal relationship among eleven tools (5S, Jidoka, Heijunka, JIT, Kaizen, Kanban, OEE, Poka-Yoke, SMED, TPM, Value Stream Mapping—a to k) using DEMATEL method is depicted below. The inputs are collected from ten respondents (in-terms of integer score of 0,1,2,.......,10).

$x_1 =$

	a	b	c	d	e	f	g	h	I	j	k
a	0	2	3	4	6	1	5	7	8	9	10
b	2	0	1	7	4	3	5	6	8	10	9
c	3	1	0	2	4	6	5	7	9	8	10
d	4	7	2	0	1	9	6	8	10	5	3
e	6	4	4	1	0	2	7	3	9	5	8
f	1	3	6	9	2	0	10	4	5	7	8
g	5	5	5	6	7	10	0	8	9	1	2
h	7	6	7	8	3	4	8	0	1	2	5
i	8	8	9	10	9	5	9	1	0	3	4
j	9	10	8	5	5	7	1	2	3	0	6
k	10	9	10	3	8	8	2	5	4	6	0

$x_2 =$

	a	b	c	d	e	f	g	h	I	j	k
a	0	3	2	5	4	1	6	8	7	10	9
b	3	0	7	1	3	4	6	5	9	10	8
c	2	7	0	2	4	6	5	7	9	8	10
d	5	1	2	0	1	9	6	8	10	5	3
e	4	3	4	1	0	2	7	3	9	5	8
f	1	1	6	9	2	0	10	4	5	7	8
g	6	6	5	6	7	10	0	8	9	1	2
h	8	8	7	8	3	4	8	0	1	2	5
i	7	7	9	10	9	5	9	1	0	3	4
j	10	10	8	5	5	7	1	2	3	0	6
k	9	9	10	3	8	8	2	5	4	6	0

$x_3 =$

	a	b	c	d	e	f	g	h	I	j	k
a	0	2	3	4	6	1	5	7	8	9	10
b	2	0	1	7	4	3	5	6	8	10	9
c	3	1	0	4	2	5	6	9	7	10	8
d	4	7	2	0	1	9	6	8	10	5	3
e	6	4	2	1	0	2	7	3	9	5	8
f	1	3	5	9	2	0	10	4	5	7	8
g	5	5	6	6	7	10	0	8	9	1	2
h	7	6	9	8	3	4	8	0	1	2	5
i	8	8	7	10	9	5	9	1	0	3	4
j	9	10	10	5	5	7	1	2	3	0	6
k	10	9	8	3	8	8	2	5	4	6	0

$x_4 =$

	a	b	c	d	e	f	g	h	I	j	k
a	0	9	8	3	8	8	2	5	4	6	10
b	9	0	7	1	3	4	6	5	9	10	8
c	8	7	0	2	4	6	5	7	9	8	10
d	3	1	2	0	1	9	6	8	10	5	3
e	8	3	4	1	0	2	7	3	9	5	8
f	8	1	6	9	2	0	10	4	5	7	8
g	2	6	5	6	7	10	0	8	9	1	2
h	5	8	7	8	3	4	8	0	1	2	5
i	4	7	9	10	9	5	9	1	0	3	4
j	6	10	8	5	5	7	1	2	3	0	6
k	10	9	10	3	8	8	2	5	4	6	0

$x_5 =$

	a	b	c	d	e	f	g	h	I	j	k
a	0	10	8	5	5	7	1	2	3	5	6
k	10	0	1	7	4	3	5	6	8	2	9
c	8	1	0	4	2	5	6	9	7	10	8
d	5	7	2	0	1	9	6	8	10	5	3
e	5	4	2	1	0	2	7	3	9	5	8
f	7	3	5	9	2	0	10	4	5	7	8
g	1	5	6	6	7	10	0	8	9	1	2
h	2	6	9	8	3	4	8	0	1	2	5
i	3	8	7	10	9	5	9	1	0	3	4
j	5	2	10	5	5	7	1	2	3	0	6
k	6	9	8	3	8	8	2	5	4	6	0

$x_6 =$

	a	b	c	d	e	f	g	h	I	j	k
a	0	9	8	1	8	8	2	5	4	6	10
b	9	0	7	1	3	4	6	5	9	1	8
c	8	7	0	2	4	6	5	7	9	8	10
d	1	1	2	0	1	9	6	8	10	5	3
e	8	3	4	1	0	2	7	3	9	5	8
f	8	1	6	9	2	0	10	4	5	7	8
g	2	6	5	6	7	10	0	8	9	1	2
h	5	8	7	8	3	4	8	0	1	2	5
i	4	7	9	10	9	5	9	1	0	3	4
j	6	10	8	5	5	7	1	2	3	0	6
k	10	9	10	3	8	8	2	5	4	6	0

$x_7 =$

	a	b	c	d	e	f	g	h	I	j	k
a	0	2	3	4	6	1	5	7	8	9	10
b	2	0	1	7	4	3	5	6	8	10	9
c	3	1	0	2	4	6	5	7	9	8	10
d	4	7	2	0	1	9	6	8	10	5	3
e	6	4	4	1	0	2	7	3	9	5	8
f	1	3	6	9	2	0	10	4	5	7	8
g	5	5	5	6	7	10	0	8	9	1	2
h	7	6	7	8	3	4	8	0	1	2	5
i	8	8	9	10	9	5	9	1	0	3	4
j	9	10	8	5	5	7	1	2	3	0	6
k	10	9	10	3	8	8	2	5	4	6	0

$x_8 =$

	a	b	c	d	e	f	g	h	I	j	k
a	0	10	8	5	5	7	1	2	3	5	6
b	10	0	1	7	4	3	5	6	8	10	9
c	8	1	0	4	2	5	6	9	7	10	8
d	5	7	4	0	1	9	6	8	10	5	3
e	5	4	2	1	0	2	7	3	9	5	8
f	7	3	5	9	2	0	10	4	5	7	8
g	1	5	6	6	7	10	0	8	9	1	2
h	2	6	9	8	3	4	8	0	1	2	5
i	3	8	7	10	9	5	9	1	0	3	4
j	5	10	10	5	5	7	1	2	3	0	6
k	6	9	8	3	8	8	2	5	4	6	0

$x_9 =$

	a	b	c	d	e	f	g	h	I	j	k
a	0	9	8	1	8	8	2	5	4	6	1
b	9	0	7	1	3	4	6	5	9	10	5
c	8	7	0	2	4	6	5	7	9	8	6
d	1	1	2	0	1	9	6	8	10	5	6
e	8	3	4	1	0	2	7	3	9	5	7
f	8	1	6	9	2	0	10	4	5	7	10
g	2	6	5	6	7	10	0	8	9	1	0
h	5	8	7	8	3	4	8	0	1	2	8
i	4	7	9	10	9	5	9	1	0	3	9
i	6	10	8	5	5	7	1	2	3	0	1
k	1	5	6	6	7	10	0	8	9	1	0

$x_{10} =$

	a	b	c	d	e	f	g	h	I	j	k
a	0	10	8	5	5	7	1	2	3	5	4
b	10	0	1	7	4	3	5	6	8	10	9
c	8	1	0	4	2	5	6	9	7	10	8
d	5	7	2	0	1	9	6	8	10	5	3
e	5	4	2	1	0	2	7	3	9	5	8
f	7	3	5	9	2	0	10	4	5	7	8
g	1	5	6	6	7	10	0	8	9	1	2
h	2	6	9	8	3	4	8	0	1	2	5
i	3	8	7	10	9	5	9	1	0	3	4
j	5	10	10	5	5	7	1	2	3	0	6
k	4	9	8	3	8	8	2	5	4	6	0

An average matrix "A" is calculated as shown in Table 9.1.

Step 2: normalizing the direct relation of the matrix Though the step 1, the matrix was normalized as Table 9.2.

Calculate the normalized initial direct-relation matrix D by,

$$D = A * S,$$

where

Table 9.1 Average matrix A by aggregating ten respondents

MATRIX *A*											
	a	b	c	d	e	f	g	h	I	j	k
a	0	6.6	5.9	3.7	6.1	4.9	3.0	5.0	5.2	7.0	7.6
b	6.6	0	3.4	4.6	3.6	3.4	5.4	5.6	7.4	9.5	4.8
c	5.9	3.4	0	2.8	3.4	4.5	4.9	7.1	7.5	7.5	7.8
d	3.7	4.7	2.2	0	1.0	9.0	6.0	8.0	10	5.1	2.0
e	6.1	3.7	3.2	1.0	0	2.0	7.0	3.0	9.0	5.1	8.0
f	4.9	2.2	4.1	9.0	2.0	0	10	4.0	5.0	7.0	8.2
g	2.9	5.4	5.2	5.8	7.0	10	0	8.0	9	1	1.8
h	5.0	6.8	8.3	8.0	3.0	4.0	8.0	0	1.0	2.0	5.8
i	5.2	7.2	7.0	9.8	9.0	5.0	9.0	1.0	0	3	4.5
j	7.0	9.2	9.8	4.9	5.0	7.0	1.0	2.0	3.0	0	5.5
k	7.6	8.5	8.8	3.6	7.7	8.2	1.8	5.3	4.5	5.5	0

Table 9.2 Dimension normalization of the direct relation of matrix, MATRIX *D*

	a	b	c	d	e	f	g	h	i	j	k
a	0	0.12	0.1	0.06	0.13	0.08	0.05	0.1	0.08	0.13	0.14
b	0.12	0	0.07	0.08	0.07	0.06	0.09	0.13	0.11	0.17	0.08
c	0.107	0.06	0	0.04	0.05	0.07	0.07	0.16	0.12	0.14	0.13
d	0.07	0.09	0.03	0	0.02	0.14	0.11	0.17	0.17	0.09	0.04
e	0.11	0.05	0.05	0.02	0	0.03	0.13	0.05	0.16	0.09	0.16
f	0.089	0.03	0.09	0.16	0.04	0	0.18	0.06	0.09	0.13	0.16
g	0.05	0.11	0.09	0.11	0.14	0.17	0	0.14	0.16	0.02	0.04
h	0.09	0.12	0.14	0.14	0.06	0.07	0.14	0	0.02	0.04	0.12
i	0.094	0.17	0.12	0.18	0.16	0.08	0.16	0.02	0	0.06	0.08
j	0.13	0.18	0.15	0.08	0.11	0.13	0.02	0.04	0.06	0	0.09
k	0.14	0.16	0.13	0.06	0.14	0.15	0.13	0.07	0.09	0.11	0

$$S = \frac{1}{\max_{1 \le i \le n} \sum_{j=1}^{n} a_{ij}}$$

Each element in matrix *D* falls between zero and one.

Step 3: Building the relation matrix of total influence after the normalization of the direct relation of the matrix, the relation matrix of the total influence (see Table 9.3) was gained from the formula of the relation matrix of total influence.

Compute the total relation matrix *T* by,

$$T = D(I - D)^{-1}$$

where *I* is the identity matrix (Table 9.3).

Step 4: Let r and c be $n * 1$ and $1 * n$ vectors representing the sum of rows and sum of columns of the total relation matrix *T*, respectively. Calculating all of the tier numbers, prominence $(r + c)$ and relation $(r - c)$ It tried to calculate the relation matrix of the total influence (T) to have r and c, and then it owned the prominence $(r + c)$, and the relation $(r - c)$ as Table 9.4.

Also determine the influential weights of criteria.

The relative importance of the criteria is calculated by using the following equation.

$$w_j = \left[\left(r_j + c_j\right)^2 + \left(r_j - c_j\right)^2 \right]^{\frac{1}{2}}$$

Table 9.3 Relation matrix of total influence (T), MATRIX T

	a	b	c	d	e	f	g	h	i	j	k
a	−0.9008	0.3561	−0.0270	0.1335	0.1170	−0.0511	0.1116	0.0812	0.2288	0.1141	−0.0658
b	0.0992	−0.6439	0.0270	0.1335	0.1170	−0.0511	0.1116	0.0812	0.2288	0.1141	−0.0658
c	0.0992	0.3561	−1.0270	0.1335	0.1170	−0.0511	0.1116	0.0812	0.2288	0.1141	−0.0658
d	0.0992	0.3561	−0.0270	−0.8665	0.1170	−0.0511	0.1116	0.0812	0.2288	0.1141	−0.0658
e	0.0992	0.3561	−0.0270	0.1335	−0.8830	−0.0511	0.1116	0.0812	0.2288	0.1141	−0.0658
f	0.0992	0.3561	−0.0270	0.1335	0.1170	−1.0511	0.1116	0.0812	0.2288	0.1141	−0.0658
g	0.0992	0.3561	−0.0270	0.1335	0.1170	−0.0511	−0.8884	0.0812	0.2288	0.1141	−0.0658
h	0.0992	0.3561	−0.0270	0.1335	0.1170	−0.0511	0.1116	−0.9188	0.2288	0.1141	−0.0658
i	0.0992	0.3561	−0.0270	0.1335	0.1170	−0.0511	0.1116	0.0812	−0.7712	0.1141	−0.0658
j	0.0992	0.3561	−0.0270	0.1335	0.1170	−0.0511	0.1116	0.0812	0.2288	0.8859	−0.0658
k	0.0992	0.3561	−0.0270	0.1335	0.1170	−0.0511	0.1116	0.0812	0.2288	0.1141	−1.0658

Table 9.4 Calculation of matrix, prominence $(r + d)$, relation $(r - d)$, weight, final criteria weight, and rank 5S, Jidoka, Heijunka, JIT, Kaizen, Kanban, OEE, Poka-Yoke, SMED, TPM, value stream mapping

r	0.0975	0.0975	0.0975	0.0975	0.0975	0.0975	0.0975	0.0975	0.0975	0.0975	0.0975
c	0.0907	2.9175	−1.2975	0.4689	0.2872	−1.5619	0.2273	−0.1065	1.5163	0.2547	−1.7238
$r + c$	0.1882	3.0150	−1.2000	0.5664	0.3847	−1.4644	0.3248	−0.0089	1.6138	0.3522	−1.6263
$r - c$	0.0069	−2.8200	1.3950	−0.3713	−0.1896	1.6594	−0.1297	0.2040	−1.4188	−0.1571	1.8213
W	0.188	4.13	1.84	0.677	0.43	2.21	0.35	0.2042	2.15	0.3856	2.45
$\hat{W}$	0.0125	0.275	0.123	0.045	0.028	0.147	0.023	0.0136	0.1433	0.0257	0.163
Rank	11	1	5	6	7	3	9	10	4	8	2

The weight of any criterion could be normalized as follows:

$$\overline{w_j} = \frac{w_j}{\sum_{j=1}^{n} w_j}$$

where $\overline{w_j}$ denotes the final criteria weights that would be required in the decision-making process. Consequently, we could obtain the influential weight for each criterion (i.e., global influential weight) by utilizing the modified 2-tuple DEMATEL approach.

Step 5: Result analysis and causal diagram. The prominence $(r + c)$ showed the influence range of factors, including influencing and influenced. In addition, the score of relation $(r - c)$ was more, and the factor influenced others more. The study organized an order for the r, c, $r +$ c, and $r - c$, and it also caught the mean score to be viewed as limitation whether the influences range large or not.

From Table 9.4, 5S, Heijunka, Kanban, Poka-Yoke, value stream mapping are net causes due to positive $(r - c)$ values, while, Heijunka, JIT, Kaizen, OEE, SMED, TPM are net receivers because of negative $(r - c)$ values. value stream mapping has the highest $(r + c)$ value indicating that this tools is the most essential tools.

Any actions taken on net causes (either tools or factor) will have direct/indirect effects on the corresponding net receivers (either tools or factor). In fact, the decision maker can pay much attention and allocate resources to cause tool(s) and factor(s) to effectively select lean tools.

9.3.2 Application 2: E-Commerce

Challenges—Company Divisions/ Focus Areas.

Criteria—Customers' Decision-Making Attributes.

E-commerce has transformed the way business is done in India. This industry is expected to form largest part of the Indian Internet market with a value of around $ 100 Billion by 2021. An increase in the smartphone penetration and Internet services drives 11% CAGR of e-commerce in India. The last 4 years have witnessed a rise in the number of unicorns primarily in the B2C segment of the E-commerce Market. While Amazon entered India with its deeply funded pockets, the Indian success stories of Flipkart and Snapdeal captured the minds of the Indian customers alike. They were able to do so with their user-friendly websites and Cash on Delivery options for the semi-literate and risk-averse Indian populace.

However, the recent regulations in e-commerce and the decline in ease of raising funding have increased the competition among these major players. In order to sustain themselves in the long run in the market, these companies must now focus on customer loyalty. For this, the major attributes that influence the attitude of customers to buy from a particular e-commerce website must be studied.

Problem Structure

Running an e-commerce company is not an easy task, with a highly diversified product and vendor portfolio. Logistics and Customer Service need to be interconnected in providing the best experience to the customers with a hope of winning their loyalty. It is essential to prioritize among the various dimensions/functions so as to positively influence the customer's buying attitude.

The main dimensions/functions to be focused upon in the e-commerce company are given in Table 9.5 and Fig. 9.2. Fifteen major criteria that influence the buyer's decision have been identified and classified among the 5 dimensions.

Table 9.5 Dimensions/functions for the e-commerce company

Dimension		Criteria	Criteria description
Website experience (A)	A1	Simplicity in navigation	Smooth Flow, Ease of placing and cancelling orders
	A2	User interface	Attractive UI, User Friendliness
	A3	Minimum server crashes	No technical disruptions or 404 errors
Customer eervice (B)	B1	Ease of returns	Quick refunds and hassle free reverse logistics
	B2	One touch resolution	One Call or Email should resolve the queries
	B3	Gracious communications	Politeness from the Company's executives
Shipping & payments (C)	C1	On time delivery	Options for Quick or scheduled delivery
	C2	Payment alternatives	Availability of CoD and easy online payment methods
	C3	Delivery reliability	Location of Vendors and its Ratings
Product management (D)	D1	Product details	Accurate Description and details along with images
	D2	Product variety	In stock Availability & diversity in product range
	D3	Useful recommendations	Effective Product Recommendation Engine
Pricing (E)	E1	Cashback offers and discounts	Heavy discounts and Low Price compared to others
	E2	Fluctuations in prices	Variations in prices with time and quantity
	E3	Mega discount sales	Higher Frequency of sales for wide range of products

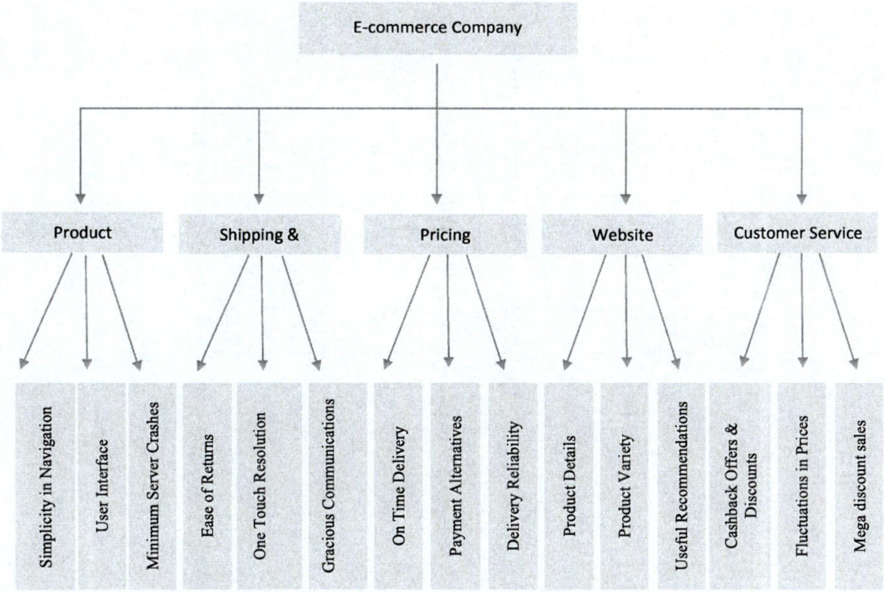

Fig. 9.2 Dimensions and Criteria

The DEMATEL approach has been applied for the case situation.

Step1: Initial Relation matrix.

The initial relation matrix $\boldsymbol{A}$ is developed using the average score for influence of 15 criteria in Table 9.6. The scores are obtained using pairwise comparisons.

Step 2: Normalization of the A matrix.

The normalizing factor is $\boldsymbol{S} = \mathbf{0.0288}$ (Table 9.7)

$$\boldsymbol{D} = \boldsymbol{AXS}$$

Step 3: The $\boldsymbol{T}$ matrix or the Total Relationship Matrix is calculated using the formula:

$$T = (I - D)^{-1}$$

We thus obtain the T matrix as (Table 9.8).

Step 4: The Prominence and Relation are mapped as presented in Table 9.9.

The average values of Prominence and Relation (R + C & R - C) for the dimensions (A to E) are represented in Table 9.10.

The cause and effect diagram of the case is shown in Fig. 9.3.

Table 9.6 Initial relation matrix

Criteria	A1	A2	A3	B1	B2	B3	C1	C2	C3	D1	D2	D3	E1	E2	E3
A1	0.00	3.33	4.00	3.67	3.67	2.33	1.33	1.00	1.00	2.33	1.33	2.33	2.67	2.33	1.67
A2	0.67	0.00	2.33	3.67	1.67	2.00	2.33	1.33	1.67	2.33	2.00	3.00	1.33	2.00	1.67
A3	2.67	1.67	0.00	1.33	3.33	1.67	0.33	1.33	2.00	1.67	2.33	1.67	2.00	1.33	1.33
B1	1.67	3.00	3.00	0.00	1.67	3.00	1.67	1.33	0.67	2.67	3.33	2.00	0.67	1.00	1.33
B2	2.33	3.00	1.67	1.67	0.00	1.67	2.33	2.67	2.67	2.33	1.33	2.33	2.33	2.00	3.00
B3	2.33	2.33	2.67	2.00	3.00	0.00	1.00	1.67	3.00	3.33	2.67	2.33	2.67	1.33	2.67
C1	1.00	3.00	2.33	2.00	2.00	3.33	0.00	2.33	1.67	2.67	1.33	1.00	2.00	3.00	1.67
C2	2.33	2.67	3.00	0.67	1.00	2.00	2.67	0.00	0.33	2.67	1.67	1.33	3.00	3.00	0.67
C3	3.00	1.33	2.00	3.00	2.33	1.67	2.33	0.00	0.00	2.33	0.67	1.33	0.67	0.67	2.00
D1	2.00	1.00	2.00	0.67	0.33	3.00	0.67	1.33	2.00	0.00	0.33	1.33	0.33	2.33	1.00
D2	3.33	1.67	2.33	1.67	2.67	2.00	2.33	1.67	0.67	2.33	0.00	2.33	2.33	2.67	2.00
D3	2.67	0.33	3.00	2.67	2.33	2.33	1.67	1.33	2.33	1.67	2.00	0.00	0.67	1.67	1.67
E1	1.33	1.67	2.00	2.00	2.00	1.67	2.33	2.00	2.00	1.33	1.67	2.00	0.00	1.00	2.33
E2	3.67	1.67	2.00	1.67	3.00	0.67	0.67	2.33	2.33	1.00	1.33	1.33	2.00	0.00	1.67
E3	2.00	3.00	2.33	1.33	0.67	1.33	1.00	3.00	2.33	1.67	3.00	1.00	2.00	1.00	0.00

Table 9.7 Normalized matrix

Criteria	A1	A2	A3	B1	B2	B3	C1	C2	C3	D1	D2	D3	E1	E2	E3
A1	0.00	0.10	0.12	0.11	0.11	0.07	0.04	0.03	0.03	0.07	0.04	0.07	0.08	0.07	0.05
A2	0.02	0.00	0.07	0.11	0.05	0.06	0.07	0.04	0.05	0.07	0.06	0.09	0.04	0.06	0.05
A3	0.08	0.05	0.00	0.04	0.10	0.05	0.01	0.04	0.06	0.05	0.07	0.05	0.06	0.04	0.04
B1	0.05	0.09	0.09	0.00	0.05	0.09	0.05	0.04	0.02	0.08	0.10	0.06	0.02	0.03	0.04
B2	0.07	0.09	0.05	0.05	0.00	0.05	0.07	0.08	0.08	0.07	0.04	0.07	0.07	0.06	0.09
B3	0.07	0.07	0.08	0.06	0.09	0.00	0.03	0.05	0.09	0.10	0.08	0.07	0.08	0.04	0.08
C1	0.03	0.09	0.07	0.06	0.06	0.10	0.00	0.07	0.05	0.08	0.04	0.03	0.06	0.09	0.05
C2	0.07	0.08	0.09	0.02	0.03	0.06	0.08	0.00	0.01	0.08	0.05	0.04	0.09	0.09	0.02
C3	0.09	0.04	0.06	0.09	0.07	0.05	0.07	0.00	0.00	0.07	0.02	0.04	0.02	0.02	0.06
D1	0.06	0.03	0.06	0.02	0.01	0.09	0.02	0.04	0.06	0.00	0.01	0.04	0.01	0.07	0.03
D2	0.10	0.05	0.07	0.05	0.08	0.06	0.07	0.05	0.02	0.07	0.00	0.07	0.07	0.08	0.06
D3	0.08	0.01	0.09	0.08	0.07	0.07	0.05	0.04	0.07	0.05	0.06	0.00	0.02	0.05	0.05
E1	0.04	0.05	0.06	0.06	0.06	0.05	0.07	0.06	0.06	0.04	0.05	0.06	0.00	0.03	0.07
E2	0.11	0.05	0.06	0.05	0.09	0.02	0.02	0.07	0.07	0.03	0.04	0.04	0.06	0.00	0.05
E3	0.06	0.09	0.07	0.04	0.02	0.04	0.03	0.09	0.07	0.05	0.09	0.03	0.06	0.03	0.00

Table 9.8 Total relationship matrix

Criteria	A1	A2	A3	B1	B2	B3	C1	C2	C3	D1	D2	D3	E1	E2	E3
A1	0.94	0.25	0.30	0.25	0.26	0.23	0.17	0.17	0.18	0.24	0.19	0.21	0.21	0.21	0.19
A2	-0.04	1.14	0.22	0.22	0.18	0.19	0.17	0.15	0.17	0.21	0.18	0.20	0.15	0.17	0.17
A3	0.06	0.18	1.16	0.17	0.23	0.18	0.12	0.15	0.18	0.19	0.18	0.17	0.17	0.16	0.16
B1	-0.87	-0.02	-0.03	0.90	-0.06	0.01	0.00	0.00	-0.02	0.00	0.04	-0.02	-0.06	-0.04	-0.02
B2	0.05	0.25	0.24	0.20	1.17	0.21	0.20	0.21	0.22	0.24	0.19	0.21	0.20	0.20	0.23
B3	0.05	0.24	0.27	0.22	0.25	1.17	0.17	0.19	0.24	0.27	0.23	0.22	0.22	0.19	0.23
C1	0.01	0.24	0.24	0.20	0.21	0.24	1.12	0.20	0.19	0.24	0.18	0.17	0.19	0.22	0.18
C2	0.07	0.22	0.26	0.16	0.18	0.21	0.19	1.13	0.15	0.23	0.18	0.17	0.21	0.22	0.15
C3	0.03	0.16	0.19	0.19	0.18	0.17	0.16	0.10	1.11	0.19	0.13	0.14	0.12	0.12	0.16
D1	0.06	0.13	0.18	0.12	0.12	0.18	0.10	0.12	0.15	1.11	0.10	0.13	0.10	0.15	0.12
D2	0.08	0.21	0.25	0.20	0.24	0.22	0.19	0.19	0.17	0.24	1.15	0.21	0.20	0.22	0.20
D3	0.04	0.15	0.24	0.20	0.20	0.20	0.15	0.15	0.19	0.19	0.18	1.12	0.14	0.17	0.17
E1	0.01	0.18	0.21	0.18	0.19	0.18	0.17	0.17	0.18	0.18	0.17	0.17	1.11	0.15	0.18
E2	0.08	0.19	0.22	0.18	0.22	0.16	0.13	0.18	0.19	0.17	0.16	0.16	0.17	1.12	0.17
E3	0.05	0.22	0.23	0.17	0.16	0.18	0.14	0.20	0.19	0.19	0.21	0.15	0.17	0.15	1.12

Table 9.9 $R - C$ matrix

Dimensions	Criteria	R	C	$R + C$	Rank	$R - C$	Impact
A	A1	4.02	0.63	4.65	3	3.39	Cause
	A2	3.50	3.74	7.24	2	-0.23	Effect
	A3	3.46	4.17	7.63	1	-0.71	Effect
B	B1	-0.18	3.55	3.38	3	-3.73	Effect
	B2	4.03	3.74	7.77	2	0.29	Cause
	B3	4.15	3.73	7.89	1	0.42	Cause
C	C1	3.82	3.19	7.01	2	0.63	Cause
	C2	3.73	3.30	7.03	1	0.42	Cause
	C3	3.13	3.47	6.61	3	-0.34	Effect
D	D1	2.87	3.90	6.76	3	-1.03	Effect
	D2	3.95	3.46	7.41	1	0.49	Cause
	D3	3.49	3.43	6.91	2	0.06	Cause
E	E1	3.44	3.30	6.74	3	0.13	Cause
	E2	3.48	3.40	6.88	2	0.09	Cause
	E3	3.52	3.40	6.92	1	0.12	Cause

Table 9.10 Ranking of dimensions

Dimensions	$R + C$	Rank	$R - C$	Impact
A	6.503	**4**	0.813	Cause
B	6.345	5	−1.005	Effect
C	6.881	**3**	0.237	Cause
D	7.031	**1**	0.160	Cause
E	6.846	2	-0.115	Effect

Key Inferences:

DEMATEL technique of multi-criteria decision making has been used to prioritize from among 5 different dimensions of an e-commerce industry, with 3 criteria for each of the dimension. From the results obtained, we can conclude that dimension D has the highest rank with the highest Prominence value ($R + C$) of 7.031. The priority order of the dimensions is D > E > C > A > B. From the graphs, it can be concluded that A, C and D are the causal factors while B and E comprise of the effects groups. Therefore, focusing on A, C and D will result in positive impact on the five dimensions. Taking the perspective of the e-commerce company, the major divisions that the company should focus on are Product Management followed by Shipping and payments, and then the website experience.

Cause group: Product Management > Shipping and Payments > Website Experience.

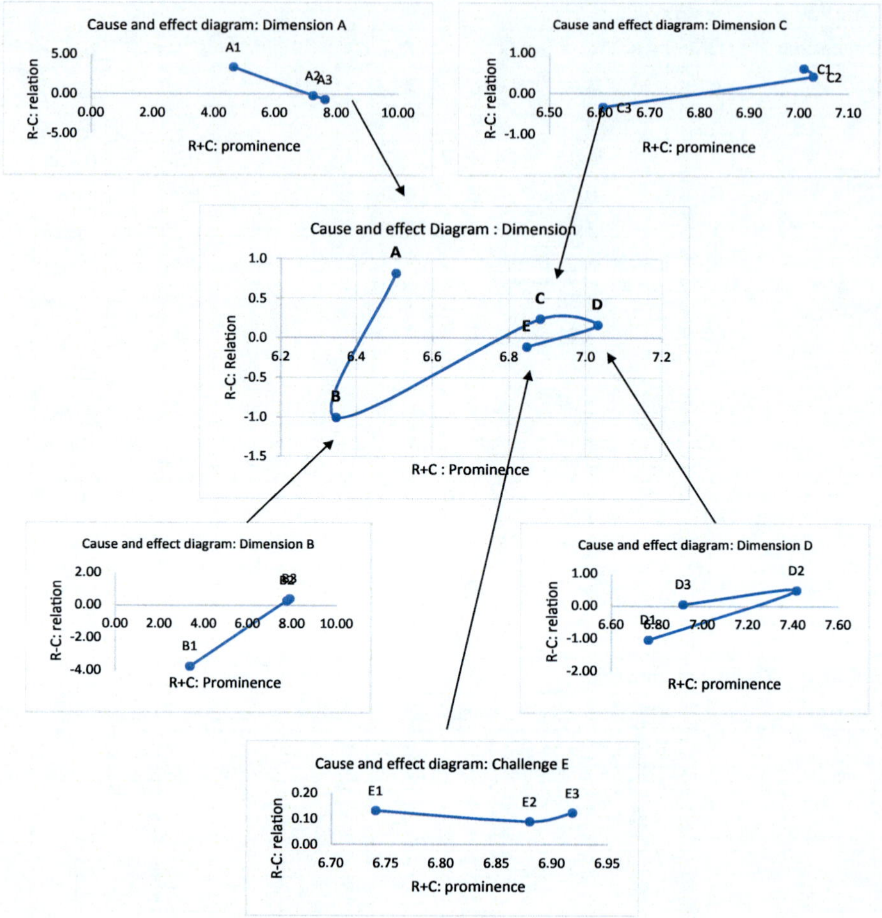

Fig. 9.3 Cause and effect diagram

9.4 Advantages of DEMATEL

- DEMATEL method is used while dealing with causality problems of a complex nature that are difficult to articulate or comprehend.
- The DEMATEL model is one such approach, allowing reliance on credibly delineable objective factors to guide analysis, to determine interdependence among the variables and constraint relations, reflecting the complex system's features in detail and their transformative trends.
- DEMETAL is one of the well appreciated techniques for its ability to accommodate causality, comparative strength, and network structure.

9.5 Limitations of DEMATEL

- DEMATEL technique is not suitable for determining the factors like hierarchy and relative factor importance.
- The subjective judgement of decision maker is presented as crisp value which inadequately reflects vagueness of the real world.

This chapter has demonstrated an application of MCDM technique—DEMATEL. The technique is widely used for its capability to accommodate cause and effect relationships. However, many real-life problems are difficult to solve using MCDM approaches because they can not be defined adequately in the absence of complete and accurate information or data. This problem can be addressed with the use of fuzzy set theory. The next chapter will demonstrate an application of fuzzy integral and grey relational technique. Grey relation analysis supports decision analysis with incomplete Information/small sample size/unknown distribution type.

Selected Bibliography and Further Reading

1. Wu, W. W., & Lee, Y. T. (2007). Developing global managers' competencies using the fuzzy DEMATEL method. *Expert Systems with Applications, 32*(2), 499–507.
2. Gabus, A., & Fontela, E. (1972). *World problems, an invitation to further thought within the framework of DEMATEL* (pp. 1–8). Geneva, Switzerland: Battelle Geneva Research Center.
3. Chang, B., Chang, C. W., & Wu, C. H. (2011). Fuzzy DEMATEL method for developing supplier selection criteria. *Expert Systems with Applications, 38*(3), 1850–1858.
4. Shieh, J. I., Wu, H. H., & Huang, K. K. (2010). A DEMATEL method in identifying key success factors of hospital service quality. *Knowledge-Based Systems, 23*(3), 277–282.
5. Hsu, C. W., Kuo, T. C., Chen, S. H., & Hu, A. H. (2013). Using DEMATEL to develop a carbon management model of supplier selection in green supply chain management. *Journal of Cleaner Production, 56,* 164–172.
6. Tseng, M. L. (2009). A causal and effect decision making model of service quality expectation using grey-fuzzy DEMATEL approach. *Expert Systems with Applications, 36*(4), 7738–7748.
7. Lin, R. J. (2013). Using fuzzy DEMATEL to evaluate the green supply chain management practices. *Journal of Cleaner Production, 40,* 32–39.
8. Zhou, Q., Huang, W., & Zhang, Y. (2011). Identifying critical success factors in emergency management using a fuzzy DEMATEL method. *Safety Science, 49*(2), 243–252.
9. Lin, C. L., & Tzeng, G. H. (2009). A value-created system of science (technology) park by using DEMATEL. *Expert Systems with Applications, 36*(6), 9683–9697.
10. Jassbi, J., Mohamadnejad, F., & Nasrollahzadeh, H. (2011). A Fuzzy DEMATEL framework for modeling cause and effect relationships of strategy map. *Expert Systems with Applications, 38*(5), 5967–5973.
11. Li, C. W., & Tzeng, G. H. (2009). Identification of a threshold value for the DEMATEL method using the maximum mean de-entropy algorithm to find critical services provided by a semiconductor intellectual property mall. *Expert Systems with Applications, 36*(6), 9891–9898.

Chapter 10
Fuzzy Integral and Grey Relation

10.1 Background

The issues dominating a multiple decision-making problem are the effective evaluation of weights of criteria and correctness of alternatives versus criteria as well as aggregate decision maker's estimated ratings.

A typical MCDM problem which is not defined properly with the required parameters will always create uncertainty in the mind of the Decision Maker. To deal with this type of ambiguity developed by non-availability of proper information or data of a MCDM problem fuzzy set theory can be best used.

Grey relation analysis is an effective means for tackling decision analysis with incomplete Information/small sample size/unknown distribution type.

10.2 Fuzzy Set Theory

Fuzzy set theory was formalized by Professor Lofti Zadeh at the University of California in 1965. What Zadeh proposed is very much a paradigm shift that first gained acceptance in the Far East and its successful application has ensured its adoption around the world. A Fuzzy set admits gradation such as all tones between black and white. A fuzzy set has a graphical description that expresses how the transition from one to another takes place. This graphical description is called a membership function (Fig. 10.1).

J. J. Thakkar, *Multi-Criteria Decision Making*, Studies in Systems, Decision and Control 336, https://doi.org/10.1007/978-981-33-4745-8_10

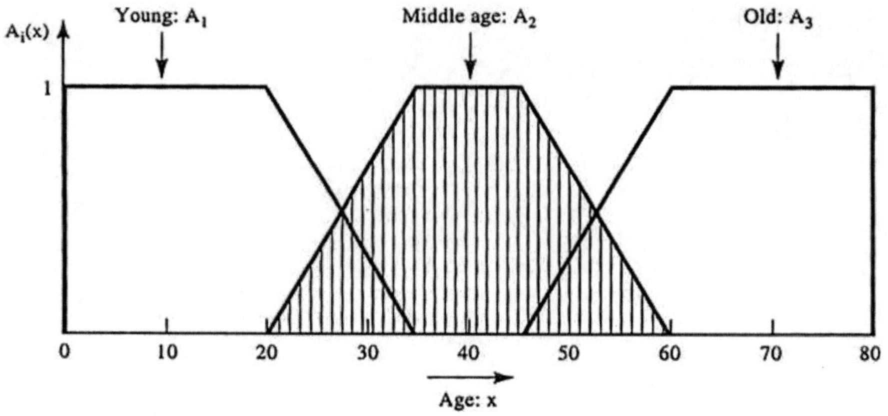

Fig. 10.1 Membership functions representing the concepts of a young, middle-aged and old person

10.2.1 Triangular Fuzzy Numbers

In a universe of discourse X, a fuzzy subset A of X is defined by a membership function $fA(x)$, which maps each element x in X to a real number in the interval [0, 1]. The function value $fA(x)$ represents the grade of membership of x in A.

A fuzzy number A (Dubois and Prade 1978) in real line k is a triangular fuzzy number if its membership function fA: $k \to [0, 1]$ is

$$fA(x) = \begin{cases} \frac{x-c}{a-c}, & c \leq x \leq a \\ \frac{x-b}{a-b}, & a \leq x \leq b \\ 0, & \text{otherwise} \end{cases}$$

with $-\infty < c \leq a \leq b < \infty$.

The triangular fuzzy number can be denoted by (c, a, b).

The parameter a gives the maximal grade of $fA(x)$, i.e. $fA(a) = 1$; it is the most probable value of the evaluation data. In addition, "c" and "b" are the lower and upper bounds of the available area for the evaluation data. They are used to reflect the fuzziness of the evaluation data. The narrower the interval $[c, b]$, the lower the fuzziness of the evaluation data.

The triangular fuzzy numbers are easy to use and easy to interpret. For example, "approximately equal to 300" can be represented by (295, 300, 306); and it can be represented more blurred by (290, 300, 313). In addition, the non-fuzzy number, an exact number, "a" can be represented by (a, a, a). For example, "300" can be represented by (300, 300, 300).

Let $A_1 = (c_1, a_1, b_1)$ and $A_2 = (c_2, a_2, b_2)$ be fuzzy numbers. According to the extension principle (Zadeh 1965), the algebraic operations of any two fuzzy numbers A_1 and A_2 can be expressed as.

- Fuzzy addition, $\oplus$:

$$A_1 \oplus A_2 = (c_1 + c_2, a_1 + a_2, b_1 + b_2),$$

- Fuzzy subtraction, $\ominus$:

$$A_1 - A_2 = (c_1 - b_2, a_1 - a_2, b_1 - c_2),$$

- Fuzzy multiplication, $\otimes$:

$$k \otimes A_2 = (kc_2, ka_2, kb_2), \quad k \epsilon K, k \geq 0,$$

$$A_1 \otimes A_2 \cong (c_1 c_2, a_1 a_2, b_1 b_2), \quad c_1 \geq 0, c_2 \geq 0,$$

- Fuzzy division, $\emptyset$:

$$A_1 \emptyset A_2 \cong (c_1/b_2, a_1/a_2, b_1/c_2), \quad c_1 \geq 0, c_2 > 0.$$

10.2.2 Linguistic Values

In fuzzy decision environments, two preference ratings can be used. They are fuzzy numbers and linguistic values characterized by fuzzy numbers (Zadeh 1975a, b, 1976). Depending on practical needs, DMs may apply one or both of them. In this paper, linguistic values characterized by triangular fuzzy numbers are used to evaluate the importance weights of all criteria and the appropriateness of alternatives versus various subjective criteria above the alternative level.

10.2.3 Ranking of Triangular Fuzzy Numbers

In fuzzy decision-making environment, ranking the alternatives under consideration is important and essential. For matching the fuzzy MCDM algorithm developed in this paper, and powerfulness in problem-solving, the graded mean integration representation method proposed by Chen and Hsieh (2000) is used to rank the final ratings of alternatives.

Let $A_i = (c_i, a_i, b_i)$, $i = 1, 2, \ldots, n$, be n triangular fuzzy numbers. By the graded mean integration representation method, the graded mean integration representation $R(A_i)$ of A_i is

$$R(A_i) = \frac{c_i + 4a_i + b_i}{6} \tag{10.1}$$

Suppose $R(A_i)$ and $R(A_j)$ are the graded mean integration representations of the triangular fuzzy numbers A_i and A_j, respectively.

Define that

$$\begin{aligned} A_i > A_j &\Leftrightarrow R(A_i) > R(A_j), \\ A_i < A_j &\Leftrightarrow R(A_i) < R(A_j), \\ A_i = A_j &\Leftrightarrow R(A_{i)} = R(A_j). \end{aligned}$$

10.2.4 Distance Between Two Triangular Fuzzy Numbers

There are many methods that can be used to characterize the inter-objects similarity (Liang et al. 2005). By considering the easy implementation and intuition, the distance measures based on triangular fuzzy numbers are utilized to build up the similarity between two objects. Because the Chen and Hsieh (2000) modified geometrical distance with distance parameter $p = 2$ can meet the concept of classical distance, and by considering the easy implementation and powerful, we utilize it to solve the distance between two triangular fuzzy numbers.

Let $A_i = (c_i, a_i, b_i)$ and $A_j = (c_j, a_j, b_j)$ be two triangular fuzzy numbers. Based on Chen and Hsieh (2000) method, the distance between A_i and A_j, denoted by d(A_i, Aj), is

$$\mathrm{d}(A_i, A_j) = \sqrt{\left\{\frac{\left[(c_i - c_j)^2 + 2(a_i - a_j)^2 + (b_i - b_j)^2\right]}{4}\right\}} \tag{10.2}$$

Entropy concept

Entropy is a measure of *unpredictability* of *information content.* To get an informal, intuitive understanding of the connection between these three English terms, consider the example of a poll on some political issue. Usually, such polls happen because the outcome of the poll isn't already known. In other words, the outcome of the poll is relatively *unpredictable*, and actually performing the poll and learning the results gives some new *information*; these are just different ways of saying that the *entropy* of the poll results is large. Now, consider the case that the same poll is performed a second time shortly after the first poll. Since the result of the first poll is already known, the outcome of the second poll can be predicted well and the results should not contain much new information; in this case the entropy of the second poll result is small relative to the first.

10.2.5 Grey Relationship Analysis

Grey system theory was developed by Deng in 1982. In grey system, all messages can be divided into three categories—the white, grey and black parts. The white part shows clear messages in a system completely, the black part has totally unknown characteristics, and the grey part happens in between and covers both known and unknown messages. This theory includes four parts (Deng 1982, 1989). It has been used in many fields, such as economy, decision, optimization, management, navigation safety, operational performance, failure model, critical success factors, customer relationship management, and effect analysis. Herein, this paper briefly reviews the mathematical equation and calculation process for the grey relation analysis.

Let $X = \{x_i | x_i = (x_{i1}, x_{i2}, \ldots, x_{in}), n \geq 3, 0 \leq i \leq m, m \geq 2\}$ be grey relation factor space, x_0 represents the referential sequence, and x_i $(i = 1, 2, \ldots, m)$ is the comparative sequence. Let x_{0k} and x_{ik} denote the respective numerals at point k ($k = 1, 2, \ldots, n$), for x_0 and x_i. Then the grey relation coefficient γ (x_{0k}, x_{ik}) of these elements at point k can be calculated by

$$\gamma(x_{0k}, x_{ik}) = \left(\min_i \min_k \Delta_{0i}(k) + \zeta \max_i \max_k \Delta_{0i}(k)\right) / \left(\Delta_{0i}(k) + \zeta \max_i \max_k \Delta_{0i}(k)\right)$$

where _0*i* (*k*) = $|x_{0k} - x_{ik}|$, and $\zeta(\zeta \in [0, 1])$ is the distinguished coefficient. The coefficient ζ can be used to change the dimension of relative value of γ (x_{0k}, x_{ik}) (Wong and Lai 2000; Wen andWu 1996). After obtaining all grey relation coefficients, the grey relation grade γ (x_0, x_i) between x_0 and x_i can be calculated by

$$\gamma(x_0, x_i) = \frac{1}{n} \sum_{k=1}^{n} \gamma(x_{0k}, x_{ik})$$

For the referential sequence x_0 and all comparative sequences x_i $(i = 1, 2, \ldots, m)$, if $\gamma(x_0, x_p)$, $\gamma(x_0, x_t)$, …, and $\gamma(x_0, x_q)$ satisfy $\gamma(x_0, x_p) > \gamma(x_0, x_t) > \cdots > \gamma(x_0, x_q)$, then the grey relational order is *x p xt … xq*.

10.3 Steps Involved in Fuzzy Integral and Grey Analysis

Step 1 Form a committee of DMs, then select the evaluation criteria and identify the perspective alternatives.
Step 2 Divide evaluation criteria into subjective and objective categories.
Step 3 Choose appropriate preference ratings for criteria weights and the superiority of alternatives versus various criteria above alternative level.
Step 4 Solve the subjective weights of all criteria above the alternative level.
Step 5 Solve the superiority of all alternatives versus all criteria above the alternative level.
Step 6 Use entropy weighting method to solve the weights of all objective criteria above the alternative level.
Step 7 Calculate the integration weights of all criteria above alternative level.
Step 8 Calculate the fuzzy grey ration grade (GRG) of all compared alternatives to reference alternative and select the optimal alternative.

10.3.1 Select Criteria

Various criteria can be considered in a multi-criteria evaluation problem. Criteria used should be identified by considering the specific requirements of the problem. The criteria can be classified into two categories: (1) subjective criteria, which have linguistic/qualitative definition, e.g. prevent noise grade; (2) objective criteria, which are defined in monetary/quantitative terms, e.g. building cost.

10.3.2 Choose Preference Rating System

Two preference ratings are used. They are triangular fuzzy numbers and linguistic values characterized by triangular fuzzy numbers. Based on the practical needs, the DMs may apply one or both of them. In this paper, weighting set $W = \{VL, L, M, H, VH\}$ and rating set $S = \{VP, P, F, G, VG\}$, where VL = Very Low, L = Low, M = Medium, H = High, VH = Very High, VP = Very Poor, P = Poor, F = Fair,G = Good, andVG = Very Good, are used to evaluate the importance weights of all criteria and the fuzzy ratings of alternatives versus various subjective criteria above the alternative level, respectively. In here, define $VL = (0, 0, 0.3)$, $L = (0, 0.3, 0.5)$, $M = (0.2, 0.5, 0.8)$, $H = (0.5, 0.7, 1)$, VH $= (0.7, 1, 1)$,$VP = (0, 0, 0.2)$, $P = (0, 0.2, 0.4)$, $F = (0.3, 0.5, 0.7)$, $G = (0.6, 0.8, 1)$, and $VG = (0.8, 1, 1)$.

10.3.3 Solve the Subjective Weights of All Criteria Above the Alternative Level

Let $w_{kj} = \_c_{kj}, a_{kj}, b_{kj}\_$, $0 \le c_{kj} \le a_{kj} \le b_{kj} \le 1$, $k = 1, 2, \ldots, n$; $j = 1, 2, \ldots, r$, be the importance degrees assigned to criterion C_k by DM D_j. Then, the weight w_k of C_k can be calculated by

$$
\begin{aligned}
w_k &= (1/r) \otimes \left(w_{k1} \oplus \cdots \oplus w_{kj} \oplus \cdots \oplus w_{kr}\right) \\
c_k &= \sum_{j=1}^{r} c_{kj}/r, \quad a_k = \sum_{j=1}^{r} a_{kj}/r, \quad b_k = \sum_{j=1}^{r} b_{kj}/r \\
w_k &= (c_k, a_k, b_k).
\end{aligned}
\tag{10.3}
$$

10.3.4 Solve the Superiority Ratings of All Alternatives Versus All Criteria Above the Alternative Level

Let $M_{ikj} = (q_{ikj}, o_{ikj}, f_{ikj})$, $0 \le q_{ikj} \le o_{ikj} \le f_{ikj} \le 1$; $i = 1, 2, \ldots, m$; $k = p + 1, p + 2, \ldots, n$; $j = 1, 2, \ldots, r$, be the linguistic rating assigned to alternative A_i by DM B_j for the subjective criterion C_k. Then, the linguistic rating M_{ik} of alternative A_i for the subjective criterion C_k can be calculated by

$$
\begin{aligned}
M_{ik} &= (1/r) \otimes \left(M_{ik1} \oplus \cdots \oplus M_{ikj} \oplus \cdots \oplus M_{ikr}\right), \quad i = 1, 2, \ldots, m; \\
&\quad k = p + 1, p + 2, \ldots, n. \\
q_{ik} &= \sum_{j=1}^{r} q_{ikj}/r, \quad o_{ik} = \sum_{j=1}^{r} o_{ikj}/r, \quad f_{ik} = \sum_{j=1}^{r} f_{ikj}/r, \\
M_{ik} &= (q_{ik}, o_{ik}, f_{ik}), \quad i = 1, 2, \ldots, m; \quad k = p + 1, p + 2, \ldots, n.
\end{aligned}
\tag{10.4}
$$

Allow $N_{ik} = (h_{ik}, e_{ik}, g_{ik})$, $i = 1, 2, \ldots, m$; $k = 1, 2, \ldots, p$, to be the fuzzy evaluation value assigned to alternative Ai for objective criterion C_k. To ensure compatibility between fuzzy evaluation value of objective criterion and linguistic rating of subjective criteria, fuzzy evaluation value has to be converted to dimensionless scales. The rating of alternative A_i versus criterion C_k can be defined as:

(1) For the benefit criterion

$$
T_{ik} = (h_{ik}/t_k, e_{ik}/t_k, g_{ik}/t_k), \tag{10.5}
$$

where

$$t_k = \max_i\{g_{ik}\}, \tag{10.6}$$

(2) For the cost criterion

$$T_{ik} = (t_k/g_{ik}, t_k/e_{ik}, t_k/h_{ik}), \tag{10.7}$$

where

$$t_k = \min_i\{h_{ik}\}. \tag{10.8}$$

Define

$$X_{ik} = \begin{cases} T_{ik}, & i = 1, 2, \ldots, m; \quad k = 1, 2, \ldots, p; \\ M_{ik}, & i = 1, 2, \ldots, m; \quad k = p + 1, p + 2, \ldots, n \end{cases} \tag{10.9}$$

where X_{ik}, $i = 1, 2, \ldots, m$; $k = 1, 2, \ldots, n$, denotes the superiority rating of alternative A_i for criterion C_k.

10.3.5 Use Entropy Weighting Method to Solve the Weights of All Objective Criteria Above the Alternative Level

The entropy weighting method (Zeleny 1982) can effectively measure the average essence of information quantity. And the larger the entropy value, the lower the information express quantity (Zeleny 1982; Feng and Chen 1992; Ding and Liang 2005). This paper tries to solve the objective weight of objective criteria, e.g. return on assets, above the alternative level. Thus, it can represent actual conditions of decision making and express the explanation ability and reliability of criteria. Let m and n denote the numbers of alternatives and the criteria above the alternative level, respectively. The steps utilized to find the objective weight of objective criteria can be summarized as follows.

Step 1 Allow X_{ik}, $i = 1, 2, \ldots, m$; $k = 1, 2, \ldots, p$; $p < n$, to be the triangular fuzzy number evaluation value of alternative A_i under objective criterion C_k above the alternative level. Let $d_{ik} = R(X_{ik})$ $i = 1, 2, \ldots, m$; $k = 1, 2, \ldots, p$; $p < n$, denote the graded mean integration representation of X_{ik}, then $0 \leq d_{ik} \leq 1$. Define

$$D = [d_{ik}]_{m \times p}, \quad i = 1, 2, \ldots, m; \quad k = 1, 2, \ldots, p,$$

$$D_k = \sum_{i=1}^{m} d_{ik}, \quad k = 1, 2, \ldots, p.$$

Step 2 The entropy measure of the *t*th objective evaluation criterion contrast intensity is

$$e_k = -q \sum_{i=1}^{m} \frac{d_{ik}}{D_k} \ln \frac{d_{ik}}{D_k}$$

where $q = 1/\ln m > 0$, and $e_k \geq 0$.
Step 3 Compute the total entropy:

$$e = \sum_{k=1}^{p} e_k.$$

Step 4 The objective weight λ_k of the *t*th objective evaluation criterion can be calculated by

$$\lambda_k = \frac{1 - e_k}{\sum_{t=1}^{p}(1 - e_t)} = \frac{1 - e_k}{p - e}, \quad 0 \leq \lambda_k \leq 1, \ \sum_{k=1}^{p} \lambda_k = 1. \tag{10.10}$$

10.3.6 Calculate the Integration Weights of All Criteria Above Alternative Level

Let *wk* = (*ck*, *ak*, *bk*), *k* = 1, 2, …, *n*, denote the subjective weights of *n* criteria above alternative level. Allow *uk*, *k* = 1, 2, …, *n*, to be the normalized subjective weight of all criteria above the alternative level. Define

$$u_k = R(w_k) / \sum_{k=1}^{n} R(w_k), \quad k = 1, 2, \ldots, n. \tag{10.11}$$

In addition, suppose that there are $p (p \leq n)$ objective criteria. Allow λ_k, $k = 1, 2, \ldots, p$, to be the objective weights of *p* criteria obtained by Eq. (10.10). Define

$$s_k = u_k / \sum_{k=1}^{p} u_k, \quad k = 1, 2, \ldots, p.$$

Then the integrated weight v_k of the *k*th objective criterion can be obtained:

$$v_k = \left[(\lambda_k \times s_k) / \sum_{k=1}^{p} (\lambda_k \times s_k) \right] \times \left(\sum_{k=1}^{p} u_k \right), \quad k = 1, 2, \ldots, p. \tag{10.12}$$

Thus, the integration weights β_k, $k = 1, 2, 3, \ldots, n$, of all criteria above alternative level can be calculated by

$$\beta_k = \begin{cases} v_k, & k = 1, 2, \ldots, p; \\ u_k, & k = p+1, p+2, \ldots, n \end{cases} \tag{10.13}$$

10.3.7 *Calculate the Grey Rational Grade of All Compared Alternatives to Reference Alternative*

Let X_i, $i = 1, 2, \ldots, m$ be the superiority ratings of m alternatives described by triangular fuzzy numbers or linguistic values characterized by triangular fuzzy numbers. Let $X_0 = (X_{01}, X_{02}, \ldots, X_{0n})$ and $X_i = (X_{i1}, X_{i2}, \ldots, X_{in})$ $(i = 1, 2, \ldots, m)$ be the referential sequence and comparative sequences, respectively. In addition, allow d_{0i} (k) to be the distance of fuzzy difference between the referential pattern X_{0k} and a comparative pattern X_{ik}, where $X0k$ is the fuzzy message of X_0 and X_{ik} is the fuzzy message of X_i at point (criterion) k. Define d_{0i} (k) as

$$d_{0i}(k) = d(X_{0k}, X_{ik})$$

Define the grey relation coefficient (GRC) of X_0 and X_i at point (criterion) k as

$$\gamma(X_{0k}, X_{ik}) = \left(\min_i \min_k d_{0i}(k) + \zeta \max_i \max_k d_{0i}(k)\right) / \left(d_{0i}(k) + \zeta \max_i \max_k d_{0i}(k)\right) \tag{10.14}$$

The GRC can be utilized to reflect the grey relation of X_i compared to X_0 at point (criterion) k.

In Eq. 10.14, max max d_{oi} (k) and min min d_{oi} (k) denote the maximum and minimum elements of the d_{oi} (k), respectively. The distinguishing coefficient ζ, which is between 0 and 1, can be used to change the dimension of relative values of λ (X_{ok}, X_{ik}). In general, $\zeta = 0.5$ is better when the relative conditions among series and elements are uncertain (Deng 1989).

Define the GRG of X_i compared to X_0 as:

$$\gamma(X_0, X_i) = \sum_{k=1}^{n} \beta_k \times \gamma(X_{0k}, X_{ik}) \tag{10.15}$$

where β_k is the integration weight of criterion C_k. When the number of GRC is too much and messages are too discrete, the GRG is used to characterize the grey relational grade of X_i compared to X_0. When the GRG is larger indicates that the series X_i and X_0 are highly related. On the contrary, these two series are lowly related when the GRG is littler.

10.4 Illustrative Application of Fuzzy Integral and Grey Analysis for Stock Selection Problem

The selection of stock portfolio is a case of an ill-defined MCDM problem. In the past, the decision as to which stocks to include within an investment portfolio was generally made on a qualitative or intuitive basis. However, in recent decades, many applications have been proposed for predicting future stock market trends and selecting suitable stock portfolios. Typical mechanisms include the use of genetic algorithms to choose optimal portfolios, the application of neural networks to predict real-world stock trends, the integration of fuzzy logic and forecasting techniques to create artificial intelligence systems for market tracking and forecasting purposes, the use of statistical approaches for the forecasting of economic indicators, the application of rough set (RS) theory to predict the S&P 100 index, fuzzy grey relation method for multiple criteria decision making and so on.

The case study is to select a most appropriate stock for investment. After primary screening, 10 companies are selected for further evaluation. The evaluation is done by three decision makers. Eight selection criteria are considered. The list of the shortlisted ten organization and eight criteria out of which seven are of objective type and one is subjective type and their respective values (collected from various data source) are mentioned in Table 10.1.

Market Capitalization

Market capitalization (market cap) is the total market value of the shares outstanding of a publicly traded company; it is equal to the share price times the number of shares outstanding. As outstanding stock is bought and sold in public markets, capitalization could be used as a proxy for the public opinion of a company's net worth and is a determining factor in some forms of stock valuation. The investment community uses this figure to determine a company's size, as opposed to sales or total asset figures.

Market cap is given by the formula $MC = N \times P$, where MC is the market capitalization, N is the number of shares outstanding, and P is the current price per share.

For example, if some company has 4 million shares outstanding and the closing price per share is $20, its market cap is then $80 million. If the closing price per share rises to $21, the market cap becomes $84 million. If it drops to $19 per share, the market cap falls to $76 million.

Table 10.1 Definition of criteria

Stocks	Criteria							
	Mkt cap	Profit growth (%) (5-year avg.)	Dividend payout (%) (5--year avg.)	Return on equity (%) (5--year avg.)	Dividend yield (%)	EPS	PE	Company management
TCS	4,837,475	23.2	43.8	37.3	3.2	110	22.1	Good
Infosys Ltd.	2,809,182	14.6	39.3	25.2	2.4	56.55	20.64	Very good
ITC Ltd,	2,643,518	18.3	57	31.7	1.9	12.11	26.53	Good
Hind. Unilever	1,881,669	15.1	73.9	102.7	1.7	18.9	46.99	Fair
Coal India	1,824,165	7.4	66.6	34.9	7.2	22.53	12.45	Fair
HDFC	1,757,354	22	28.5	21.1	1.3	56.59	19.07	Good
Wipro	1,394,121	13.3	28.6	22.8	2.1	36.17	15.12	Very good
HCL Tech.	1,279,185	42.2	19.7	28.1	3.3	38.95	21.35	Good
Axis Bank	1,065,667	24.6	27.6	17	1	34.8	12.11	Good
Hindustan Zinc	784,642	15.1	18	20.2	2.4	18.97	8.45	Fair

Why It's Important

A common misconception is that the higher the stock price, the larger the company. Stock price, however, may misrepresent a company's actual worth. If we look at two fairly large companies, IBM and Microsoft, on 15 February 2013 stock prices were \$199.98 and \$28.05, respectively. Although IBM's stock price was higher, we can see that MSFT's market cap of \$234.6 billion was actually larger than IBM's \$225.1 billion. If we compare the two companies by solely looking at their stock prices, we would not be comparing their true values, which are affected by the number of outstanding shares each company has.

The classification of companies into different caps also allows investors to gauge the growth versus risk potential. Historically, large caps have experienced slower growth with lower risk. Meanwhile, small caps have experienced higher growth potential, but with higher risk.

Profit Growth

Companies normally want profits to grow. To calculate profit growth, analysts use a percent change as measure. This shows the percentage the profit grew from one period to another. Analysts can use any period to determine the profit growth, such as weekly, monthly, quarterly, semi-annually or annually.

Why It's Important.

Investors want to know whether a company's stock will provide a good return, as measured by increase in stock price and dividend payments. Earnings and net income drive stock prices, and hence investors will have a concern for any drop in gross or net profit margins. Growth-stock investor's pay higher prices for stocks of companies with expanding profit margins and sell companies that report dropping margins. Income investors look for stocks that offer good value and steady or increasing dividends. A drop in profit margins could threaten a company's ability to increase dividends.

Dividend Payout Ratio:

Dividend payout ratio is the fraction of net income a firm pays to its stockholders in dividends:

$$\text{Dividend payout ratio} = \frac{\text{Dividends}}{\text{Net Income for the same period}}$$

According to Financial Accounting by Walter T. Harrison, the calculation for the payout ratio is as follows:

$$\text{Payout Ratio} = (\text{Dividends} - \text{Preferred Stock Dividends})/\text{Net Income}$$

Why It's Important

The part of the earnings not paid to investors is left for investment to provide for future earnings growth. Investors seeking high current income and limited capital growth prefer companies with high dividend payout ratio. However, investors seeking capital growth may prefer lower payout ratio because capital gains are taxed at a lower rate. High growth firms in early life generally have low or zero payout ratios. As they mature, they tend to return more of the earnings back to investors.

Return on Equity (ROE)

Return on equity (ROE) is the amount of net income returned as a percentage of shareholders equity. Return on equity measures a corporation's profitability by revealing how much profit a company generates with the money shareholders have invested.

ROE is expressed as a percentage and calculated as:

$$\text{Return on Equity} = \text{Net Income}/\text{Shareholder's Equity}$$

Why It's Important

Investors want to see a high return on equity ratio because this indicates that the company is using its investors' funds effectively. Higher ratios are almost always better than lower ratios, but have to be compared to other companies' ratios in the industry. Since every industry has different levels of investors and income, ROE can't be used to compare companies outside of their industries very effectively.

Many investors also choose to calculate the return on equity at the beginning of a period and the end of a period to see the change in return. This helps track a company's progress and ability to maintain a positive earnings trend.

DIVIDEND YIELD:

A financial ratio that indicates how much a company pays out in dividends each year relative to its share price. Dividend yield is represented as a percentage and can be calculated by dividing the dollar value of dividends paid in a given year per share of stock held by the dollar value of one share of stock.

The formula for calculating dividend yield may be represented as follows:

$$= \frac{\textbf{Annual Dividends Per share}}{\textbf{Price Per Share}}$$

Why It's Important

Investors who require a minimum stream of cash flow from their investment portfolio can secure this cash flow by investing in stocks paying relatively high, stable dividend yields. Yet, high dividends may often come at the cost of growth potential. Every dollar a company is paying in dividends to its shareholders is a dollar that company is not reinvesting in itself in an effort to make capital gains. While being paid for holding a stock is attractive to many, and for good reason, shareholders can earn high returns if the value of their stock increases while they hold it. In other words, when companies pay high dividends it may come at a cost.

10.4.1 Step-by-Step Application of the Technique

Stock	Mkt cap in crore	Profit growth (%)(5-yr avg)	Dividend payout (%)(5-yr avg)	Return on equity (%)(5-yr avg)	Dividend yield (%)	EPS	PE	Company management evaluation by 3 decision makers		
TCS	4,837,475	23.2	43.8	37.3	3.2	110	22.1	Good	very good	Good
Infosys Ltd.	2,809,182	14.6	39.3	25.2	2.4	56.55	20.64	Very good	Very good	Very good
ITC Ltd.	2,643,518	13.3	57	31.7	1.9	12.11	26.53	Good	Poor	Fair
Hind. Unilever	1,881,669	15.1	73.9	102.7	1.7	18.9	46.99	Fair	Good	Fair
Coal India	1,824,165	7.4	66.6	34.9	7.2	22.53	12.45	Fair	Very poor	Fair
HDFC	1,757,354	22	28.5	21.1	1.3	56.59	19.07	Good	Good	Good
Wipro	1,394,121	13.3	28.6	22.8	2.1	36.17	15.12	Very good	Poor	Poor
HCL Tech.	1,279,185	42.2	19.7	28.1	3.3	38.95	21.35	Good	Good	Poor
Axis Bank	1,065,667	24.6	27.6	17	1	34.8	12.11	Good	Poor	Good
Hindustan Zinc	784,642	15.1	18	20.2	2.4	18.97	8.45	Fair	Fair	Good

Criteria Evaluation by Three Experts

	Mkt cap	Profit growth (%) (5-year avg)	Dividend payout (%) (5-year avg)	Return on equity (%) (5-yr avg)	Dividend yield (%)	EPS	PE	Company management
D1	High	Very high	Medium	Medium	Low	Medium	Medium	Low
D2	Low	Medium	High	Medium	Medium	Low	Very high	High
D3	Medium	High	Medium	Low	Very low	Very high	Low	Very high

Weighing set assumed based on triangular fuzzy theory

Weighting set			
Very low	0	0	0.3
Low	0	0.3	0.5
Medium	0.2	0.5	0.8
High	0.5	0.7	1
Very high	0.7	1	1

Rating set assumed based on triangular fuzzy theory

Rating set			
Very poor	0	0	0.2
Poor	0	0.2	0.4
Fair	0.3	0.5	0.7
Good	0.6	0.8	1
Very good	0.8	1	1

Converting Data into Fuzzy Triangleular Form

Mkt cap Lakh crore			Profit growth (%)(5-year avg)			Dividend payout (%)(5-year avg)			Return on equity (%)(5-year avg)		
46	48.37475	50	22	23.2	25	42	43.8	45	35	37.3	39
25	28.09182	30	13	14.6	16	36	39.3	42	24	25.2	26
23	26.43518	29	16	18.3	20	55	57	60	30	31.7	35
17	18.81669	20	14	15.1	16	71	73.9	75	100	102.7	105
17.5	18.24165	19.5	6	7.4	8	63	66.6	69	32	34.9	36
15	17.57354	19	20	22	24	25	28.5	30	20	21.1	23
12	13.94121	16	12	13.3	15	26	28.6	29	20	22.8	26
11	12.79185	15	40	42.2	45	16	19.7	21	26	28.1	30
8	10.65667	12	22	24.6	26	25	27.6	30	15	17	19
6	7.84642	9	14	15.1	18	16	18	19	18	20.2	22

Dividend yield (%)			EPS			PE		
3	3.2	4	105	110	118	20	22.1	23
2	2.4	3	50	56.55	60	19	20.64	22
1	1.9	2	10	12.11	14	23	26.53	28
1	1.7	2	15	18.9	20	45	46.99	49
6	7.2	8	20	22.53	25	10	12.45	16
1	1.3	2	50	56.59	60	15	19.07	23
2	2.1	3	33	36.17	39	14	15.12	16
3	3.3	4	35	38.95	40	20	21.35	24
0.5	1	1.8	32	34.8	37	10	12.11	15
2	2.4	3	15	18.97	20	7	8.45	9

Company management	Decision maker 1			Decision maker 2				Decision maker 3			
Good	0.6	0.8	1	Very good	0.8	1	1	Good	0.6	0.8	1
Very good	0.8	1	1	Very good	0.8	1	1	Very good	0.8	1	1
Good	0.6	0.8	1	Poor	0	0.2	0.4	Fair	0.3	0.5	0.7
Fair	0.3	0.5	0.7	Good	0.6	0.8	1	Fair	0.3	0.5	0.7
Fair	0.3	0.5	0.7	Very poor	0	0	0.2	Fair	0.3	0.5	0.7
Good	0.6	0.8	1	Good	0.6	0.8	1	Good	0.6	0.8	1
Very good	0.8	1	1	Poor	0	0.2	0.4	Poor	0	0.2	0.4
Good	0.6	0.8	1	Good	0.6	0.8	1	Poor	0	0.2	0.4
Good	0.6	0.8	1	Poor	0	0.2	0.4	Good	0.6	0.8	1
Fair	0.3	0.5	0.7	Fair	0.3	0.5	0.7	Good	0.6	0.8	1

In the subsequent phase, now we will solve the superiority ratings.
Finding X_{ik}.

(1) For the benefit criterion

$$T_{ik} = (h_{ik}/t_k, e_{ik}/t_k, g_{ik}/t_k),$$

where

$$t_k = \max_i \{g_{ik}\},$$

(2) For the cost criterion

$$T_{ik} = (t_k/g_{ik}, t_k/e_{ik}, t_k/h_{ik}),$$

where

$$t_k = \min_i \{h_{ik}\}.$$

Define

$$X_{ik} = \begin{cases} T_{ik}, & i = 1, 2, \dots, m; \quad k = 1, 2, \dots, p; \\ M_{ik}, & i = 1, 2, \dots, m; \quad k = p+1, p+2, \dots, n \end{cases}$$

X_{11}	0.92	0.967495	1	X_{12}	0.488889	0.515556	0.555556	X_{13}	0.56	0.584	0.6
X_{21}	0.5	0.561836	0.6	X_{22}	0.288889	0.324444	0.355556	X_{23}	0.48	0.524	0.56
X_{31}	0.46	0.528704	0.58	X_{32}	0.355556	0.406667	0.444444	X_{33}	0.733333	0.76	0.8
X_{41}	0.34	0.376334	0.4	X_{42}	0.311111	0.335556	0.355556	X_{43}	0.946667	0.985333	1
X_{51}	0.35	0.364833	0.39	X_{52}	0.133333	0.164444	0.177778	X_{53}	0.84	0.888	0.92
X_{61}	0.3	0.351471	0.38	X_{62}	0.444444	0.488889	0.533333	X_{63}	0.333333	0.38	0.4
X_{71}	0.24	0.278824	0.32	X_{72}	0.266667	0.295556	0.333333	X_{73}	0.346667	0.381333	0.386667
X_{81}	0.22	0.255837	0.3	X_{82}	0.888889	0.937778	1	X_{83}	0.213333	0.262667	0.28
X_{91}	0.16	0.213133	0.24	X_{92}	0.488889	0.546667	0.577778	X_{93}	0.333333	0.368	0.4
X_{101}	0.12	0.156928	0.18	X_{102}	0.311111	0.335556	0.4	X_{103}	0.213333	0.24	0.253333

X_{14}	0.333333	0.355238	0.371429	X_{15}	0.375	0.4	0.5	X_{16}	0.889831	0.932203	1
X_{24}	0.228571	0.24	0.247619	X_{25}	0.25	0.3	0.375	X_{26}	0.423729	0.479237	0.508475
X_{34}	0.285714	0.301905	0.333333	X_{35}	0.125	0.2375	0.25	X_{36}	0.084746	0.102627	0.118644
X_{44}	0.952381	0.978095	1	X_{45}	0.125	0.2125	0.25	X_{46}	0.127119	0.160169	0.169492
X_{54}	0.304762	0.332381	0.342857	X_{55}	0.75	0.9	1	X_{56}	0.169492	0.190932	0.211864
X_{64}	0.190476	0.200952	0.219048	X_{65}	0.125	0.1625	0.25	X_{66}	0.423729	0.479576	0.508475
X_{74}	0.190476	0.217143	0.247619	X_{75}	0.25	0.2625	0.375	X_{76}	0.279661	0.306525	0.330508
X_{84}	0.247619	0.267619	0.285714	X_{85}	0.375	0.4125	0.5	X_{86}	0.29661	0.330085	0.338983
X_{94}	0.142857	0.161905	0.180952	X_{95}	0.0625	0.125	0.225	X_{96}	0.271186	0.294915	0.313559
X_{104}	0.171429	0.192381	0.209524	X_{105}	0.25	0.3	0.375	X_{106}	0.127119	0.160763	0.169492

X_{17}	0.408163	0.45102	0.469388	X_{18}	0.666667	0.866667	1
X_{27}	0.387755	0.421224	0.44898	X_{28}	0.8	1	1
X_{37}	0.469388	0.541429	0.571429	X_{38}	0.3	0.5	0.7
X_{47}	0.918367	0.95898	1	X_{48}	0.4	0.6	0.8
X_{57}	0.204082	0.254082	0.326531	X_{58}	0.2	0.333333	0.533333
X_{67}	0.306122	0.389184	0.469388	X_{68}	0.6	0.8	1
X_{77}	0.285714	0.308571	0.326531	X_{78}	0.266667	0.466667	0.6
X_{87}	0.408163	0.435714	0.489796	X_{88}	0.4	0.6	0.8
X_{97}	0.204082	0.247143	0.306122	X_{98}	0.4	0.6	0.8
X_{107}	0.142857	0.172449	0.183673	X_{108}	0.4	0.6	0.8

After finding X_{ij} we need to find E_k that is entropy calculation which will be done by following formula

$$e_k = -q \sum_{i=1}^{m} \frac{d_{ik}}{D_k} \ln \frac{d_{ik}}{D_k}$$

where $q = 1/\ln m > 0$, and $e_k \geq 0$. Here $m = 10$ from the data we have 10 companies

$$D_k = \sum_{i=1}^{m} d_{ik}, \quad k = 1, 2, \ldots, p.$$

$$d_{ik} = R(X_{ik}).$$

For example.

Let $A_i = (c_i, a_i, b_i), i = 1, 2, \ldots, n$, be n triangular fuzzy numbers.

$$R(A_i) = \frac{c_i + 4a_i + b_i}{6}$$

Then we will compute total entropy

$$e = \sum_{k=1}^{p} e_k.$$

e_k	
e_1	0.941931
e_2	0.958065
e_3	0.954096
e_4	0.923346
e_5	0.931811
e_6	0.9132
e_7	0.952171

And total entropy $(e) = 6.574621$.

Next step is to find lambda(k) objective weight

$$\lambda_k = \frac{1 - e_k}{\sum_{t=1}^{p} (1 - e_t)} = \frac{1 - e_k}{p - e}, \quad 0 \leq \lambda_k \leq 1, \sum_{k=1}^{p} \lambda_k = 1.$$

lamda(k)	
lamda(l)	0.040739
lamda(2)	0.02942
lamda(3)	0.032205
lamda(4)	0.053778
lamda(5)	0.047839
lamda(6)	0.060896
lamda(7)	0.033555

u_k		s_k		v_k		B_k	
u_1	0.116129	s_1	0.136778	v_1	0.1146	B_1	0.1146
u_2	0.167742	s_2	0.197568	v_2	0.119541	B_2	0.119541
u_3	0.132903	s_3	0.156535	v_3	0.103678	B_3	0.103678
u_4	0.099355	s_4	0.117021	v_4	0.129428	B_4	0.129428
u_5	0.064516	s_5	0.075988	v_5	0.074763	B_5	0.074763
u_6	0.134194	s_6	0.158055	v_6	0.197948	B_6	0.197948
u_7	0.134194	s_7	0.158055	v_7	0.109075	B_7	0.109075
u_8	0.150968					B_8	0.150968

Next calculate the grey rational grade of all compared alternatives to reference alternative.

Let $X_0 = (X_{01}, X_{02}, \ldots, X_{0n})$ and $X_i = (X_{i1}, X_{i2}, \ldots, X_{in})$ ($i = 1, 2, 3, \ldots, m$) be the referential sequence and comparative sequences, respectively. In addition, allow $d_{0i}(k)$ to be the distance of fuzzy difference between the referential pattern X_{0k} and a comparative pattern X_{ik}, where X_{0k} is the fuzzy message of X_0 and X_{ik} is the fuzzy message of X_i at point (criterion) k. Define d_{0i} as.

$$D_{0i} = d(X_{0k}, X_{ik})$$

$d_{0i}(k)$	$i=1$	$i=2$	$i=3$	$i=4$	$i=5$	$i=6$	$i=7$	$i=8$
$k=1$	0	0.422515	0.397381	0.623343	0.471865	0	0.514272	0.11547
k=2	0.407896	0.617994	0.457448	0.738164	0.583229	0.466156	0.539336	0
$k=3$	0.439624	0.537876	0.216253	0.671445	0.676561	0.836861	0.428356	0.458258
$k=4$	0.590623	0.607141	0	0	0.688919	0.784784	0	0.360555
$k=5$	0.596532	0.781505	0.095819	0.649065	0	0.748166	0.699558	0.60553
$k=6$	0.618015	0.452303	0.606019	0.77432	0.714307	0.465991	0.571342	0.173205
$k=7$	0.684349	0.643526	0.605353	0.759057	0.602793	0.633141	0.651896	0.503322
$k=8$	0.705853	0	0.724685	0.710008	0.465279	0.615213	0.516776	0.360555

(continued)

(continued)

$d_{0i}(k)$	$i=1$	$i=2$	$i=3$	$i=4$	$i=5$	$i=6$	$i=7$	$i=8$
$k=9$	0.757186	0.401313	0.612041	0.815246	0.754077	0.645406	0.708007	0.360555
$k=10$	0.310314	0.595645	0.742686	0.785722	0.583229	0.784493	0.79137	0.360555

Define the grey relation coefficient (GRC) of X_0 and X_i at point (criterion) k *as*

$$\gamma(X_{0k}, X_{ik}) = \left(\min_i \min_k d_{0i}(k) + \zeta \max_i \max_k d_{0i}(k)\right) / \left(d_{0i}(k) + \zeta \max_i \max_k d_{0i}(k)\right)$$

$$\zeta = 0.5$$

$\max_i \max_k d_{0i}(k)$	0.836861
$\min_i \min_k d_{0i}(k)$	0

x_{01}, x_{11}	1	X_{02}, X_{12}	0.497572	X_{03}, X_{13}	0.512901	X_{04}, X_{14}	0.401652	X_{05}, X_{15}	0.469991	X_{06}, X_{16}	1	X_{07}, X_{17}	0.448622	X_{08}, X_{18}	0.783724
x_{01}, x_{21}	0.506374	X_{02}, X_{22}	0.403725	X_{03}, X_{23}	0.477727	X_{04}, X_{24}	0.361778	X_{05}, X_{25}	0.417737	X_{06}, X_{26}	0.473024	X_{07}, X_{27}	0.436881	X_{08}, X_{28}	1
x_{01}, x_{31}	0.48765	X_{02}, X_{32}	0.437548	X_{03}, X_{33}	0.659274	X_{04}, X_{34}	0.383925	X_{05}, X_{35}	0.382131	X_{06}, X_{36}	0.333333	X_{07}, X_{37}	0.494139	X_{08}, X_{38}	0.477285
x_{01}, x_{41}	0.414676	X_{02}, X_{42}	0.407997	X_{03}, X_{43}	1	X_{04}, X_{44}	1	X_{05}, X45	0.377867	X_{06}, X_{46}	0.34776	X_{07}, X_{47}	1	X_{08}, X_{48}	0.537148
x_{01}, x_{51}	0.412262	X_{02}, X_{52}	0.348711	X_{03}, X_{53}	0.813672	X_{04}, X_{54}	0.391974	X_{05}, X_{55}	1	X_{06}, X_{56}	0.358676	X_{07}, X_{57}	0.374271	X_{08}, X_{58}	0.408639
x_{01}, x_{61}	0.403717	X_{02}, X_{62}	0.480549	X_{03}, X_{63}	0.408444	X_{04}, X_{64}	0.350811	X_{05}, X_{65}	0.369398	X_{06}, X_{66}	0.473112	X_{07}, X_{67}	0.422754	X_{08}, X_{68}	0.707244
x_{01}, x_{71}	0.379432	X_{02}, X_{72}	0.394018	X_{03}, X_{73}	0.40871	X_{04}, X_{74}	0.355359	X_{05}, X_{75}	0.409734	X_{06}, X_{76}	0.39791	X_{07}, X_{77}	0.390937	X_{08}, X_{78}	0.453951
x_{01}, x_{81}	0.372175	X_{02}, X_{82}	1	X_{03}, X_{83}	0.366044	X_{04}, X_{84}	0.370805	X_{05}, X_{85}	0.473493	X_{06}, X_{86}	0.404811	X_{07}, X_{87}	0.44742	X_{08}, X_{88}	0.537148
x_{01}, x_{91}	0.355924	X_{02}, X_{92}	0.510441	X_{03}, X_{93}	0.406057	X_{04}, X_{94}	0.339174	X_{05}, X_{95}	0.356868	X_{06}, X_{96}	0.393322	X_{07}, X_{97}	0.371464	X_{08}, X_{98}	0.537148
x_{01}, x_{101}	0.340535	X_{02}, X102	0.412623	X_{03}, X_{103}	0.360369	X_{04}, X_{104}	0.34749	X_{05}, X_{105}	0.417737	X_{06}, X_{106}	0.347845	X_{07}, X_{107}	0.345867	X_{08}, X_{108}	0.537148

Calculation of Final GRG Value for all Companies

$$\gamma(X_0, X_i) = \sum_{k=1}^{n} \beta_k \times \gamma(X_{0k}, X_{ik})$$

GRG1	0.1146	0.05948	0.053177	0.051985	0.035138	0.197948	0.048933	0.118317	0.679578
GRG2	0.05803	0.048262	0.04953	0.046824	0.031231	0.093634	0.047653	0.150968	0.526132
GRG3	0.055885	0.052305	0.068352	0.04969	0.028569	0.065983	0.053898	0.072055	0.446737
GRG4	0.047522	0.048772	0.103678	0.129428	0.02825	0.068838	0.109075	0.081092	0.616656
GRG5	0.047245	0.041685	0.08436	0.050732	0.074763	0.070999	0.040824	0.061691	0.472299
GRG5	0.046266	0.057445	0.042347	0.045405	0.027617	0.093652	0.046112	0.106771	0.465614
GRG7	0.043483	0.047101	0.042374	0.045993	0.030633	0.078765	0.042642	0.068532	0.399523
GRG8	0.042651	0.119541	0.037951	0.047992	0.0354	0.080132	0.048802	0.081092	0.493561
GRG9	0.040789	0.061019	0.042099	0.043898	0.02668	0.077857	0.040517	0.081092	0.413952
GRG10	0.039025	0.049325	0.037362	0.044975	0.031231	0.068855	0.037726	0.081092	0.389591

By comparing various values of GRG for 10 companies, we get the following order

$$\text{GRG1} > \text{GRG4} > \text{GRG2} > \text{GRG8} > \text{GRG5} > \text{GRG6} > \text{GRG3} > \text{GRG9} > \text{GRG7} > \text{GRG10}$$

The most appropriate company is TCS (GRG1).

10.5 Advantages of Fuzzy Integral and Grey Analysis

- The use of triangular fuzzy numbers and linguistic values characterized by triangular fuzzy numbers accommodates the subjectivity of human opinions.
- The entropy weighting method can efficiently grasp the actual conditions of decision making.
- The process of decision making becomes clear and definite by using the concept of ideal, since the concept establishes an ideal to stimulate the creativity and invention of new ways and directions for selecting alternatives.
- The fuzzy grey relation method can overcome the problems of small size and unknown distribution of samples in decision process.

10.6 Limitation of Fuzzy Integral and Grey Analysis

- Outcome of this method is very much dependent on the criteria selected for evaluation. A careful consideration is necessary while selecting the criteria.
- One frequently encountered problem in selection process is the halo effect, which implies a decision maker's tendency to let one attribute of the candidate or organization influence their overall assessment.

This chapter has demonstrated the use of Fuzzy Integral and Grey relation analysis which combines the use of Grey and Fuzzy theory. The next chapter will discuss the application of one of the recent techniques in the domain of MCDM—Multi-Objective Optimization on the Basis of Ratio Analysis Method (MOORA). It employs separate mathematical models to benefit the non-benefit criteria and accommodates graphical qualitative criteria of the decision matrix. Due to this added advantage, in MOORA method, the chance of losing information is very small.

Selected Bibliography and Further Reading

1. Liao, M.-S., Liang, G.-S., & Chen, C.-Y. (2013). Fuzzy grey relation method for multiple criteria decision-making problems. *Quality & Quantity, 47*, 3065–3077.

Chapter 11
Multi-objective Optimization on the Basis of Ratio Analysis Method (MOORA)

11.1 Background

Multi-Objective Optimization on the Basis of Ratio Analysis Method (MOORA) is developed by Brauers and Zavadskas in 2006. The method has been successfully employed for the various problems such as material selection, performance evaluation in real estate sector and contractor selection, design selection, robot selection, personnel selection, quality control, dynamic scheduling in production systems, location selection, firm selection, healthcare waste management, etc. MOORA uses a ratio system in which each response of an alternative on an objective is compared to a denominator, which is representative for all alternatives concerning that objective. The procedure of MOORA is illustrated in Fig. 11.1.

11.2 Step-by-Step Procedure of MOORA

Step 1. Decision matrix which shows values of different alternatives for different objective is created

$$X = \begin{bmatrix} x_{11} \; \ldots \; x_{1j} \; \ldots \; x_{1n} \\ \ldots \; \ldots \; \ldots \; \ldots \; \ldots \\ x_{i1} \; \ldots \; x_{ij} \; \ldots \; \ldots \\ \ldots \; \ldots \; \ldots \; \ldots \; \ldots \\ x_{m1} \; \ldots \; \ldots \; \ldots \; x_{mn} \end{bmatrix}$$

In this, matrix x_{ij} shows value of alternative i according to objective or criteria j. m is number of alternatives and n is number of objective or criteria.

Step 2. Due to differences in criteria, values of measurements have different units. Normalization is required so that values of measurements transform into dimensionless numbers. These dimensionless numbers are in [0, 1]

J. J. Thakkar, *Multi-Criteria Decision Making*, Studies in Systems, Decision and Control 336, https://doi.org/10.1007/978-981-33-4745-8_11

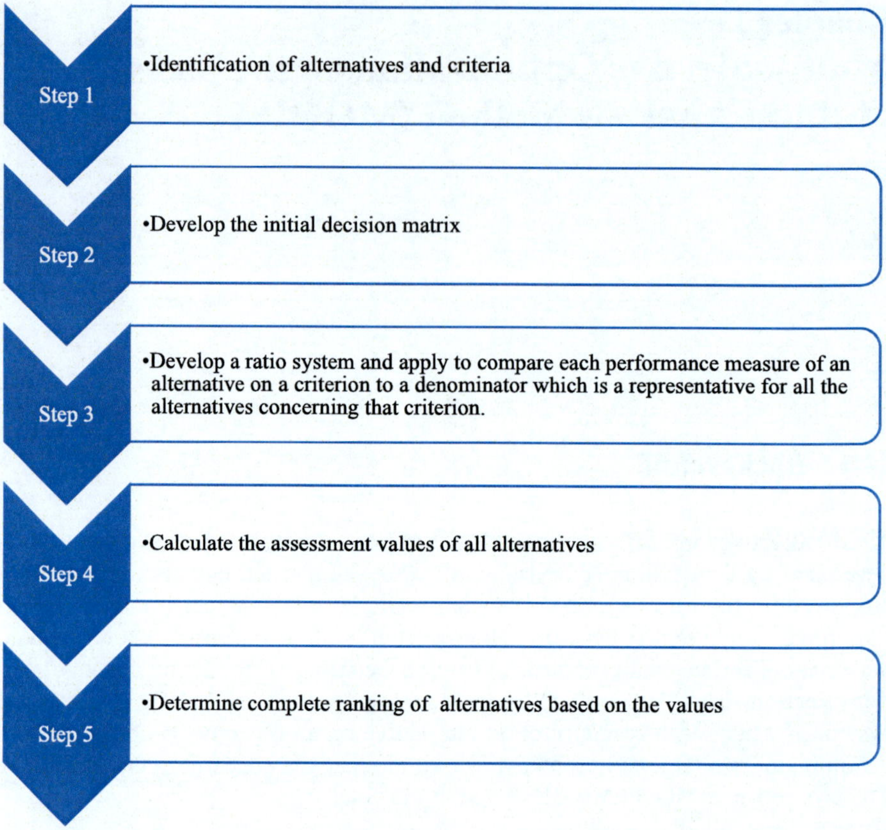

Fig. 11.1 MOORA procedure

interval so that commenting is very easy. Normalized values are showed with x_{ij}^*

$$x_{ij}^* = \frac{x_{ij}}{\sqrt{\sum_{j=1}^{m} x_{ij}^2}}$$

According to Eq. (2.2), normalized value is obtained dividing alternative's value by squared root of sum of alternative's squares.

MOORA method has two different components: ratio system and reference point approach.

Step 3. According to ratio system, ranking value can be determined as below.

$$y_i^* = \sum_{i=1}^{g} x_{ij}^* - \sum_{i=g+1}^{n} x_{ij}^*$$

This equation determines difference between sum of to maximize criteria and sum of to minimize criteria. Attained y_i^* values are ranked and the alternative has biggest value is selected as the best alternative.

Step 4. In reference point approach, reference point equals difference between maximal objective reference point and normalized value found in step 2. Maximal objective reference point is realistic and non-subjective approach. Selected reference point between the alternatives is shown with r_j and it means reference point of jth objective (criteria).

$$r_j - x_{ij}^*$$

Step 5. In this step, distance of values is calculated and attained results are written in matrix. This matrix is analyzed for Tchebyheff's min–max metric operation to determine the ranking.

$$\underset{j}{\text{Min}}\left\{\max_i\left(r_j - x_{ij}^*\right)\right\}$$

11.3 Illustrative Application of MOORA for the Selection of Manufacturing System

An application of MOORA has been illustrated for a hypothetical case of selecting an appropriate manufacturing system for a textile industry. SM Mills, one of the largest producers of yarns for the knitting and weaving industries in the world. Offering a diverse product line to fulfil demands of customers worldwide, the company uses automated spinning processes and innovations to every phase of yarn manufacturing process.

Starting with one open-end spinning facility in Guntur, the company, since its inception, has grown into one of the largest producers of 100% cotton and cotton-blend yarns in India. SM Mills Ltd has attained a target of 500 crores of turnover in year 2020.

The company is targeting capacity expansion and wishes to explore the suitability of various manufacturing systems. The board of directors are interested in evaluating various manufacturing systems such as Single Cell Manufacturing System (SCM), Machine Cluster (MC), Manual Line Assembly (MAL), Automated Transfer Line (ATL), Automated Assembly System (AAS), Machine Cells (MC) and Flexible Manufacturing Cells (FMS) on the basis of eleven criteria as listed in Table 11.1.

Step 1.

Formation of the comparative DECISION Matrix with the help of Expert support. Experts have compared seven various manufacturing systems, i.e. Single Cell Manufacturing System (SCM), Machine Cluster (MC), Manual Line Assembly (MAL),

Automated Transfer Line (ATL), Automated Assembly System (AAS), Machine Cells (MC) and Flexible Manufacturing Cells (FMS) on the basis of eleven criteria (Table 11.1).

The Comparative Matrix is calculated in Table 11.2.

Step 2.

Calculation of Normalized Dimensionless Decision Matrix as indicated in Table 11.3.

Step 3.

The Ranking Values are obtained (in Table 11.4) using step-by-step procedure of MOORA provided in Sect. 11.2.

Table 11.1 Definition of criteria

S. No.	Criteria	Definition	Allocated weight out of 5
1	In process inventory (PI)	The volume of inventory required for unfinished products (lesser the better)	1
2	Labour cost (LC)	The cost of labour involved (lesser the better)	4
3	Infrastructure cost (IC)	The total cost of infrastructure set-up (lesser the better)	3
4	Annual production volume (APV)	Maximum annual capacity (higher the better)	5
5	Material handling cost (MHC)	Cost related to material handling (lesser the better)	2
6	Finished product inventory (FPI)	Total inventory related to the finished product (lesser the better)	1
7	Automation level (AL)	The level of automation (higher the better)	4
8	Setup time (ST)	The setup time of the job (lesser the better)	3
9	Loading/unloading time (LUT)	The total loading/unloading time in unit operation (lesser the better)	2
10	Accuracy (*A*)	The level of accuracy of the finished products, as per desired (higher the better)	4
11	Work transport b/w stations (WTS)	The time related to work transferring b/w different stations (lesser the better)	2

Table 11.2 Comparative matrix

Process/criteria	PI w.r.t 10	LC (Lac.)	IC (Cr.)	APV (ton.)	MHC (lac)	FPI (lac.)	AL wrt 10	ST (min)	LUT (min)	A wrt 10	WTS (min)
Single station cell	10	12	50	14	20	7	1	10	9	6	10
Machine cluster	7	7	60	25	12	13	5	8	7	7	7
Manual assembly line	3	9	65	45	27	23	4	6	7	7	2
Automated transfer line	4	3	110	60	30	30	7	4	3	8	4
Automated assembly system	2	11	800	100	45	50	10	1	2	10	2
Machine cells	8	7	70	35	34	17	5	7	6	7	6
FMS	7	8	90	20	36	10	4	5	4	10	8
Normalized weight	0.0322	0.1290	0.0967	0.1612	0.0645	0.03225	0.1290	0.0967	0.0645	0.1290	0.0645

Table. 11.3 Normalized dimensionless decision matrix

Process/criteria	PI w.r.t 10	LC (lac.)	IC (Cr.)	APV (ton.)	MHC (lac.)	FPI (lac.)	AL wrt 10	ST (min)	LUT (min)	A wrt 10	WTS (min)
SSC	0.586210382	0.5277594	0.244484893	0.10414464	0.245255736	0.103934927	0.065653216	0.58621038	0.57616596	0.28379	0.605227533
MC	0.410347267	0.30785965	0.293381872	0.18597258	0.147153441	0.193022008	0.328266082	0.46896831	0.44812908	0.331089	0.423659273
MAL	0.175863115	0.39581955	0.317830361	0.33475064	0.331095243	0.341500476	0.262612866	0.35172623	0.44812908	0.331089	0.121045507
ATL	0.234484153	0.13193985	0.537866766	0.44633419	0.367883604	0.445435403	0.459572515	0.23448415	0.19205532	0.378387	0.242091013
AAS	0.117242076	0.48377945	0.39117583	0.74389032	0.551825406	0.742392339	0.656532164	0.05862104	0.12803688	0.472984	0.121045507
MC	0.468968305	0.30785965	0.342278851	0.26036161	0.416934751	0.252413395	0.328266082	0.41034727	0.38411064	0.331089	0.36313652
FMS	0.410347267	0.3518396	0.440072808	0.14877806	0.441460324	0.148478468	0.262612866	0.29310519	0.25607376	0.472984	0.484182026

Table. 11.4 Ranking of criteria

Process/criteria	Rank
Single station cell	−0.20091
Machine cluster	−0.08363
Manual assembly line	−0.06004
Automated transfer line	0.014668
Automated assembly system	0.080373
Machine cells	−0.08387
FMS	−0.09171

11.4 Advantages of MOORA

- The decision maker can easily apply this method to evaluate the alternatives and select the most suitable one, while being completely unaware of the physical meaning of the decision-making process.
- Unlike the other MCDM methods, it employs separate mathematical models to benefit the non-benefit criteria and accommodates graphical qualitative criteria of the decision matrix. Due to this added advantage, in MOORA method, the chance of losing information is very small.

11.5 Limitations of MOORA

- Specially while assigning weight to the alternatives, it's quite difficult to predict weights given to various criteria. In select situations, unmanageable number of pairwise comparisons of attributes and alternatives with respect to each of the attributes makes the process highly complex.
- As the number of alternatives increases, the amount of calculations rises quite rapidly and computational procedures become quite elaborate.

This chapter has demonstrated an application of a recent MCDM technique—MOORA. The next chapter will discuss another recently developed MCDM technique called—LINMAP which is an extension of PROMETHEE. Linear Programming Technique for Multidimensional Analysis of Preference (LINMAP) is a computationally efficient and intuitive tool for visualizing and exploring multidimensional parameter spaces.

Selected Bibliography and Further Reading

1. Ertuğrul, I. & Öztaş, T. (2015). The application of sewing machine selection with the multi-objective optimization on the basis of ratio analysis method (MOORA) in apparel sector. *Journal of Textile & Apparel/Tekstil ve Konfeksiyon, 25*(1).
2. Brauers, W. K. M., Zavadskas, E. K., Peldschus, F., et al. (2008). Multi-objective decision-making for road design. *Transport, 23*(3), 183–193.
3. Chakraborty, S. (2011). Applications of the MOORA method for decision making in manufacturing environment. *The International Journal of Advanced Manufacturing Technology, 54*(9–12), 1155–1166.

Chapter 12
Linear Programming Techniques for Multidimensional Analysis of Preference (LINMAP)

12.1 Background

Multi-attribute decision-making (MADM) problems are an important type of multi-criteria decision-making (MCDM) problems and are widespread in real-life decision situations. They can be dealt with using several existing methods such as the technique for order preference by similarity to ideal solution (TOPSIS) developed by Hwang and Yoon (1981), **the linear programming technique for multidimensional analysis of preference (LINMAP)** developed by Srinivasan and Shocker.

Uni-Criterion Method:

$$\text{Optimize}\{f(a)|a \in A\}$$

Uni-criterion model is based on mathematically well-stated problems either maximization or minimization problem having.

1. Optimal solution;
2. Complete ranking of the actions.

However, these criteria do not accommodate the need of multi-criteria decision making problem. Thus, there is a need for multi-criteria decision making method.

Multi-Criteria Method:

$$\text{Optimize}\{f1(a), f2(a), f3(a)\ldots|a \in A\}$$

Multi-criteria model is mathematically ill-stated problem as there exists no alternative optimizing all the criteria at the same time. However, most (nearly all) human problems have a multi-criteria nature. In most of the real life problems, it is necessary to consider technological, economic, environmental and social criteria.

J. J. Thakkar, *Multi-Criteria Decision Making*, Studies in Systems, Decision and Control 336, https://doi.org/10.1007/978-981-33-4745-8_12

PROMETHEE (Preference Ranking Organisation Method for Enrichment Evaluations technique for project scheduling) is used to convert the MCDM problem to linear programming problem when there are constraints. We solve the case using LINMAP technique which is an extension to PROMETHEE with constraints.

The PROMETHEE method is a multi-criteria decision aid system that permits the building of an outranking between different alternatives.

The basic idea was to build an outranking relation from pairwise comparisons of the actions. **PROMETHEE** method utilizes preferences and priorities of decision maker as follows:

1. Preference functions are used to model the perception of the decision maker.
2. Weights are allocated to the criteria to reflect the priorities of the decision maker

In the LINMAP method, all the decision data are known precisely or given as crisp values. However, under many conditions, crisp data is inadequate or insufficient to model real-life decision problems.

Linear programming technique for multidimensional analysis of preference (LINMAP), LINMAP is a computationally efficient and intuitive tool for visualizing and exploring multidimensional parameter spaces.

LINMAP technique has got a wide range of applications including environmental management, distribution sector, hydrology and water management, finance, chemistry, logistics and transportation, energy management, health care, manufacturing and assembly, sports, etc.

Srinivasan and Shocker (1973) developed LINMAP (linear programming techniques for multidimensional analysis of preference) for assessing the weights as well as locating the ideal solution. In this method, m alternatives composed of n attributes are represented as m points in the n-dimensional space. A decision maker (DM) is assumed to have his ideal point that denotes the most preferred alternative location. Once the location of the ideal solution is decided, we can choose an alternative which has the shortest distance from the ideal solution. Given two alternatives, a DM is presumed to prefer an alternative which is closer to his ideal point. Then, the weighted Euclidean distance, di is calculated.

In the LINMAP method, pairwise comparisons of alternatives given by the decision maker are evaluated and the best alternative that has the shortest distance to positive-ideal solution (PIS) is selected. Each alternative is assessed on the basis of its distance to the PIS, which is unknown. The PIS and the weights of attributes are then estimated using a new linear programming model using the consistency and inconsistency indices defined. The lower value of the distance for an alternative indicates that the alternative is closer to PIS. A procedure of LINMAP is illustrated in Fig. 12.1.

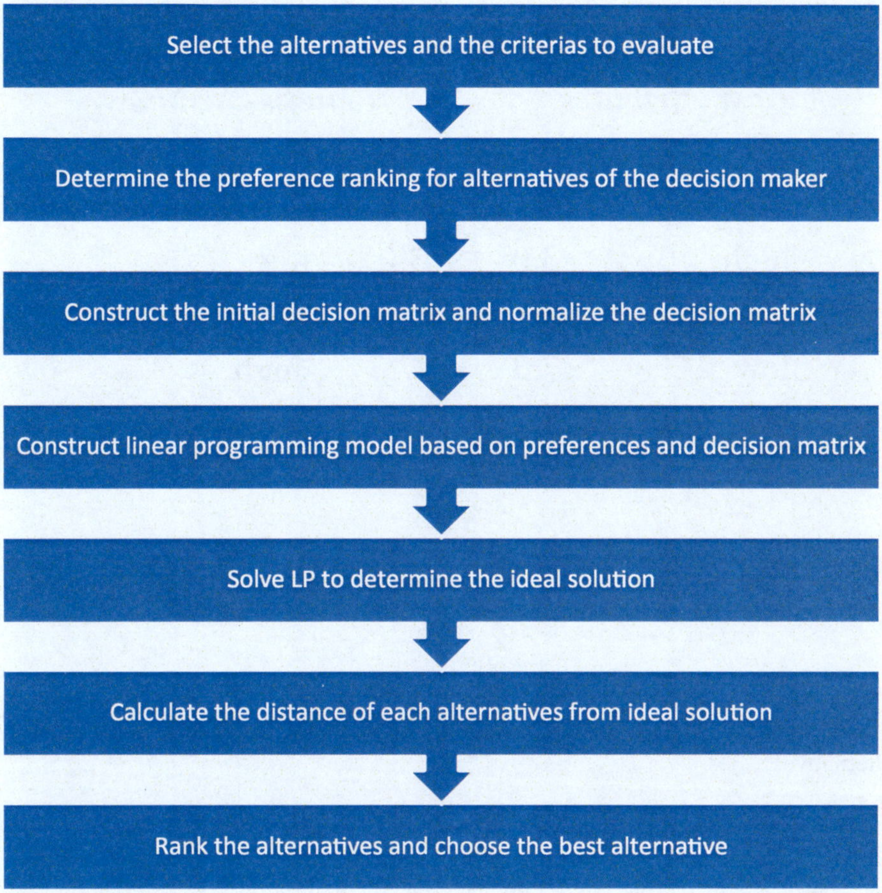

Fig. 12.1 LINMAP procedure

12.2 Step-By-Step Procedure of LINMAP

Step 1: Formulation of the multi-criterion table

Let us consider *n* projects/alternatives/challenges and *k* criteria. Criteria can be tangible or intangible.

Intangible criteria are not measurable thus scale is used to score the intangible criteria. Selection of scale is based on cost attribute and benefits attribute.

Scale for Intangibles

Cost attributes			*Benefit attributes*
	0		
very high	1.0	1.0	very low
high	3.0	3.0	low
average	5.0	5.0	average
low	7.0	7.0	high
very low	9.0	9.0	very high
	10		

Each criteria is provided with certain weightage w_k such that

$$\sum_{k=0}^{q} w_k = 1$$

Step 2: Calculate the evaluative differences of *i*th alternative with respect to other alternatives.

Step 3: Preference function selection

Although there are various preference functions such as U Shape, V Shape, Linear and Gaussian considering the preference threshold and indifference threshold.

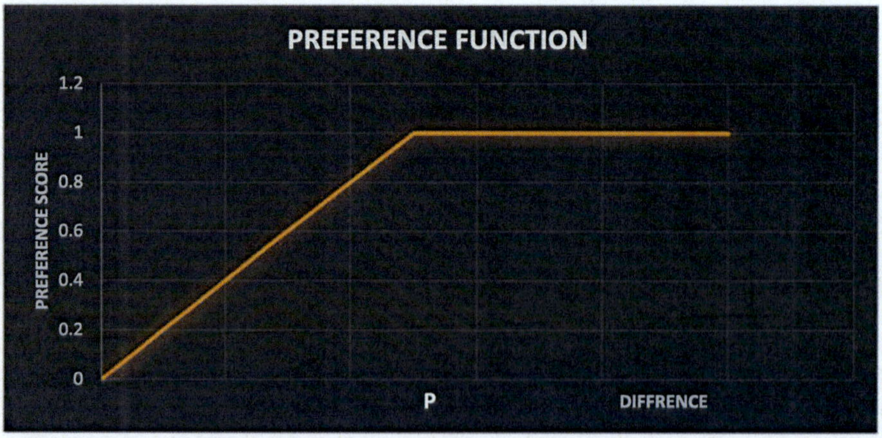

where P = preference threshold.

Definition:

$$P(d) = \begin{cases} 0 & d \le 0 \\ \frac{d}{p} & 0 \le d \le p \\ 1 & d > p \end{cases}$$

Parameter to fix = Preference (P).

Step 4: Computation of preference matrix

$$\forall a_i, a_j \in A : \pi\left(a_i, a_j\right) = \sum_{k=1}^{q} w_k \pi_k(a_i, a_j)$$

Here

$$\pi(a_i, a_i) = 0$$
$$\pi\left(a_i, a_j\right) \ge 0$$
$$\pi\left(a_i, a_j\right) + \pi\left(a_j, a_i\right) \le 1$$

Step 5: Determination of positive and negative outranking flows as follows:

$$\text{Positive flow score: } \varphi^+(a_i) = \frac{1}{n-1} \sum_{b \in A} \pi(a_i, b)$$

$$\text{Negative flow score: } \varphi^-(a_i) = \frac{1}{n-1} \sum_{b \in A} \pi(b, a_i)$$

$$\text{Net flow score: } \varphi(a_i) = \varphi^+(a_i) - \varphi^-(a_i)$$

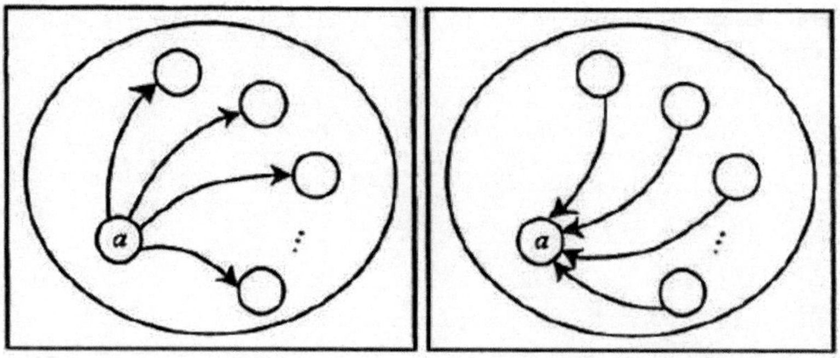

(a) The $\phi^+(a)$ outranking flow. (b) The $\phi^-(a)$ outranking flow.

Step 6: Determine the ranking of all the considered alternatives depending on the values of $\varphi(a_i)$.

For PROMETHEE I: Highest values of flow score are considered.

For PROMETHEE II: Net flow score is considered for the ranking of the alternatives.

Step 7: Convert the MCDM problem to linear programming problem using PROMETHEE V.

The PROMETHEE ranking is obtained for which the net flows $\varphi(a_i)$ have been computed.

The following $\{0, 1\}$ linear programme is then considered in order to take into account the additional constraints

$$\max\left\{\sum_{i=1}^{k} \varphi(a_i)x_i\right\}$$

s.t.

$$Ax_i \sim B$$
$$x_i = \{0, 1\}$$

where $\sim$ holds for $=$, $\leq$ or $\geq$.

The coefficients of the objective function are the net outranking flows. The higher the net flow, the better the alternative. The purpose of the $\{0, 1\}$ linear programme is to select alternatives collecting as much net flow as possible and taking the constraints into account.

How do we Evaluate Attributes in LINMAP?

Attributes matrix is calculated by decision makers (DM). For each attribute, various factors are considered as represented in Table 12.1. Each attribute is scaled to the same unit.

Table 12.1 Decision matrix

Preferences	Attributes				
	1	2	.	.	n
1	x_{11}	x_{12}	.	.	x_{1n}
.	.	.	.	.	.
.	.	.	.	.	.
M	x_{m1}	x_{m2}	.	.	x_{mn}

Preference Relation.

Preference relation between alternatives provided by DM is represented as follows:

$$\Omega = \{(k, j)|x_k \geq x_j \quad k, j = 1, 2, \ldots\ldots\ldots n\} \tag{12.1}$$

Linear Programming Formulation.

12.2.1 Defining the Parameters and Variables

(i) x_{ij} =Value of jth attributes for kth alternative.
(ii) z_{kl} = Preference variable
(iii) w_j = Weight for the jth attributes
(iv) s_i = Euclidean distance of ith alternative

Constraints formulation.

(i) Error constraints between two alternatives

$$\sum_{j=1}^{n} w_j (x_{kj} - x_{lj}) + z_{kl} \leq 0 \quad \forall (k, l) \in \Omega \tag{12.2}$$

(ii) Weighted Euclidean distance, di

$$\sum_{j=1}^{n} w_j \sum_{(k,l)\in\Omega} (x_{kj} - x_{lj}) = h \tag{12.3}$$

(iii) Non-negativity constraints

$$w_j \geq 0, z_{kl} \geq 0 \forall (k, l) \in \Omega,$$

Define the objective function

$$\max \sum_{(k,l)\in\Omega} z_{kl}$$

Solution of LPP

After solving the LPP, we get the value of z^* and w^*

Calculation of Euclidean distance

$$s_i = -\sum_{j=1}^{n} w_J \times x_{ij}$$

Analysis of Result

Analysis is done by comparing and interpreting s_i value with preference relation.

Higher S_i value is given high rank, i.e. the alternative having high s_i value is more preferable than the lower one.

12.3 Illustrative Application of LINMAP

Application 1: Ranking of cities for smart city criteria

Cities are engines of growth for the economy of every nation, including India. Nearly 31% of India's current population lives in urban areas and contributes 63% of India's GDP (Census 2011). With increasing urbanization, urban areas are expected to house 40% of India's population and contribute 75% of India's GDP by 2030. This requires comprehensive development of physical, institutional, social and economic infrastructure. All are important in improving the quality of life and attracting people and investments to the city, setting in motion a virtuous cycle of growth and development. Development of smart cities is a step in that direction.

A smart city would be the one which plans judiciously to meet its aspirations and challenges in a sustainable manner while fostering principles of good governance. These are achieved in a smart city by utilizing the enhanced power of technology, engaging with a more aware and informed citizenry and creating a more competent and capacitated set of people working within an accountable framework.

There is no one way of defining smart city. Smart city concept varies from city to city depending up on the level of development, willingness to change and reform, resources and aspirations of the city residents.

There is no one way of defining smart city. Smart city concept varies from city to city depending on the level of development, willingness to change and reform, resources and aspirations of the city residents. This concept is urban development strategy and mostly technology-driven future vision. Smart city is about how citizens are shaping the city, and how citizens are empowered to contribute to urban development.

The objective is to promote cities that provide core infrastructure and give a decent quality of life to its citizens, a clean and sustainable environment and application of "Smart" solutions. The focus is on sustainable and inclusive development, and the idea is to look at compact areas, create a replicable model which will act like a lighthouse to other aspiring cities. The Smart Cities Mission of the Government is a bold, new initiative. It is meant to set examples that can be replicated both within and outside the smart city, catalyzing the creation of similar Smart Cities in various regions and parts of the country. Table 12.2 presents SWOT analysis for smart city problem.

Critical Challenge.

Some typical challenges of comprehensive development in smart cities are described below.

- Mixed land use in area-based developments;
- Housing and inclusiveness opportunities for all;
- Preserving and developing open spaces;
- Transport options to reduce traffic;
- Provide an identity to the city;
- Reducing the pollution and make city ecofriendly;
- Built smart infrastructure.

For prioritization of smart city, various criteria have been taken as mentioned in the guidelines of smart city project. Each criterion is given some weightage in Table 12.3

For the ranking of city, we take five cities for the evaluation purpose.

The ten criteria listed in Table 12.3 are taken from the Government of India Smart City Project. For each criterion, a numerical score is provided on GOI website. We analyze all the criteria and then evaluate value of each by giving some feasible number. Score of each city and their validation is given in Table 12.4.

Decision matrix is scaled to 1–10 in Table 12.5

The forced-choice ordered paired comparison judgments are:

$$\Omega = \{(1, 2), (1_a, 3), (4, 1), (5, 1), (2, 3), (4, 2), (5, 2), (4, 3), (5, 3), (4, 5)\}$$

Table 12.2 SWOT analysis for smart city problem

Strength	Opportunities
• Development strategy planning • Cultural heritage • Major development initiative • Major industry situated in the city Connectivity of city • Educational institutes	• Exploiting service innovation • Opportunities towards new business • Water level in city • Competitiveness of local cluster • Opportunities to use renewable energy source
Weakness	**Threats**
• Political condition • Environmental sustainability • Digital gap • Weak policy and funding instruments • Illiteracy rate	• Economic crisis • Lack of resources • Vulnerable business model for sustainability of public sector initiatives • Low level of private investment in R&D and research • Weak institutional environment for technology and innovation

Table 12.3 Criteria for ranking smart city

S. No.	Criteria	Description
1	Cleanliness	Swachh Bharat baseline on number of household sanitary latrines
2	Digital connectivity	Online grievance redresses system
3	Monthly e-newsletter	Publication of monthly e-newsletter which describes about the city, various opportunity and complaints resolution
4	e-placed municipal budget	Electronically place project-wise municipal budget expenditure information for the last two financial years on the website
5	Excess of internet	% of households with access to computer/laptop with internet without internet
6	Excess of bank	% of households with access to banking facilities
7	Payment of salary	Payment of salaries by ULB up-to last month
8	Contribution of revenue	Percentage contribution of tax revenue, fees and user charges, rents and other internal revenue sources to the ULB Budgeted receipts
9	Water supply	Percentage of operation and maintenance cost of water supply
10	City-level JnNURM reform	Percentage of city-level JnNURM (Jawaharlal Nehru National Urban Renewal Mission) reforms achieved

Table 12.4 Criteria evaluation for smart city problem

City	Cleanliness (Swachh Bharat ranking)	Digital connectivity of municipal	Monthly e-newsletter	e-placed municipal budget	Excess of Internet (%)
Bhopal	106	9	No	7	22
Bhubaneswar	331	9	No	7	30
Guwahati	51	1	No	8	33
Pune	31	9	Yes	9	38
Surat	63	10	Yes	5	13
City	**Excess of bank (%)**	**Payment of salary**	**Contribution of revenue**	**Water supply (%)**	**JnNURM**
Bhopal	67.92	2 month	3	75.58	62.59
Bhubaneswar	72.54	2 month	10	62.96	46.85
Guwahati	84.46	1 month	1	32.64	100
Pune	86.36	3 month	6	99.20	90.12
Surat	53.51	3 month	10	80.5	90.08

Table 12.5 Decision matrix for smart city problem

City	Criteria									
	1	2	3	4	5	6	7	8	9	10
Bhopal	8	9	5	7	6	10	6	3	8	6
Bhubaneswar	3	9	5	7	8	6	6	10	6	5
Guwahati	9	1	5	8	9	8	3	1	3	10
Pune	10	9	10	9	10	3	10	6	10	9
Surat	9	10	10	5	4	10	10	10	9	9

That is, City Bhopal is preferred to City Bhubaneswar; City Pune is preferred to Bhopal, etc.

The LP formulation of LINMAP is given below.

$$\max z = z_{12} + z_{13} + z_{41} + z_{51} + z_{23} + z_{42} + z_{52} + z_{43} + z_{53} + z_{45}$$

Subject to

$$5w_1 + 0w_2 + 0w_3 + 0w_4 - 2w_5 + 4w_6 + 0w_7 - 7w_8 + 2w_9 + 1w_{10} + z_{12} \leq 0$$
$$-1w_1 + 8w_2 + 0w_3 - 1w_4 - 3w_5 + 2w_6 + 3w_7 + 2w_8 + 5w_9 + 4w_{10} + z_{13} \leq 0$$
$$2w_1 + 0w_2 + 5w_3 + 2w_4 + 4w_5 - 7w_6 + 4w_7 + 3w_8 + 2w_9 + 3w_{10} + z_{41} \leq 0$$
$$1w_1 + 1w_2 + 5w_3 - 2w_4 - 2w_5 + 0w_6 + 4w_7 + 7w_8 + 1w_9 + 3w_{10} + z_{51} \leq 0$$
$$-6w_1 + 8w_2 + 0w_3 - 1w_4 - 1w_5 - 2w_6 + 3w_7 + 9w_8 + 3w_9 - 5w_{10} + z_{23} \leq 0$$
$$7w_1 + 0w_2 + 5w_3 + 2w_4 + 2w_5 - 3w_6 + 4w_7 - 4w_8 + 4w_9 + 4w_{10} + z_{42} \leq 0$$
$$6w_1 + 1w_2 + 5w_3 - 2w_4 - 4w_5 + 4w_6 + 4w_7 + 0w_8 + 3w_9 + 4w_{10} + z_{52} \leq 0$$
$$1w_1 + 8w_2 + 5w_3 + 1w_4 + 1w_5 - 5w_6 + 7w_7 + 5w_8 + 7w_9 + 1w_{10} + z_{43} \leq 0$$
$$0w_1 + 9w_2 + 5w_3 - 3w_4 - 5w_5 + 2w_6 + 7w_7 + 9w_8 + 6w_9 - 1w_{10} + z_{53} \leq 0$$
$$1w_1 - 1w_2 + 0w_3 + 4w_4 + 6w_5 - 7w_6 + 0w_7 - 4w_8 + 1w_9 + 0w_{10} + z_{45} \leq 0$$
$$16w_1 + 34w_2 + 30w_3 + 0w_4 - 4w_5 - 12w_6 + 36w_7 + 20w_8 + 34w_9 + 14w_{10} = 1$$

The above linear programming is solved in MATLAB. (Code is given in Appendix) and the results are as follows: Table 12.6

Calculation of Euclidean Distance.

For City Bhopal

$$\begin{aligned} S_i &= (8 * 0.1113 + 9 * 0.2109 + 5 * 0.1373 + 7 * 0.2174 + 6 * 2.4866 \\ &\quad + 10 * 1.9608 + 6 * 0.1387 + 3 * 0.5295 + 8 * 0.1421 + 0 * 6) \\ &= 43.0819 \end{aligned}$$

Table 12.6 Criteria weightages for smart city problem

w_1	0.1113
w_2	0.2109
w_3	0.1373
w_4	0.2174
w_5	2.4866
w_6	1.9608
w_7	0.1387
w_8	0.5295
w_9	0.1421
w_{10}	0.0000

For City Bhubaneswar

$$\begin{aligned} S_i &= (3 * 0.1113 + 9 * 0.2109 + 5 * 0.1373 + 7 * 0.2174 + 8 * 2.4866 \\ &\quad +6 * 1.9608 + 6 * 0.1387 + 10 * 0.5295 + 6 * 0.1421 + 0 * 5) \\ &= 43.0777 \end{aligned}$$

For City Guwahati

$$\begin{aligned} S_i &= (9 * 0.1113 + 1 * 0.2109 + 5 * 0.1373 + 8 * 0.2174 \\ &\quad +9 * 2.4866 + 8 * 1.9608 + 3 * 0.1387 + 1 * 0.5295 + 3 * 0.1421 + 0 * 6) \\ &= 43.076 \end{aligned}$$

For City Pune

$$\begin{aligned} S_i &= (10 * 0.1113 + 9 * 0.2109 + 10 * 0.1373 + 9 * 0.2174 + 10 * 2.4866 \\ &\quad +3 * 1.9608 + 10 * 0.1387 + 6 * 0.5295 + 10 * 0.1421 + 0 * 9) \\ &= 43.0741 \end{aligned}$$

For City Surat

$$\begin{aligned} S_i &= (9 * 0.1113 + 10 * 0.2109 + 10 * 0.1373 + 5 * 0.2174 + 4 * 2.4866 \\ &\quad +10 * 1.9608 + 10 * 0.1387 + 10 * 0.5295 + 9 * 0.1421 + 0 * 9) \\ &= 43.086 \end{aligned}$$

Based on the Calculated S_i Euclidean distance, the rank of city is given as in Table 12.7.

Table 12.7 Ranking of the cities for smart city problem

Rank	City
1	Surat
2	Bhopal
3	Bhubaneswar
4	Guwahati
5	Pune

Interpretation of results

- For all city, S_i value is nearly 43 this happens because there is no intransitivity between preferences.
- For the ranking, we consider value after first decimal point.
- If there is transitivity between preferences exist than the Euclidean values are nearly same.
- In the above problem, all criteria are converted into benefit criteria and hence the objective function is maximized. If the criteria are cost related, then we minimize the objective function.
- If there exist some criteria which are qualitative in nature such as low, minimum and high, then we can apply Fuzzy LINMAP techniques.

Application 2: Public Distribution System Challenges

The public distribution system (PDS) in the country facilitates the supply of food grains and distribution of essential commodities to a large number of poor people through a network of fair price shops at a subsidized price on a recurring basis.

With a network of more than 4 lakh fair price shops claiming to distribute annually, commodities worth more than Rs 15,000 crore to about 16 crore families, the PDS in India is perhaps the largest distribution network of its type in the world.

With a coverage of around 400 million below poverty line (BPL) families, a review of the PDS has discovered the following structural shortcomings and disturbances:

Growing instances of the consumers receiving inferior quality food grains in ration shops:

1. Deceitful dealers replace good supplies received from the Food Corporation of India (FCI) with inferior stock and sell FCI stock in the black market.
2. Illicit fair price shop owners have been found to create large number of bogus cards to sell food grains in the open market.
3. Many FPS dealers resort to malpractice, illegal diversions of commodities, holding and black marketing.
4. Numerous malpractices make safe and nutritious food inaccessible and unaffordable to many poor thus resulting in their food insecurity

5. Identification of households to be denoted status and distribution to granted PDS services has been highly irregular and diverse in various states. The recent development of aadhar UIDAI cards has taken up the challenge of solving the problem of identification and distribution of PDs services along with direct cash transfers.
6. Regional allocation and coverage of FPS are unsatisfactory and the core objective of price stabilization of essential commodities has not met.

Problem Definition

LINMAP technique is used to model the challenges faced by the PDS. The economic, time and quantity constraints are modelled to prioritize the PDS challenges.

The challenges (P1, P2, P3, P4, P5) and criteria (C1, C2, C3, C4, C5, C6, C7, C8) are defined in Tables 12.8 and 12.9.

In addition to the above factors, there are following three constraints which restrict the selection of challenges.

(1) Challenge 1 and Challenge 5 cannot be chosen simultaneously.
(2) Challenge 3 and challenge 4 cannot be chosen simultaneously.
(3) Only one challenge can be chosen from challenges 3, 4 and 5.

Step-by-Step Application of LINMAP for PDS Problem.

1. Multi-criterion Table: Hypothetical intangible criteria are used in the case of public distribution system in Table 12.10.
2. Compute uni-criterion preference degree for every pair of alternatives as shown in Table 12.11. The intangible criteria are evaluated on numerical subjective

Table 12.8 Challenge definition for public distribution system (PDS) problem

S. No.	Challenge		Definition
1	No ration cards	*P*1	A lot of people below poverty line do not have ration or aadhar cards, which excludes them from taking benefits
2	Lack of transparency and accountability	*P*2	There is no proper account nor accurate official records for reference, which is a major challenge to PDS
3	Reduced scale & quality issue	*P*3	The amount of distribution of utilities and its quality has reduced to a very low level
4	Inclusion and exclusion errors	*P*4	During distribution, there is no proper allocation to distributing agencies which leads to further which result in erroneous and misdistribution
5	Bogus cards	*P*5	There are a lot of fake ration and aadhar cardholders which is another major problem

Table 12.9 Criteria definition and priority for public distribution system (PDS) problem

S. No.	Criteria		Definition	Priority
1	Tackling capacity	*C*1	How much is the existing system capable of overcoming the challenge	Maximize
2	Tackling time	*C*2	How much time will be needed to overcome the challenge	Minimize
3	IT and HR requirements	*C*3	IT infrastructure and human resources required for particular challenge	Minimize
4	Distribution	*C*4	How much will it help in effective distribution of utilities if challenge is resolved	Maximize
5	Effectiveness on losses	*C*5	How much will it help in reducing the losses	Maximize
6	Investment	*C*6	How much initial investment is required for particular challenge	Minimize
7	Effectiveness on corruption	*C*7	How effectively corruption can be reduced after tackling the challenge	Maximize
8	Customer satisfaction	*C*8	Increase in customer satisfaction expected after removal of problem	Maximize

Table 12.10 Criteria definition for public distribution system (PDS) problem

*C*1		*C*2	*C*3	*C*4	*C*5	*C*6	*C*7	*C*8
*P*1	High	Low	V high	V high	Poor	Average	V Low	Low
*P*2	V Poor	High	High	Average	Good	High	V High	V Good
*P*3	V Poor	Average	Average	V Good	Good	Low	Low	Low
*P*4	Good	Low	Low	Average	V High	Low	Good	High
*P*5	Poor	High	High	V Good	Average	Average	V High	V High

Table 12.11 Uni-criterion preference degree for alternatives

*C*1		*C*2	*C*3	*C*4	*C*5	*C*6	*C*7	*C*8
*P*1	7	4	9	7	8	5	1	2
*P*2	3	8	7	5	4	3	9	6
*P*3	2	5	5	6	5	8	3	3
*P*4	4	3	3	4	4	6	4	7
*P*5	3	7	8	8	1	5	6	9
W_K (Weights)	0.0277	0.111	0.166	0.194	0.055	0.1388	0.0833	0.222

scale. The weights are allocated to the criteria.

3. 'V'-shape preference function has been selected for formulation of preference matrix.
4. Maximum difference table for the criteria in Table 12.12.
5. Compute global preference degree for every pair of alternatives and construct the preference matrix as shown in Table 12.13.

Calculations:

$$\pi(P1, P2) = 0.277 * 0.8 + 0.111 * 0.8 + 0.194 * 0.667 + 0.055 * 0.571 = 0.240$$
$$\pi(P2, P1) = 0.166 * 0.33 + 0.1388 * 0.4 + 0.0833 * 1 + 0.222 * 0.571 = 0.321$$

Table 12.14

$$\pi(P2, P3) = 0.277 * 0.2 + 0.055 * 1 + 0.222 * 0.428 = 0.24$$
$$\pi(P3, P2) = 0.111 * 0.6 + 0.166 * 0.33 + 0.194 * 0.25 + 0.055 * 0.142 + 0.0833 + 0.75 = 0.262$$

Similarly, all other pairs are evaluated and preference matrix obtained is shown in Table 12.15.

Table 12.12 Maximum difference table

Maximum difference in criteria	Difference
*C*1	5
*C*2	5
*C*3	6
*C*4	4
*C*5	7
*C*6	5
*C*7	8
*C*8	7

Table 12.13 Global preference degree

Preference	0.8	0.8	0	0.5	0.571	0	0	0
(*P*1–*P*2)	+4	−4	+2	+2	+4	+2	−8	−4
*P*1	7	4	9	7	8	5	1	2
	*C*1	*C*2	*C*3	*C*4	*C*5	*C*6	*C*7	*C*8
*P*2	3	8	7	5	4	3	9	6
(*P*2–*P*1)	−4	+4	−2	−2	−4	−2	+8	+4
Preference	0	0	0.33	0	0	0.4	1	0.571

Table 12.14 Global preference degree

Preference	0.2	0	0	0	1	0	0	0.428
(*P*2–*P*3)	−1	−3	−2	−1	−1	−5	−6	+3
*P*2	3	8	7	5	4	3	9	6
	*C*1	*C*2	*C*3	*C*4	*C*5	*C*6	*C*7	*C*8
*P*3	2	5	5	6	5	8	3	3
(*P*3–*P*2)	−1	−3	−2	+1	+1	+5	+6	−3
Preference	0	0.6	0.33	0.25	0.142	0	0.75	0

Table 12.15 Global preference degree

*P*1		*P*2	*P*3	*P*4	*P*5
*P*1	0	0.240	0.228	0.246	0.104
*P*2	0.321	0	0.24	0.207	0.115
*P*3	0.165	0.262	0	0.129	0.1277
*P*4	0.3788	0.259	0.303	0	0.233
*P*5	0.351	0.279	0.416	0.3465	0

Preference Matrix

6. Computing positive, negative and net flow score Table 12.16
7. The following {0, 1} linear programme is then considered in order to take into account the additional constraints.

Objective function is to maximize the net flow score to choose the best alternative for the public distribution system.

$$\text{Max} Z = -0.00994x_1 - 0.0392x_2 - 0.1253x_3 + 0.06135x_4 + 0.203x_5$$

Table 12.16 Global preference degree

Positive flow score		Negative flow score	Net flow score	Ranking
*P*1	0.2045	0.3039	−0.00994	3
*P*2	0.22075	0.26	−0.0392	4
*P*3	0.171	0.2965	−0.1253	5
*P*4	0.29345	0.2321	0.06135	2
*P*5	0.348	0.1449	0.203	1

s.t.

$$x_1 + x_2 + x_3 + x_4 + x_5 \leq 2$$
$$x_1 + x_5 \leq 1$$
$$x_3 + x_4 \leq 1$$
$$x_3 + x_4 + x_5 \leq 1$$
$$x_{i=}\{0, 1\}$$

where

x_i = Decision variable corresponding to selection of particular challenge

Solving the above linear programming model, the results are

$$x_2 = 1, x_5 = 1$$

Inferences

Considering all the economic, political and technological constraints, it is found that P2 and P5 are the most appropriate set of challenges which need to be tackled for most effective results on working in public distribution system.

12.4 Advantages of LINMAP

- The LINMAP procedure does not require that the set Ω consist of all
- $m(m - 1)/2$ paired comparison judgments from the DM (Decision Maker).
- The set of weights obtained by LINMAP are more reliable if the number of pairs in Ω is large, i.e. when the number of alternatives is greater than the number of attributes.
- The method does not require that the paired comparison judgment be transitive.
- It supports group-level decision making because it constitutes a useful platform for debate and consensus building.
- It (as all outranking methods) can simultaneously deal with qualitative and quantitative criteria.
- Criteria scores can be expressed in our own scale.
- It can deal with uncertain and fuzzy information.

12.5 Limitation of LINMAP

- When the number of attributes exceeds the number of alternatives (i.e. $n > m$), it is not easy for the DM to assess the partial orders, and it is hard to obtain reliable weights by the LINMAP; therefore, in this case it is better to use ELECTRE.
- In the LINMAP method, all the decision data are known precisely or given as crisp values. However, under many conditions, crisp data is inadequate or insufficient to model real-life decision problems.
- It suffers from the rank reversal problem when a new alternative is introduced.
- It does not provide the possibility to really structure a decision problem. In the case of many criteria and options, it thus may become difficult for the decision maker to obtain a clear view of the problem and to evaluate the results.
- It does not provide any formal guidelines for weighing, but assumes that the decision maker is able to weigh the criteria appropriately.
- The way in which the preference information is processed is complicated and hard to explain to non-specialists.

This chapter has discussed the key features and use of MCDM technique—LINMAP. The next chapter will discuss the application of one more recent advancement in the field of MCDM—Complex Proportion Assessment Method (COPRAS). It is a method of normalizing, in which we combine all indicators into a single quantitative account, that is the value of method criterion. This method takes account of direct and proportional dependence of priority.

Selected Bibliography and Further Reading

1. Deshmukh, S.C. Preference Ranking Organization Method of Enrichment Evaluation (Promethee), BVDU, COE Pune
2. Jean-Pierre Brans International Journal of Engineering Science Invention ISSN, PROMETHEE Methods, Centrum voor Statistiek en Operationeel Onderzoek, Vrije Universiteit Brussel Pleinlaan 2, B-1050 Brussels, Belgium jpbrans@vub.ac.be
3. Facility Location Selection using PROMETHEE II Method by Vijay Manikrao Athawale Department of Mechanical Engineering, Government Polytechnic Amravati—444603, India and Shankar Chakraborty Department of Production Engineering, Jadavpur University Kolkata—700032, India Proceedings of the 2010, International Conference on Industrial Engineering and Operations Management Dhaka, Bangladesh
4. Maheswaran, K., & Loganathan, T. (2013). A novel approach for prioritization of failure modes in FMEA using MCDM. International Journal of Engineering Research and Applications (IJERA) ISSN: 2248-9622 www.ijera.com *3*(4), 733–739 Jul-Aug 2013.
5. Srinivasan, V., & Shocker, A. D. (1973). Linear programming techniques for multidimensional analysis of preference. *Psychometrika, 38,* 337–342.
6. Xia, H.-C., Li, D.-F., Zhou, J.-Y., & Wang, J.-M. (2006). Fuzzy LINMAP method for multiattribute decision making under fuzzy environments. *Journal of Computer and System Sciences, 72,* 741–759.

7. Multiple Attribute Decision Making by Ching-Lai Hwang.
8. Multicriterion Decision in Management By Pomerol.
9. https://smartcities.gov.in/.
10. https://www.punecorporation.org/.
11. https://www.bhopalmunicipal.com/.
12. https://www.suratmunicipal.gov.in/.
13. https://bmc.gov.in/.

Chapter 13
Complex Proportion Assessment Method (COPRAS)

13.1 Background

COPRAS method was developed in 1996 by Vilnius Gediminas Technical University scientists Zavadskas and Kaklauskas. It is actually a method of normalizing, in which we combine all indicators into a single quantitative account, that is the value of method criterion. This method accommodates the direct and proportional dependence of priority.

Complex Proportional Assessment (COPRAS) is one of the most commonly applied approaches in MCDM which is utilized to rank various alternatives based on different criteria with the help of associated criteria weights and utility degree of alternatives. The selection of the best alternative is performed by considering the ideal and anti-ideal solutions. COPRAS assumes direct and proportional dependence of the importance and utility degree of investigated versions on a system of criteria efficiently defining the alternatives. COPRAS has several advantages as having less computational time, very simple and transparent computation process, etc. over the other MCDM methods such as EVAMIX, TOPSIS, VIKOR and AHP. COPRAS can more easily be adopted for majority of the decision problems. Also, decisions provided by COPRAS are shown to be more efficient and less biased than those provided by TOPSIS and Simple Additive Weighting. Figure 13.1 represents a flow diagram of COPRAS procedure.

J. J. Thakkar, *Multi-Criteria Decision Making*, Studies in Systems, Decision and Control 336, https://doi.org/10.1007/978-981-33-4745-8_13

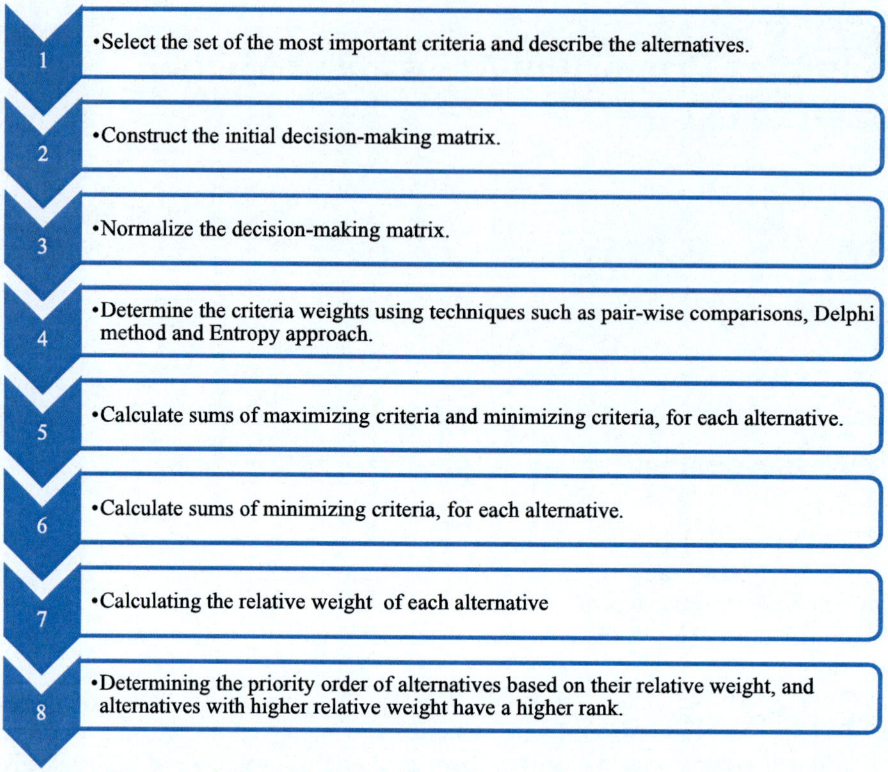

Fig. 13.1 CORPAS procedure

13.2 Step-by-Step Procedure of CORPAS-G Method

Ranking alternatives by the COPRAS method assumes direct and proportional dependence of significance and priority of investigated alternatives on a system of criteria. COPRAS-G uses criterion values determined in intervals. Replacement of the crisp x_{ij} with the interval value $[\underline{x}_{ij}, \bar{x}_{ij}]$, where $\underline{x}_{ij}$ is the lower limit and $\bar{x}_{ij}$ is the upper limit of interval, requires some modifications in the ranking procedure

Step 1: Construction and normalization of decision-making matrix

- Selecting the set of the most important attributes, describing the alternatives
- Constructing the decision-making matrix X
- For example:

$$X = \begin{bmatrix} [w_{11}; b_{11}] & [w_{12}; b_{12}] & \ldots & [w_{1m}; b_{1m}] \\ [w_{21}; b_{21}] & [w_{22}; b_{22}] & \ldots & [w_{2m}; b_{2m}] \\ \vdots & \vdots & \ldots & \vdots \\ [w_{n1}; b_{n1}] & [w_{n2}; b_{n2}] & \ldots & [w_{nm}; b_{nn}] \end{bmatrix}; j = \overline{1, n}; \quad i = \overline{1, m},$$

Here w_{ij} is smallest value (lower limit) and b_{ij} is biggest value (upper limit).

- Normalizing decision-making matrix X

$$\overline{w_{ij}} = \frac{w_{ij}}{\frac{1}{2}\left(\sum_{j=1}^{n} w_{ij} + \sum_{j=1}^{n} b_{ij}\right)} = \frac{2w_{ij}}{\sum_{j=1}^{n} w_{ij} + \sum_{j=1}^{n} b_{ij}}$$

$$\overline{b_{ij}} = \frac{b_{ij}}{\frac{1}{2}\left(\sum_{j=1}^{n} w_{ij} + \sum_{j=1}^{n} b_{ij}\right)} = \frac{2b_{ij}}{\sum_{j=1}^{n} (w_{ij} + b_{ij})}$$

w_{ij} is the lower value of the j attribute in the i alternative of the solution; b_{ij} is the upper value of the j attribute in the i alternative of the solution.

m is the number of attributes

n is the number of the alternatives compared

The normalized decision-making matrix is given as:

$$\bar{X} = \begin{bmatrix} [\overline{w_{11}}, \overline{b_{11}}] & [\overline{w_{12}}; \overline{b_{12}}] & \ldots & [\overline{w_{1m}}; \overline{b_{1m}}] \\ [\overline{w_{21}}, \overline{b_{21}}] & [\overline{w_{21}}; \overline{b_{22}}] & \ldots & [\overline{w_{2m}}; \overline{b_{2m}}] \\ \vdots & \vdots & \ldots & \vdots \\ [\overline{w_{n1}}, \overline{b_{n1}}] & [\overline{w_{n2}}, \overline{b_{n2}}] & \ldots & [\overline{w_{nm}}; \overline{b_{nm}} \end{bmatrix}$$

Step 2: Weighted normalized decision-making matrix

- Determining weighted attributes q_j

$$\hat{w}_{ij} = \bar{w}_{ij} \cdot q_j;$$
$$\hat{b}_{ij} = \overline{b_{ij}} \cdot q_j.$$

- So weighted normalized matrix becomes:

$$\hat{X} = \begin{bmatrix} \left[\hat{w}_{11}; \hat{b}_{11}\right] & \left[\hat{w}_{12}; \hat{b}_{12}\right] & \ldots & \left[\hat{w}_{1m}; \hat{b}_{1m}\right] \\ \left[\hat{w}_{21}; \hat{b}_{21}\right] & \left[\hat{w}_{21}; \hat{b}_{22}\right] & \ldots & \left[\hat{w}_{2m}; \hat{b}_{2m}\right] \\ \vdots & \vdots & \ldots & \vdots \\ \left[\hat{w}_{n1}; \hat{b}_{n1}\right] & \left[\hat{w}_{n2}; \hat{b}_{n2}\right] & \ldots & \hat{w}_{nm}; \hat{b}_{nm} \end{bmatrix}$$

Step 3: Calculating the sums P_j of the attribute values, whose larger values are more preferable (optimization direction is maximization), for each alternative (each row of the decision-making matrix)

$$P_j = \frac{1}{2} \sum_{i=1}^{k} \left(\hat{w}_{ij} + \hat{b}_{ij}\right).$$

where k is the number of attributes that must be maximized

Step 4: Calculating the sums R_j of attribute values, whose smaller values are more preferable (optimization direction is minimization), for each alternative (each row of the decision-making matrix)

$$R_j = \frac{1}{2} \sum_{i=k+1}^{m} \left(\hat{w}_{ij} + \hat{b}_{ij}\right);$$

Step 5: Determining minimal value of $\boldsymbol{R_j}$

$$R_{\min} = \min_j R_j$$

Step 6: Determining the relative weights of each alternative

$$Q_j = P_j + \frac{R_{\min} \sum_{j=1}^{n} R_j}{R_j \sum_{j=1}^{n} \frac{R_{\min}}{R_j}}.$$

This can also be written as

$$Q_j = P_j + \frac{\sum_{j=1}^{n} R_j}{R_j \sum_{j=1}^{n} \frac{1}{R_j}}.$$

Step 7: Ranking alternatives according to relative significance of each alternative

- For determining the ranking or priority of the project significance (Relative weightage of alternative) Q_j.
- The greater is the significance Q_j, the higher is the priority (rank) of the alternative

13.3 Illustrative Applications of COPRAS

Application 1: Quality management
Quality management is the act of overseeing all activities and tasks needed to maintain a desired level of excellence. This includes creating and implementing quality planning and assurance, as well as quality control and quality improvement. In rapidly changing global, dynamic change have taken place in the organization's strategy. Organizations are paying more attention towards optimizing their management practices. An organization fails to implement a set of management practices that have been successful in other organizations.

A set of criteria and practices are involved in quality management for improving organizational performance. It is now attracting the attention of researchers in various fields like supply chain management, education, manufacturing, etc. There are many issues which need to be addressed and prioritized with regard to implementation of a quality management programme in any institution. To effectively implement quality management, it is essential to focus on the most important TQM practices initially and address the less important ones in the later stage. This involves ranking of the TQM practices.

In this study, we aim to prioritize the relative importance of the TQM practices for implementation so that the organization can evaluate their current practices and improve their performance. We use the Complex Proportional Assessment Method (COPRAS-G) to rank the TQM practices. We try to approach the problem through the following steps:

1. Identification of the Quality management practices to be implemented
   A literature review has helped to identify the various criteria involved in the TQM implementation as listed in Table 13.1.
2. Establishing a comparison matrix of the TQM implementation criterion
3. Scale selection: A scale of 1–10 has been selected in creation of the comparison matrix for the application of the COPRAS-G method. The decision-making time and the resources are the criteria that need to be minimized, i.e. optimization direction is minimization while the others are to be maximized, i.e. optimization direction is maximization.

Table 13.1 Description of quality management practices

Practices	Description
Strategic planning ($C1$)	Strategic planning functions as a means to integrate the requires quality of commercial activities of organization so that total quality is reflected in corporate vision, mission and strategies
Leadership ($C2$)	Associated with clear vision and guidelines in relation to the act of sharing knowledge can produce strong commitment to education, training is another sub-factor that provides high skilled employee in dealing with work and personal purpose
Top management commitment ($C3$)	Commitment of senior management for ensuring the (TQM) implementation is an essential factor
Tools and techniques ($C4$)	TQM embraces a wide range of systems, approaches, techniques and tools. Systems and techniques are also process analysis and improvement is another sub-factor that helps organizations evaluate the achievements of predicted results and monitor continuous improvement efforts moving in the right direction
Quality system ($C5$)	Organizations should develop their quality philosophy, policy, procedures and objectives, and acquire information from employees, customers, suppliers and competitors factors that have their own role in quality management
Continuous improvement ($C6$)	In order to transfer connection quality strategy throughout the organization, senior management must create an organizational environment that is focused on continuous improvement. Helps organizations evaluate the achievements of predicted results and monitor continuous improvement efforts moving in the right direction
Organizational culture ($C7$)	The behaviours and public opinion reflect a common culture in an organization. First of all, the existing organizational culture will affect TQM implementation unconsciously as a model to be guaranteed. So it seems necessary to understand what culture is and how the application of TQM affects
Employee involvement ($C8$)	Employee involvement is a process in the employee empowerment in management decision and improves the activities proportional to their levels in the organization. Supporting from senior management is essential to stabilize the availability of concrete action. Increased participation means more responsibility, which in turn requires a greater level of skill
Human resource development ($C9$)	Human resource development plays an important role to maximize the ability of individuals
Customer satisfaction ($C10$)	Represents the value derived by the customer

(continued)

Table 13.1 (continued)

Practices	Description
Supplier relationship ($C11$)	Having effective supplier chain management can contribute to the quality performance in many ways. Long-term partnerships with suppliers also help the parties involved the solving programme related to quality improvement investment
Decision-making time ($C12$)	Direct involvement of senior management allows all decisions to be taken quickly and provide the TQM path
Resources ($C13$)	Providing maximum quality output using minimum resources

Step-by-Step Application for Prioritizing TQM Practices:

1. **Creating decision-making matrix**

Hypothetical values have been chosen in Table 13.2 to demonstrate the application of the COPRAS-G method in ranking the quality management practices.

2. **Normalized decision matrix**
   See Table 13.3.
3. **Weighted normalized decision matrix**
   See Table 13.4.
4. **Calculating the sums P_j of the attribute values, whose larger values are more preferable (optimization direction is maximization), for each alternative (each row of the decision-making matrix) and calculating the sums R_j of attribute values, whose smaller values are more preferable (optimization direction is minimization), for each alternative (each row of the decision-making matrix)**
   See Table 13.5.
5. **Determining the relative weights for each alternative**
   See Table 13.6.
   Thus, we can rank the quality management practices as follows.

   1: ISO (Q1)
   2: SIX SIGMA (Q4)
   3: QFD (Q2)
   4: TQM (Q5)
   5: KAIZEN (Q3)
   6: TRIZ (Q6)

Table 13.2 Decision matrix

Quality management practices	*C*1		*C*2		*C*3		*C*4		*C*5		*C*6		*C*7		*C*8		*C*9		*C*10	
Optimization direction	Max		Max		Max		Max		Max		Max		Max		Max		Min		Min	
Attribute weights	0.120		0.056		0.205		0.110		0.060		0.090		0.057		0.090		0.017		0.195	
	Min	Max	Min	Max	Min	Max	Min	Max	Min	Max	Min	Max	Min	Max	Min	Max	Min	Max	Min	Max
ISO (Q1)	3.000	4.000	6.000	7.000	1.000	3.0000	7.000	8.000	2.000	6.000	8.000	2.000	9.000	4.000	7.000	8.000	1.000	5.000	1.000	4.000
QFD (Q2)	5.000	6.000	7.000	8.000	5.000	6.000	4.000	6.000	5.000	9.000	4.000	6.000	1.000	3.000	4.000	7.000	6.000	9.000	4.000	5.000
KAIZEN (Q3)	2.000	9.000	3.000	8.000	6.000	8.000	5.000	6.000	4.000	6.000	2.000	3.000	4.000	6.000	5.000	6.000	7.000	8.000	4.000	9.000
SIX SIGMA (Q4)	2.000	3.000	4.000	6.000	7.000	8.000	4.000	5.000	7.000	9.000	4.000	5.000	4.000	5.000	4.000	6.000	1.000	6.000	2.000	3.000
TQM (Q5)	7.000	3.000	5.000	6.000	4.000	5.000	2.000	3.000	5.000	6.000	5.000	6.000	6.000	7.000	4.000	6.000	1.000	3.000	5.000	9.000
TRIZ (Q6)	1.000	3.000	5.000	6.000	1.000	6.000	2.000	6.000	4.000	5.000	1.000	3.000	2.000	3.000	4.000	5.000	4.000	6.000	1.000	5.000

Table 13.3 Normalized decision matrix

Quality management practices	$C1$		$C2$		$C3$		$C4$		$C5$		$C6$		$C7$		$C8$		$C9$		$C10$	
Optimization direction	Max		Max		Max		Max		Max		Max		Max		Max		Min		Min	
Attribute weights	0.120		0.056		0.205		0.110		0.060		0.090		0.057		0.090		0.017		0.195	
	Min	Max	Min	Max	Min	Max	Min	Max	Min	Max	Min	Max	Min	Max	Min	Max	Min	Max	Min	Max
ISO	0.125	0.167	0.169	0.197	0.033	0.100	0.241	0.276	0.059	0.176	0.327	0.082	0.333	0.148	0.212	0.242	0.035	0.175	0.038	0.154
QFD	0.208	0.250	0.197	0.225	0.167	0.200	0.138	0.207	0.147	0.265	0.163	0.245	0.037	0.111	0.121	0.212	0.211	0.316	0.154	0.192
KAIZEN	0.083	0.375	0.085	0.225	0.200	0.267	0.172	0.207	0.118	0.176	0.082	0.122	0.148	0.222	0.152	0.182	0.246	0.281	0.154	0.346
SIX SIGMA	0.083	0.125	0.113	0.169	0.233	0.267	0.138	0.172	0.206	0.265	0.163	0.204	0.148	0.185	0.121	0.182	0.035	0.105	0.192	0.346
TQM	0.292	0.125	0.141	0.169	0.133	0.167	0.069	0.103	0.147	0.176	0.204	0.245	0.222	0.259	0.121	0.182	0.035	0.105	0.192	0.346
TRIZ	0.042	0.125	0.141	0.169	0.033	0.200	0.069	0.207	0.118	0.147	0.041	0.122	0.074	0.111	0.121	0.152	0.140	0.211	0.038	0.192

Table 13.4 Weighted normalized decision matrix

Quality management practices	*C*1		*C*2		*C*3		*C*4		*C*5		*C*6		*C*7		*C*8		*C*9		*C*10	
Optimization direction	Max		Max		Max		Max		Max		Max		Max		Max		Min		Min	
Attribute weights	0.120		0.056		0.205		0.110		0.060		0.090		0.057		0.090		0.017		0.195	
	Min	Max	Min	Max	Min	Max	Min	Max	Min	Max	Min	Max	Min	Max	Min	Max	Min	Max	Min	Max
ISO	0.015	0.020	0.020	0.024	0.004	0.012	0.029	0.033	0.007	0.021	0.039	0.010	0.040	0.018	0.025	0.029	0.004	0.021	0.005	0.018
QFD	0.025	0.030	0.024	0.027	0.020	0.024	0.017	0.025	0.018	0.032	0.020	0.029	0.004	0.013	0.015	0.025	0.025	0.038	0.018	0.023
KAIZEN	0.010	0.045	0.010	0.027	0.024	0.032	0.021	0.025	0.014	0.021	0.010	0.015	0.018	0.027	0.018	0.022	0.029	0.034	0.018	0.042
SIX SIGMA	0.010	0.015	0.014	0.020	0.028	0.032	0.017	0.021	0.025	0.032	0.020	0.024	0.018	0.022	0.015	0.022	0.004	0.025	0.009	0.014
TQM	0.035	0.015	0.017	0.020	0.016	0.020	0.008	0.012	0.018	0.021	0.024	0.029	0.027	0.031	0.015	0.022	0.004	0.013	0.023	0.042
TRIZ	0.005	0.015	0.017	0.020	0.004	0.024	0.008	0.025	0.014	0.018	0.005	0.015	0.009	0.013	0.015	0.018	0.017	0.025	0.005	0.023

Table 13.5 P_j and R_j values of attributes

*P*1	0.173
*P*2	0.174
*P*3	0.169
*P*4	0.166
*P*5	0.165
*P*6	0.112
*R*1	0.024
*R*2	0.052
*R*3	0.062
*R*4	0.026
*R*5	0.041
*R*6	0.035

Table 13.6 Alternative weightages

*ISO (Q*1)	0.232
*QFD (Q*2)	0.201
*KAIZEN (Q*3)	0.192
*SIX SIGMA (Q*4)	0.221
*TQM (Q*5)	0.200
*TRIZ (Q*6)	0.153

To effectively implement quality management, it is really essential to focus on the most important TQM practices initially and address the less important ones in the later stage. This involves implementation of the ranking of the TQM practices. Thus, it is really essential to determine the priority of the quality management criteria to achieve maximum benefits and desired goal. This analysis can benefit many types of organizations such as manufacturing, education, operations and supply chain.

Application 2: Smart cities

Cities are engine of any economy. According to 2011 census, Nearly 31% of India's current population lives in urban areas and contributes 63% of India's GDP. With increasing urbanization, urban areas are expected to house 40% of India's population and contribute 75% of India's GDP by 2030. The conceptualization of Smart City varies from city to city and country to country, depending on the level of development, willingness to change and reform, resources and aspirations of the city residents. In the approach to the Smart Cities Mission, the objective is to promote cities that provide core infrastructure and give a decent quality of life to its citizens, a clean and sustainable environment and application of "Smart" Solutions.

Smart Cities concept was formally introduced as a government initiative in June 2015 under government programme SMART CITY AWAS YOJNA. The Smart City Mission will be operated as a Centrally Sponsored Scheme (CSS) and the Central Government proposes to give financial support to the Mission to the extent of Rs. 48,000 crores over five years, i.e. on an average Rs. 100 crore per city per year. An equal amount, on a matching basis, will have to be contributed by the State/ULB; The Mission will cover 100 cities and its duration will be five years (FY2015-16 to FY2019-20). The Mission may be continued thereafter in the light of an evaluation to be done by the Ministry of Urban Development (MoUD) and incorporating the learnings into the Mission.

Table 13.7 Description of smart city criteria

S. No.	Criteria	Definition
1	Revenue generated	Collection of internally generated revenue (e.g. taxes, fees, charges) during the last three FYs (2012–15)
2	Road length	If there exists sufficient length of road then less investment in road sector is required hence it indicates the financial viability
3	JnNURM reforms	Percentage of completion of Projects sanctioned up to March, 2012 under JnNURM
4	Environmental impact	Current pollution level of the city more the pollution level more is the priority since it needs quick reforms
5	Employment generation	The employment that will be generated while implementation of the project
6	Cost-effectiveness of smart city plan	Whether the plan is financially feasible or not. Higher the cost lesser the priority
7	Population	The number of people who is going to be benefited from the smart city this is also the indicator of utility of the facilities
8	No. of educational institute(higher educations)	More the number of educational institute more number of human resource will get benefited

Table 13.8 Description of smart city challenges

S. No.	Challenges	Definitions
1	Retrofitting existing legacy city infrastructure to make it smart:	Determining the existing problem of the area that means most consideration
2	Availability of master plan or city development plan:	70–80% of cities don't have a city development plan
3	Reliability of utility services:	These project should have 24 * 7 h of electricity supply, which is not possible with current infrastructure
4	Three-tier governance	There is a lot of conflict between central govt, state govt, and local govt

Indian government is planning to develop few cities as smart city to provide better standard of life. But due to availability of limited fund and time there must be a definite plan and criteria so as smart city project proceeds in definite direction. This demands propirtization of the proposed smart city projects. Given ten cities and basic available demographic and other data like population, GDP, availability of educational institutes, etc., we need to rank the cities to be considered for smart cities. The description of smart city criteria and challenges is given in Tables 13.7 and 13.8, respectively.

Analysing

For the selection of smart cities, the quantitative data used is shown in Table 13.9. In this table, cities are considered as alternative and are placed in rows. The criteria or attributes are placed in column.

Next step is normalization of performance scores of the alternatives with respect to the considered attributes.

The normalization is done and the results are shown in Table 13.10.

Then we compute the weighted normalized performance scores d_{ij}.

All the weighted normalized performance scores determined are shown in Table 13.11.

Now the sums of weighted normalized values are computed for both the beneficial criteria (S_{i+}) and non-beneficial criteria (S_{i-}).

In the similar way, all the values of (S_{i+}), (S_{i-}) and (Q_i) are determined and shown in Table 13.12. Then, we determine the performance index (P_i) for each alternative considering $Q_{\max}$ (as shown in Fig. 13.2).

Table 13.9 Quantitative data for smart city problem

Alternatives	Criteria							
	Revenue generated (lakhs)	JnNURM Reforms	Population	Road Network km	Particulate content (level of pollution) PM2.5	Employment generation ok (in million)	Plan budget outlay (in crore)	No. of institutions
Delhi	451,154	46.18	16,753,235	32,000	153	9.8	864	1200
Bhubaneswar	310,810	62.90	648,032	11,345	93	5.6	712	400
Pune	1,686,695	57.55	2,538,000	23,453	57	4.7	854	360
Jaipur	574,549	51.86	2,233,000	12,343	88	6.5	636	413
Surat	568,504	65.50	2,434,000	2341	103	5.2	728	258
Ahmedabad	856,905	75.50	3,520,000	2365	100	7.4	855	599
Guwahati	183,798	54.85	809,895	7653	71	6.3	560	174
Chennai	976,703	64.22	4,344,000	5643	81	7.5	826	678
Bhopal	508,006	51.15	1,437,000	34,564	112	6.8	751	490
Vishakhapatnam	520,030	63.20	982,904	232	65	5.6	562	365

Table 13.10 Normalized matrix for Smart City Problem

	Revenue generated (lakhs)	JnNURM Reforms	Population	Road Network km	Particulate content (level of pollution) PM2.5	Employment generation ok (in million)	Plan budget outlay (in crore)	No. of institutions
Delhi	0.068	0.078	0.469	0.243	0.166	0.150	0.118	0.243
Bhubaneswar	0.047	0.106	0.018	0.086	0.101	0.086	0.097	0.081
Pune	0.254	0.097	0.071	0.178	0.062	0.072	0.116	0.073
Jaipur	0.087	0.087	0.063	0.094	0.095	0.099	0.087	0.084
Surat	0.086	0.110	0.068	0.018	0.112	0.080	0.099	0.052
Ahmedabad	0.129	0.127	0.099	0.018	0.108	0.113	0.116	0.121
Guwahati	0.028	0.093	0.023	0.058	0.077	0.096	0.076	0.035
Chennai	0.147	0.108	0.122	0.043	0.088	0.115	0.112	0.137
Bhopal	0.077	0.086	0.040	0.262	0.121	0.104	0.102	0.099
Vishakhapatnam	0.078	0.107	0.028	0.002	0.070	0.086	0.076	0.074

Table 13.11 Weighted normalized matrix for smart city problem

	Revenue generated (lakhs)	JnNURM Reforms	Population	Road Network km	Particulate content (level of pollution) PM2.5	Employment generation ok (in million)	Plan budget outlay (in crore)	No. of institutions
Weightage	0.3	0.15	0.10	0.12	0.075	0.1	0.11	0.045
Delhi	0.0204	0.0117	0.0469	0.0291	0.0124	0.0150	0.0129	0.0109
Bhubaneswar	0.0140	0.0159	0.0018	0.0103	0.0076	0.0086	0.0107	0.0036
Pune	0.0762	0.0146	0.0071	0.0213	0.0046	0.0072	0.0128	0.0033
Jaipur	0.0260	0.0131	0.0063	0.0112	0.0072	0.0099	0.0095	0.0038
Surat	0.0257	0.0166	0.0068	0.0021	0.0084	0.0080	0.0109	0.0024
Ahmedabad	0.0387	0.0191	0.0099	0.0022	0.0081	0.0113	0.0128	0.0055
Guwahati	0.0083	0.0139	0.0023	0.0070	0.0058	0.0096	0.0084	0.0016
Chennai	0.0441	0.0162	0.0122	0.0051	0.0066	0.0115	0.0124	0.0062
Bhopal	0.0230	0.0129	0.0040	0.0314	0.0091	0.0104	0.0112	0.0045
Vishakhapatnam	0.0235	0.0160	0.0028	0.0002	0.0053	0.0086	0.0084	0.0033

Table 13.12 Ranking of criteria for smart city problem

Alternative	S_{i+}	$S_{i}-$	Q_i	P_i	Rank
Delhi	0.1340	0.0254	0.1471	98.89	2
Bhubaneswar	0.0543	0.0182	0.0725	48.76	9
Pune	0.1297	0.0174	0.1488	100.00	1
Jaipur	0.0703	0.0167	0.0902	60.62	6
Surat	0.0615	0.0193	0.0788	52.93	7
Ahmedabad	0.0866	0.0209	0.1025	68.89	5
Guwahati	0.0426	0.0142	0.0661	44.43	10
Chennai	0.0953	0.0189	0.1129	75.87	3
Bhopal	0.0862	0.0203	0.1026	68.93	4
Vishakhapatnam	0.0543	0.0137	0.0786	52.83	8

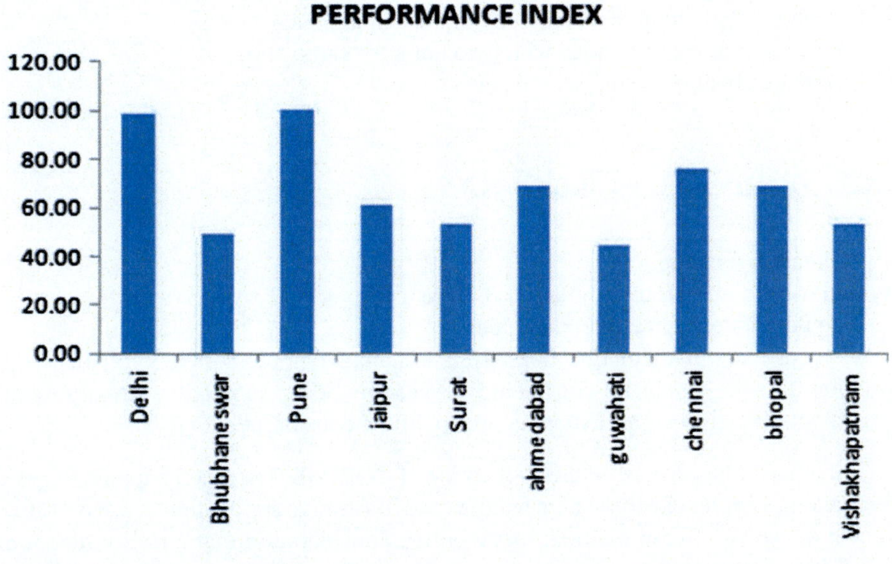

Fig. 13.2 Performance index of smart city project

Inferences

From the results, we can see that

1. Performance index of Guwahati is very less. This may be attributed to less revenue generated within the city. Therefore, the government should explore the ways by which revenue can be augmented for various smart city projects.
2. Delhi and Pune have performed very well and hence government should adopt them among smart city on priority.

13.4 Advantages of COPRAS

- A single criterion cannot give a full expression of the goals pursued by various clients. For this purpose, multi-attribute decision method COPRAS can be used.
- It can be used for multi-criteria evaluation of both maximizing and minimizing criteria.
- Allows for imprecise input; takes into account insufficient information.
- It is very simple to understand and easy to implement.
- We can normalize quantities which do not have same unit.
- Eliminate bias in the study.

13.5 Disadvantages of COPRAS

- Difficult to develop
- Can require numerous simulations before use.
- It could not determine interval values.
- Many data are mathematical approximation and are not completely reliable.
- Some factors cannot be considered because of being subjective in nature and difficult to assign numerical value like political consideration.

This chapter has discussed the application of CORPAS. The next chapter will help the readers to appreciate the salient features and illustrative application of ARAS (also known as ARM). These methods work on the consideration that a utility function value determining the complex relative efficiency of a feasible alternative is directly proportional to the relative effect of values and weights of the main criteria considered in a project.

Selected Bibliography and Further Reading

1. Podvezko, V. (2011). The comparative analysis of MCDA methods SAW and COPRAS. *Inzinerine Ekonomika-Engineering Economics, 22*(2), 134–146.

2. Popović, G., Stanujkić, D., & Stojanović, S. (2012). Investment project selection by applying COPRAS method and imprecise data. *Serbian Journal of Management, 7*(2), 257–269.
3. Bielinskas, V., Burinskienė, M., & Palevičius, V. (2015). *Assessment of neglected areas in Vilnius city using MCDM and COPRAS methods, 122,* 29–38.
4. website: http://smartcities.gov.in/.
5. https://www.ijert.org/research/prioritization-of-critical-factors-impacting-on-tqm-implementation-in-golgohar-mining-industrial-company-by-using-ahp-IJERTV3IS080561.pdf.

Chapter 14
Additive Ratio Assessment Method (ARM/ARAS)

14.1 Background

ARAS/ARM is a quantitative method developed by Zavadskas and Turskis [4]. In a mono-criterion approach, the analyst builds a unique criterion capturing all the relevant aspects of the problem. Such a one-dimensional approach is an oversimplification of the actual nature of the problem. In many real-world decision problems, a decision maker has a set of multiple conflicting objectives. All new ideas and possible variants of decisions must be compared for various criteria. The problem of a decision maker consists of evaluating a finite set of alternatives in order to find the best one, to rank them from the best to the worst, to group them into predefined homogeneous classes, or to describe how well each alternative meets all the criteria simultaneously. There are many methods for determining the ranking of a set of alternatives in terms of a set of decision criteria. In a multi-criteria approach, the analyst seeks to build several criteria using several points of view. MCDM is one of the most widely used decision methodologies in science, business and governments, which are based on the assumption of a complex world, and can help to improve the quality of decisions by making the decision-making process more explicit, rational and efficient.

The typical MCDM problem is concerned with the task of ranking a finite number of decision alternatives, each of which is explicitly described in terms of different decision criteria that need to be considered simultaneously. According to the ARAS method, a utility function value determining the complex relative efficiency of a feasible alternative is directly proportional to the relative effect of values and weights of the main criteria considered in a project.

Although it is a newly proposed method, the ARAS method is an effective and easy to use for various MCDM problems. The effectiveness and usefulness of the ARAS method is confirmed by its extensions, such as an extension of the method with grey numbers ARAS-G and an extension with the use of interval-valued triangular fuzzy numbers. The method has been successfully used for solving many MCDM problems in different areas such as the selection of the supplier, the ranking of

J. J. Thakkar, *Multi-Criteria Decision Making*, Studies in Systems, Decision and Control 336, https://doi.org/10.1007/978-981-33-4745-8_14

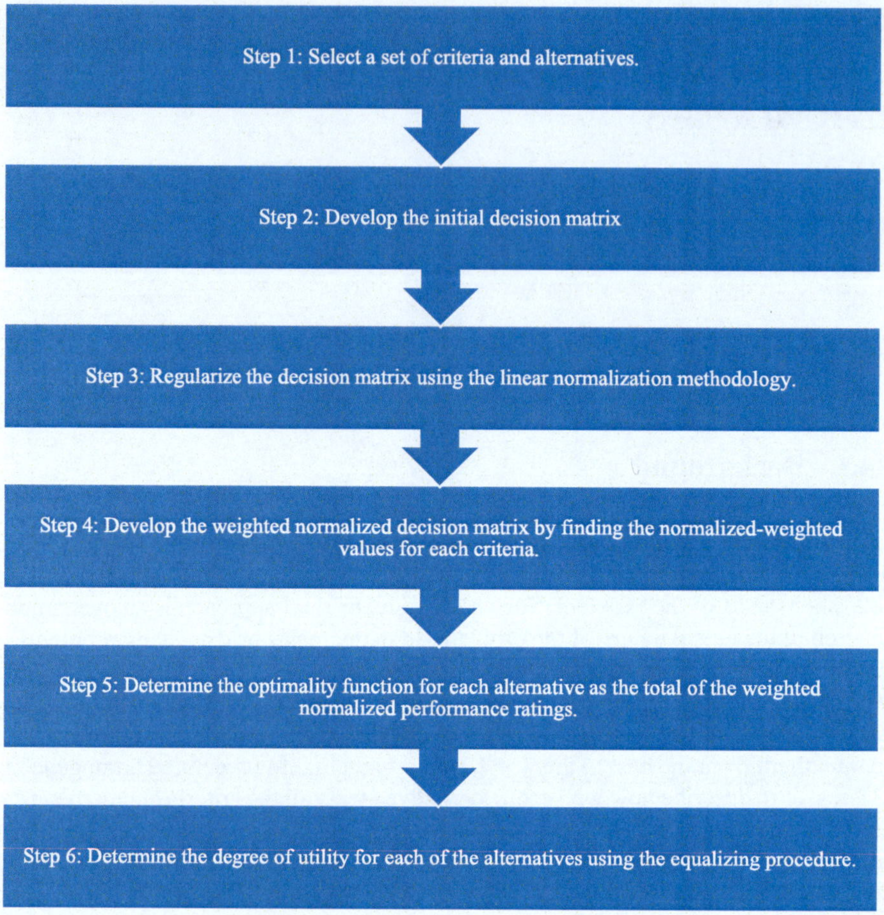

Fig. 14.1 ARAS/ARM procedure

financial institutions and problems related to construction etc. Figure 14.1 represents the step-by-step procedure of ARAS/ARM.

14.2 Step-by-Step Procedure of ARM

Step 1: Construct the decision-making matrix (DMM). In the MCDM of the discrete optimization problem, any problem to be solved is represented by the following DMM of preferences for m feasible alternatives (rows) rated on n signfull criteria (columns):

$$X = \begin{bmatrix} x_{01} \ldots x_{0j} \ldots x_{0n} \\ . \\ x_{i1} \ldots x_{ij} \ldots x_{in} \\ . \\ x_{m1} \ldots x_{mj} \ldots x_{mn} \end{bmatrix}; i = 0, 1, 2, \ldots, m; \; j = 1, 2, 3, \ldots, n$$

where m—number of alternatives, n—number of criteria describing each alternative, x_{ij}—value representing the performance value of the i alternative in terms of the j criterion and x_{0j}—optimal value of j criterion.

If optimal value of j criterion is unknown, then

$$\begin{aligned} x_{0j} &= \max x_{ij}; \text{ if maximum is preferable.} \\ &= \min x_{ij}; \text{ if minimum is preferable.} \end{aligned}$$

Step 2: The initial values of all the criteria are normalized. The values of normalized $\bar{x}_{ij}$ are defined as follows,

$$\bar{x}_{ij} = \frac{x_{ij}}{\sum_{i=0}^{m} x_{ij}}$$

The criteria, whose preferable values are minima, are normalized by applying two-stage procedure:

$$x_{ij} = \frac{1}{x_{ij}^*}; \text{ Then, } \quad \bar{x}_{ij} = \frac{x_{ij}}{\sum_{i=0}^{m} x_{ij}}$$

$$\bar{X} = \begin{bmatrix} \bar{x}_{01} \ldots \bar{x}_{0j} \ldots \bar{x}_{0n} \\ . \\ \bar{x}_{i1} \ldots \bar{x}_{ij} \ldots \bar{x}_{in} \\ . \\ \bar{x}_{m1} \ldots \bar{x}_{mj} \ldots \bar{x}_{mn} \end{bmatrix}; i = 0, 1, 2, \ldots, m; \; j = 1, 2, 3, \ldots, n$$

Step 3: Define normalized-weighted matrix X_1. The values of weight w_j are usually determined by the expert evaluation method. The sum of weights w_j would be limited as follows:

$$\sum_{j=0}^{n} w_j = 1.$$

Normalized-weighted values of all the criteria are calculated as follows:

$$\bar{x}_{ij} w_j = \hat{x}_{ij}$$

where

w_j: weight of the j criterion,
$\bar{x}_{ij}$: Normalized rating of j criterion,

$$X_1 = \begin{bmatrix} \hat{x}_{01} \ldots \hat{x}_{0j} \ldots \hat{x}_{0n} \\ . \\ \hat{x}_{i1} \ldots \hat{x}_{ij} \ldots \hat{x}_{in} \\ . \\ \hat{x}_{m1} \ldots \hat{x}_{mj} \ldots \hat{x}_{mn} \end{bmatrix}; i = 0, 1, 2, \ldots, m; \; j = 1, 2, 3, \ldots, n$$

Step 4: The following operation determines the values of optimality function:

$$S_i = \sum_{i=0}^{n} \hat{x}_{ij} \quad i = 0, 1, 2, \ldots m$$

where S_i is the value of optimality function of i alternative.

The biggest value is the best and the least one is the worst.

The degree of the alternative utility is determined by a comparison of the variant with the ideally best one S_0. The equation used for the calculation of the utility degree K_i of an alternative is given as below:

$K_i = \frac{S_i}{S_0}$ where S_i and S_0 are the optimality criterion values.

The calculated values K_i are in the interval [0, 1] and can be ordered in an increasing sequence, which is the wanted order of precedence.

14.3 Illustrative Application of ARM/ARAS

Application 1: Prioritization of banks for online banking

A person wants to avail an online banking facility. He wants to open his account from the alternatives he has, viz.,

1. AXIS Bank;
2. State Bank of India (SBI);
3. ICICI Bank;
4. HDFC Bank;
5. Punjab National Bank, based on the features (criteria) each banking website offer.

Table 14.1 Decision matrix for online banking problem

	Security	Balance enquiry	Transaction history	Bank statement	Payment transfer	SMS alert	Payment to insurance and bills
AXIS	Very high	Very high	Average	High	Very high	Very high	High
SBI	Very high	High	High	High	High	High	Very high
ICICI	High	Very high	High	Average	Very high	High	Average
HDFC	Very high	High	High	High	Very high	Very high	Average
PNB	High	Low	Average	Average	Average	Low	Average

The criteria based on which the person is going to take decision are namely,

1. **Security**: The criteria which define how secure the online facility is over Internet;
2. **Balance enquiry**: The accessibility to enquire balance of the account;
3. **Transaction history**: The past data which shows the transactions made in previous;
4. **Bank statement**: The statement which bank sends regarding the concerned account;
5. **Payment transfer**: The ease of transferring money from one to another account;
6. **SMS alert**: Short message alert regarding the change made in the account recently;
7. **Payment to insurance or electricity or telephone bills**: Payment of electricity or telephone bills or monthly/quarterly/yearly insurance money.

Table 14.1 represents the whole decision-making problem in tabular form:

To convert this linguistics quantity to numerical value, we have adopted two scales: beneficial attribute scale and non-beneficial attribute scale.

1. Beneficial attribute scale:

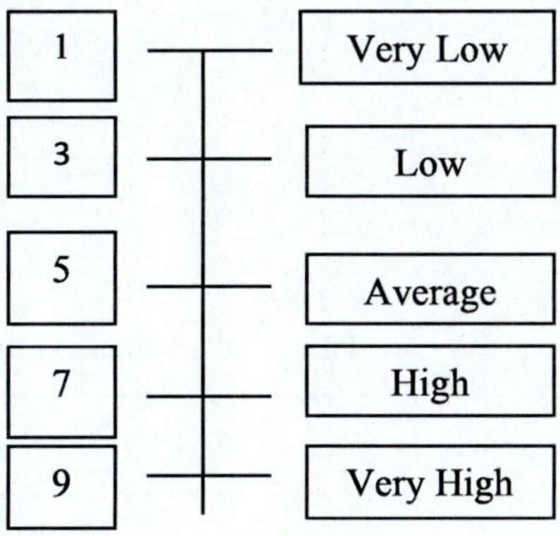

2. Non-beneficial attribute scale:

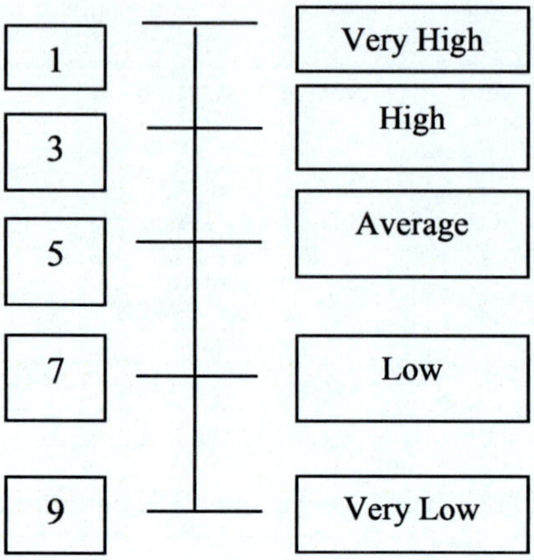

Step 1: Based on the above scales, the decision-making matrix is formed (Table 14.2).

Step 2: All the inputs are normalized and the normalized decision matrix is obtained as (Table 14.3):

Step 3: The weightage of different criterion is given in Table 14.4.

Table 14.2 Decision matrix

	Security	Balance enquiry	Transaction history	Bank statement	Payment transfer	SMS alert	Payment to insurance and bills
AXIS	9	9	5	7	9	9	7
SBI	9	7	7	7	7	7	9
ICICI	7	9	7	5	9	7	5
HDFC	9	7	7	7	9	9	5
PNB	7	3	5	5	5	3	5

Table 14.3 Normalized decision matrix

	Security	Balance enquiry	Transaction history	Bank statement	Payment transfer	SMS alert	Payment to insurance and bills
AXIS	0.219512	0.257143	0.16129	0.225806	0.230769	0.257143	0.225806
SBI	0.219512	0.2	0.225806	0.225806	0.179487	0.2	0.290323
ICICI	0.170732	0.257143	0.225806	0.16129	0.230769	0.2	0.16129
HDFC	0.219512	0.2	0.225806	0.225806	0.230769	0.257143	0.16129
PNB	0.170732	0.085714	0.16129	0.16129	0.128205	0.085714	0.16129

Table 14.4 Criteria weightages

S. No.	Criteria	Weightage
1	Security	0.6
2	Balance enquiry	0.05
3	Transaction history	0.03
4	Bank statement	0.04
5	Payment transfer	0.2
6	SMS alert	0.02
7	Payment to bills	0.06

Table 14.5 Weighted criteria matrix

	Security	Balance enquiry	Transaction history	Bank statement	Payment transfer	SMS alert	Payment to insurance and bills
AXIS	0.131707	0.012857	0.004839	0.009032	0.046154	0.005143	0.013548
SBI	0.131707	0.01	0.006774	0.009032	0.035897	0.004	0.017419
ICICI	0.102439	0.012857	0.006774	0.006452	0.046154	0.004	0.009677
HDFC	0.131707	0.01	0.006774	0.009032	0.046154	0.005143	0.009677
PNB	0.102439	0.004286	0.004839	0.006452	0.025641	0.001714	0.009677

The normalized-weighted matrix is obtained as shown in Table 14.5.
Step 4: The value of optimality function S_i for i alternative is,

$$S_i = \sum_{i=0}^{n} \hat{x}_{ij}$$

See Table 14.6.
The utility degree K_i can be calculated and put in Table 14.7.

Table 14.6 Optimality values (S_i) for alternatives

	Security	Balance enquiry	Transaction history	Bank statement	Payment transfer	SMS alert	Payment to insurance and bills	S_i
AXIS	0.131707	0.012857	0.004839	0.009032	0.046154	0.005143	0.013548	0.223
SBI	0.131707	0.01	0.006774	0.009032	0.035897	0.004	0.017419	0.214
ICICI	0.102439	0.012857	0.006774	0.006452	0.046154	0.004	0.009677	0.188
HDFC	0.131707	0.01	0.006774	0.009032	0.046154	0.005143	0.009677	0.218
PNB	0.102439	0.004286	0.004839	0.006452	0.025641	0.001714	0.009677	0.155

Table 14.7 Utility degree (K_i) for alternatives

	Security	Balance enquiry	Transaction history	Bank statement	Payment transfer	SMS alert	Payment to insurance and bills	S_i	K_i
AXIS	0.131707	0.012857	0.004839	0.009032	0.046154	0.005143	0.013548	0.223	1
SBI	0.131707	0.01	0.006774	0.009032	0.035897	0.004	0.017419	0.214	0.962155
ICICI	0.102439	0.012857	0.006774	0.006452	0.046154	0.004	0.009677	0.188	0.843572
HDFC	0.131707	0.01	0.006774	0.009032	0.046154	0.005143	0.009677	0.218	0.978535
PNB	0.102439	0.004286	0.004839	0.006452	0.025641	0.001714	0.009677	0.155	0.694408

As per ARM technique of multi-criteria decision making, as per utility degree (Si) the best alternative is AXIS Bank, and the worst alternative is PNB.

The ranking of the alternatives can be shown as,

(1) AXIS Bank;
(2) HDFC;
(3) SBI;
(4) ICICI;
(5) PNB.

Application 2: Prioritization of quality management tools for construction project in Indian scenario

A Hyderabad-based Construction Company XYZ wants to analyse and implement a quality management tool out of the following existing options:

- Plan Do Check Act (PDCA) Cycle;
- Fishbone Diagram;
- Pareto Chart;
- Scatter Diagram;
- Decision Matrix;
- Flowchart;
- Stratification;
- Control Chart;
- Histogram;
- Brainstorming;
- Tree Diagram.

Every quality management tool has its own advantages and limitations. Company head and project managers have considered some critical criteria for choosing a quality management tool as explained below.

- **Break down**: It helps in making planning consistent and provides effective project execution. The best break down will have maximum value and vice versa.
- **Link together**: It gives how does the tool links with past projects and useful for future projects. If the tool have good link up with past projects and have good scope of reuse, it will have maximum value and vice versa.
- **Buildup time**: It is the time required to represent the available database. Buildup time has to be minimum. Lesser the buildup time, more will be the value and vice versa.
- **Comparison**: It caters for how much quantum and variety of data can be compared/analyzed by the tool. Higher the comparison ability, higher the value given and vice versa.
- **Multi-functionality**: It caters for how the data can be analyzed (trend) and represented in different ways vis-à-vis requirement for a given set of inputs. Higher the multi-functionality, higher the value given and vice versa.

Table 14.8 Weights to different criteria

Criteria						
Break down ($C1$)	Link together ($C2$)	Buildup ($C3$)	Comparison ($C4$)	Multi-functionality ($C5$)	Cost ($C6$)	Ease of understanding ($C7$)
0.1	0.1	0.1	0.2	0.1	0.2	0.2

- **Cost of usage**: Cost of usage of tool should be minimum. Lesser the cost of usage, more the value given to that tool and vice versa
- **Ease of understanding**: Eventually, the ease of understanding by human resource is a critical criterion. More the ease of understanding more value is given to that and vice versa.

In line with the above discussion, weights are given to different criteria are tabulated in Table 14.8.

The scale used for this is from 0 to 100. Value 100 denotes the best and 0 denotes the worst. Then, the decision-making matrix (DMM) for the above problem with respective values and weights is tabulated in Table 14.9.

Table 14.9 Decision-making matrix with respective values

Criteria							
	Break down ($C1$)	Link together ($C2$)	Buildup ($C3$)	Comparison ($C4$)	Multi-functionality ($C5$)	Cost ($C6$)	Ease of understanding ($C7$)
Weight to criteria (w_{ij})	0.1	0.1	0.1	0.2	0.1	0.2	0.2
Optimal value	100	100	100	100	100	100	100
Plan do check act cycle	90	50	60	70	80	60	70
Fishbone diagram	80	90	70	80	80	40	50
Pareto chart	60	40	80	60	90	50	70
Scatter diagram	40	40	30	20	50	40	30
Decision matrix	70	70	50	50	80	40	30
Flowchart	80	90	40	50	90	40	60
Stratification	80	90	50	60	50	60	60
Control chart	50	30	40	90	60	50	60
Histogram	40	50	80	70	30	80	60
Brainstorming	20	50	40	60	100	20	30
Tree diagram	90	90	90	20	60	50	90

After formulating Table 14.9, the DMM is normalized. The normalized matrix is shown in Table 14.10.

Finally, the Normalized Weighted Matrix with optimum function value and ranks are as shown in Table 14.11.

14.4 Advantages of ARM/ARAS

- Good and easy to apply for quantitative measurements.
- According to goals and opportunities, we can change the weights and preference values.
- The normalization will take care of maxima and minima criteria
- Strong in single-dimensional problem.
- Can remove any unit of measure.
- Ability to compensate among criteria; intuitive to decision makers; calculation is simple does not require complex computer programs.

14.5 Limitations of ARM/ARAS

- Not effective for qualitative initial measurements.
- Not effective for comparative preference.
- Not effective for verbal decision-making analysis.
- Not effective for decision environments involving high levels of uncertainty.
- Estimates revealed do not always reflect the real situation; result obtained may not be logical.
- Weightage given may be different by the different expert. That's why result will be different if the expert is different.
- No solution with equal weight of DMs.

This chapter has detailed the use of a recently developed technique ARM. The next chapter will help the readers to understand the salient features and use of the recent advancement in MCDM typically called as weighted aggregated sum product assessment (WASPAS). This method combines two well-known methods—the weighted sum model (WSM) and the weighted product method (WPM)—to provide a method with accuracy greater than the original two methods, with an optimization of the aggregation being conducted.

Table 14.10 Normalized matrix

Criteria	Break down ($C1$)	Link together ($C2$)	Buildup ($C3$)	Comparison ($C4$)	Multi-functionality ($C5$)	Cost ($C6$)	Ease of understanding ($C7$)
Weight to criteria (w_{ij})	0.1	0.1	0.1	0.2	0.1	0.2	0.2
Optimal value	0.125	0.126582	0.136986	0.136986	0.114943	0.15873	0.140845
Plan do check act cycle	0.1125	0.063291	0.082192	0.09589	0.091954	0.095238	0.098592
Fishbone diagram	0.1	0.113924	0.09589	0.109589	0.091954	0.063492	0.070423
Pareto chart	0.075	0.050633	0.109589	0.082192	0.103448	0.079365	0.098592
Scatter diagram	0.05	0.050633	0.041096	0.027397	0.057471	0.063492	0.042254
Decision matrix	0.0875	0.088608	0.068493	0.068493	0.091954	0.063492	0.042254
Flowchart	0.1	0.113924	0.054795	0.068493	0.103448	0.063492	0.084507
Stratification	0.1	0.113924	0.068493	0.082192	0.057471	0.095238	0.084507
Control chart	0.0625	0.037975	0.054795	0.123288	0.068966	0.079365	0.084507
Histogram	0.05	0.063291	0.109589	0.09589	0.034483	0.126984	0.084507
Brainstorming	0.025	0.063291	0.054795	0.082192	0.114943	0.031746	0.042254
Tree diagram	0.1125	0.113924	0.123288	0.027397	0.068966	0.079365	0.126761

Table 14.11 Normalized weighted matrix

Criteria	Break down ($C1$)	Link together ($C2$)	Buildup ($C3$)	Comparison ($C4$)	Multi-functionality ($C5$)	Cost ($C6$)	Ease of understanding ($C7$)	Optimum function value	Rank
Weight to criteria (w_{ij})	0.1	0.1	0.1	0.2	0.1	0.2	0.2		
Optimal value	0.0125	0.012658	0.013699	0.027397	0.011494	0.031746	0.028169	0.137663	1
Plan do check act cycle	0.01125	0.006329	0.008219	0.019178	0.009195	0.019048	0.019718	0.092938	2
Fishbone Diagram	0.01	0.011392	0.009589	0.021918	0.009195	0.012698	0.014085	0.088878	3
Pareto chart	0.0075	0.005063	0.010959	0.016438	0.010345	0.015873	0.019718	0.085897	7
Scatter diagram	0.005	0.005063	0.00411	0.005479	0.005747	0.012698	0.008451	0.046549	12
Decision matrix	0.00875	0.008861	0.006849	0.013699	0.009195	0.012698	0.008451	0.068503	10
Flowchart	0.01	0.011392	0.005479	0.013699	0.010345	0.012698	0.016901	0.080515	8
Stratification	0.01	0.011392	0.006849	0.016438	0.005747	0.019048	0.016901	0.086376	6
Control chart	0.00625	0.003797	0.005479	0.024658	0.006897	0.015873	0.016901	0.079855	9
Histogram	0.005	0.006329	0.010959	0.019178	0.003448	0.025397	0.016901	0.087213	5
Brainstorming	0.0025	0.006329	0.005479	0.016438	0.011494	0.006349	0.008451	0.057041	11
Tree diagram	0.01125	0.011392	0.012329	0.005479	0.006897	0.015873	0.025352	0.088572	4

Selected Bibliography and Further Reading

1. Zavadskas, E. K., Kaklauskas, A., Peldschus, F., & Turskis, Z. (2007). Multi-attribute assessment of road design solutions by using the COPRAS method. *The Baltic Journal of Road and Bridge Engineering, 2*(4), 195–203.
2. Zavadskas, E. K., Kaklauskas Turskis, Z., & Tamošaitienė, J. (2008). Selection of the effective dwelling house walls by applying attributes values determined at intervals. *Journal of Civil Engineering and Management, 14*(2), 85–93. https://doi.org/10.3846/1392-3730.2008.14.3.
3. Zavadskas, E. K., Turskis, Z., Tamošaitienė, K., & Marina, V. (2008). Multicriteria selection of project managers by applying grey criteria. *Technological and Economic Development of Economy, 14*(4), 462–477. https://doi.org/10.3846/1392-8619.2008.14.462-477.
4. Zavadskas, E. K., & Turskis, Z. (2010). A new additive ratio assessment (ARAS) method in multi-criteria decision-making. *Technological and Economic Development of Economy, 16*(2), 159–172.

Chapter 15
Weighted Aggregated Sum Product Assessment (WASPAS)

15.1 Background

Multiple criteria decision-making (MCDM) methods can be successfully applied to rank complex alternative decisions and to achieve optimal solutions based on multiple, usually conflicting criteria. Weighted aggregated sum product assessment (WASPAS) method is preferred to the variety of available methods because of its abilities to increase the accuracy of ranking. WASPAS enables to reach the highest accuracy of estimation applying suggested methodology for optimization of weighted aggregated function. It combines two well-known methods—the weighted sum model (WSM) and the weighted product method (WPM)—to provide a method with accuracy greater than the original two methods, with an optimization of the aggregation being conducted. It has already been proved that the accuracy of an aggregated method would always be better than that of the single methods.. The effect of parameter λ on the ranking performance of WASPAS method is also studied, revealing the fact that better performance is attained at higher λ values. When the value of λ is set at 0, WASPAS method behaves like a WPM method, and when λ is 1, it is transformed into WSM method. The main advantage of this method is identified as its strong resistance against rank reversal of the considered alternatives. It is also found that this method has the unique capability of dealing with both single and multi-response optimization problems in various decision-making process. As this method involves simple and sound mathematics, and is quite comprehensive in nature, it can be successfully applied to any decision-making-related situation.

As the WASPAS model is simple to apply and yet provides accurate results compared to both the WSM and WPM methods, it is being widely accepted and used as an efficient decision-making approach. The WASPAS method tries to find optimal solution based on two joint criteria, i.e., weighted aggregation of additive and multiplicative methods. It has been applied for solving MCDM problems for increasing ranking accuracy, and it has the capability to reach the highest accuracy of estimation. A procedure of WASPAS is presented in Fig. 15.1.

J. J. Thakkar, *Multi-Criteria Decision Making*, Studies in Systems, Decision and Control 336, https://doi.org/10.1007/978-981-33-4745-8_15

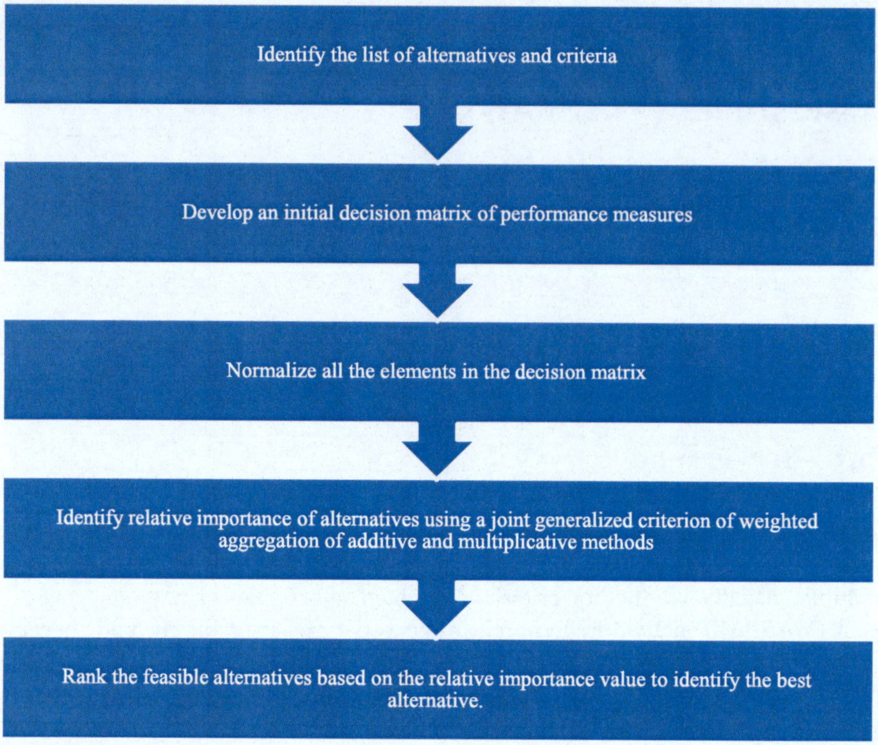

Fig. 15.1 WASPAS procedure

15.2 Step-by-Step Procedure of WASPAS

Mathematical Formulation of Decision

Suppose that a given MCDM problem is defined on m alternatives and n decision criteria. Let w_j denote the relative significance of the criterion, and x_{ij} is the performance value of alternative i when it is evaluated in terms of criterion j.

The total relative importance of alternative i as per the WSM method, denoted by $Q_i^{(1)}$, is defined as:

$$Q_i^{(1)} = \sum_{j=1}^{n} \bar{x}_{ij} w_j, \tag{15.1}$$

where

$$\bar{x}_{ij} = \begin{cases} \frac{x_{ij}}{\max_i x_{ij}} & \text{if } \max_i x_{ij} \text{ is preferable} \\ \frac{\min_i x_{ij}}{x_{ij}} & \text{if } \min_i x_{ij} \text{ is preferable} \end{cases}$$

This is defined as the linear normalization of initial criteria values.

The total relative importance of alternative i as per the WPM method, denoted by $Q_i^{(2)}$, is defined as:

$$Q_i^{(2)} = \prod_{j=1}^{n} \left(\bar{x}_{ij}\right)^{w_j}, \tag{15.2}$$

where $\bar{x}_{ij}$ is the linear normalization of initial criteria values as explained above.

The weighted aggregated sum product assessment (WASPAS) method for ranking of alternatives is defined as:

$$Q_i = \lambda \sum_{j=1}^{n} \bar{x}_{ij} w_j + (1-\lambda) \prod_{j=1}^{n} \left(\bar{x}_{ij}\right)^{w_j}, \lambda = 0, 0.1, 0.2, \ldots, 1. \tag{15.3}$$

When $\lambda = 0$, WASPAS reduces to the WPM, and it reduces to the WSM when $\lambda = 1$.

As λ is optimized for minimum variance of the initial criteria values,

$$\lambda_{\text{optimal}} = \frac{\sigma^2\left(Q_i^{(2)}\right)}{\sigma^2\left(Q_i^{(1)}\right) + \sigma^2\left(Q_i^{(2)}\right)}, \tag{15.4}$$

where

$$\sigma^2\left(Q_i^{(1)}\right) = \sum_{j=1}^{n} \left(\frac{\partial Q_i^{(1)}}{\partial \bar{x}_{ij}}\right)^2 \sigma^2\left(\bar{x}_{ij}\right), \tag{15.5}$$

$$\sigma^2\left(Q_i^{(2)}\right) = \sum_{j=1}^{n} \left(\frac{\partial Q_i^{(2)}}{\partial \bar{x}_{ij}}\right)^2 \sigma^2\left(\bar{x}_{ij}\right), \tag{15.6}$$

Optimal λ (Eq. 15.4) should be calculated for every alternative before applying WASPAS (Eq. 15.3). Optimal λ may vary depending on ratio of $\frac{\sigma^2\left(Q_i^{(1)}\right)}{\sigma^2\left(Q_i^{(2)}\right)}$ $\Im$ in every particular case.

Every MCDM problem starts with the following decision/evaluation matrix:

$$X = \begin{bmatrix} x_{11} & x_{12} & \dots & x_{1n} \\ x_{21} & x_{22} & \dots & x_{2n} \\ \dots & \dots & \dots & \dots \\ x_{m1} & x_{m2} & \dots & x_{mn}. \end{bmatrix}$$

where m is the number of candidate alternatives, n is the number of evaluation criteria and x_{ij} is the performance of i'th alternative with respect to j'th criterion.

The application of WASPAS method which is a unique combination of two well-known MCDM approaches, i.e., weighted sum model (WSM) and weighted product model (WPM) at first requires linear normalization of the decision matrix elements using the following two equations:

For beneficial criteria: $\overline{x}_{ij} = \frac{x_{ij}}{\max_i, x_{ij}}$.

And for non-beneficial criteria: $\overline{x}_{ij} = \frac{\min_i, x_{ij}}{x_{ij}}$.

In WASPAS method, a joint criterion of optimality is sought based on two criteria of optimality. The first criterion of optimality, i.e., criterion of a mean weighted success, is similar to WSM method. It is a popular and well-accepted MCDM approach applied for evaluating a number of alternatives in terms of a number of decision criteria. Based on WSM method, the total relative importance of i'th alternative is calculated as follows:

$$\left(Q_i^{(1)}\right) = \sum_{j=1}^{n} \bar{x}_{ij} w_i$$

On the other hand, according to WPM method, the total relative importance of i th alternative is computed using the following expression:

$$\left(Q_i^{(2)}\right) = \prod_{j=1}^{n} \bar{x}_{ij} w_i$$

In order to have increased ranking accuracy and effectiveness of the decision-making process, in WASPAS method, a more generalized equation for determining the total relative importance of i th alternative is developed as below

$$Q_i = \lambda Q_i^{(1)} + (1 - \lambda) Q_i^{(2)} = \lambda \sum_{j=1}^{n} \bar{x}_{ij} w_i + (1 - \lambda) \prod_{j=1}^{n} \bar{x}_{ij} w_i$$

$$\lambda = 0, 0.1, 0.2, \dots.1$$

15.3 Illustrative Applications of WASPAS

Application 1: Selection of elective course

Industrial and system engineering is an interdisciplinary department at IIT Kharagpur. It offers B.Tech, Dual Degree, M.Tech, MS, and Ph.D. programs. The objective of this department is to provide a platform for academicians, researchers, practitioners, and entrepreneurs to explore the areas such as supply chain, E-business, virtual reality, safety analytics and its current applications.

The M.Tech programme is of two years duration. The compulsory subjects include: decision modelling, work system design, production planning and inventory control, advanced decision modelling, system modelling analysis, TQM and six sigma, management information systems. In addition, a large number of elective subjects are offered in various areas of Operations Management, Information Technology, and Quality and Productivity that includes Supply Chain Management, Technology Management, Safety Engineering and Management.

This section demonstrates an application of WASPAS for prioritizing elective courses of M.Tech. program. The decision-making problems often include four basic elements: (1) the set of substitutive alternatives, (2) the set of evaluation criteria, (3) the outcome (or decision) matrix in regard to the alternatives scored based on the evaluation criteria, and (4) the preference structure of decision making about criteria importance or weights.

For demonstrating an application of WASPAS for prioritizing various electives offered by the department, list of electives and various attributes (criteria) for evaluation are listed in Tables 15.1 and 15.2 respectively.

A step-by-step procedure of WASPAS is employed for prioritizing the electives as below.

Step 1: Table 15.3 represents the linguistics assessment of the elective subjects for the set of attributes (criteria) defined in Table 15.2.

According to various literature reviews, to convert this linguistics quantity to numerical value we have adopted two scales: beneficial attributes scale and non-beneficial attributes scale.

Table 15.1 Elective subjects for MCDM problem

S. No.	Elective subject's name
1	Engineering system safety design and control (ESSDC)
2	Human factor engineering (HF)
3	Supply chain management (SCM)
4	Recommender system in e-business (RSEB)
5	Project engineering and management (PEM)
6	Statistical decision modelling (SDM)

Table 15.2 List of attributes for elective selection problem

S. No.	Name of attributes
1	Scope for Ph.D. (SPHD)
2	Scope for job (SJ)
3	Prior knowledge (PK)
4	Experience of lecturer (EL)
5	Subject of interest (SI)
6	Course content (CC)
7	Advice of guide (AG)
8	Structure of routine/Slot (SR)

Table 15.3 Linguistic assessment of the alternatives

SPHD		SJ	PK	EL	SI	CC	AG	SR
ESSDC	High	High	0.43	0.47	Very high	High	High	0.85
HF	Very high	Very low	0.12	0.95	Low	Average	Low	0.23
SCM	High	High	0.83	0.81	Average	Very high	High	0.62
RSEB	Average	Very high	0.76	0.85	High	Average	Low	0.80
PEM	Average	Average	0.32	0.89	Very high	Low	Average	0.61
SDM	High	Average	0.78	0.91	High	Average	High	0.35

1. Beneficial attributes scale:

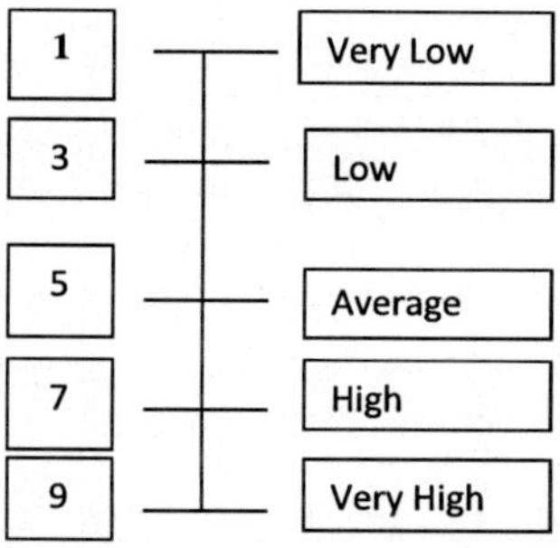

2. Non-beneficial attributes scale:

Step 2: Then, we have identified that between our eight selected attributes which attributes are beneficial and which are non-beneficial.

Table 15.4 Beneficial attributes

S. No.	Beneficial attributes
1	Scope for Ph.D. (SPHD)
2	Scope for job (SJ)
3	Subject of interest (SI)
4	Advice of guide (AG)

Table 15.5 Non-beneficial attributes

S. No.	Non-beneficial attributes
1	Course content (CC)

The beneficial attributes are: Tables 15.4 and 15.5

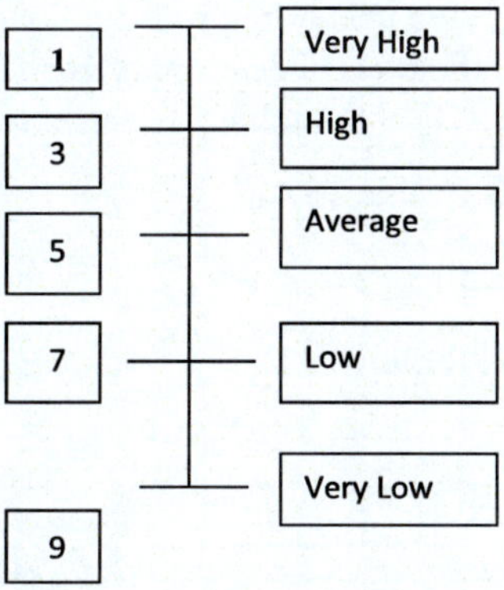

The non-beneficial attributes are:

Step 3: Now using the upper scales we have adopted the numerical table as 15.6

Step 4: Now we have derived the following weightage matrix using the following equations (Table 15.7):

$$\overline{x}_i j = \frac{x_i j}{\max i, x_i j} \text{ And } \overline{x}_i j = \frac{\min i, x_i j}{x_i j}$$

Table 15.6 Numerical table using upper scale

SPHD		SJ	PK	EL	SI	CC	AG	SR
ESSDC	7	7	0.43	0.87	9	3	7	0.85
HF	9	1	0.12	0.95	1	5	1	0.23
SCM	7	7	0.83	0.81	5	1	7	0.62
RSEB	5	9	0.76	0.85	7	5	1	0.80
PEM	5	5	0.32	0.89	9	7	5	0.61
SDM	7	5	0.78	0.91	7	5	7	0.35

Table 15.7 Weighted matrix

	SPHD	SJ	PK	EL	SI	CC	AG	SR
ESSDC	0.778	0.778	0.518	0.916	1	0.33	1	1
HF	1	0.111	0.146	1	0.111	0.2	0.143	0.271
SCM	0.778	0.778	1	0.853	0.555	1	1	0.729
RSEB	0.555	1	0.916	0.895	0.778	0.2	0.143	0.941
PEM	0.555	0.555	0.385	0.937	1	0.143	0.714	0.718
SDM	0.778	0.555	0.940	0.958	0.778	0.20	1	0.412

Step 5: The attributes are assigned weightages in Table 15.8 based on inputs received from Ph.D scholars of the department.

Step 6: Table 15.9 is derived using the procedural steps of WSM method.

Step 7: Table 15.10 is dervied using the procedural steps of WPM method.

Step 8: The WASPAS method has been employed and results for eight attributes for λ (varies from 0, 0.1, 0.2…1) are obtained as presented in Tables 15.11 to 15.18.

1. **Scope for Ph.D. (SPHD)**: Table 15.11

2. **Scope for Job (SJ)**: Table 15.12

3. **Prior Knowledge (PK)**: Table 15.13

Table 15.8 Weightages of eight attributes

S. No.	Name of attributes weightage
1	Scope for Ph.D. (SPHD) 0.20
2	Scope for job (SJ) 0.25
3	Prior knowledge (PK) 0.03
4	Experience of lecturer (EL) 0.13
5	Subject of interest (SI) 0.18
6	Course content (CC) 0.07
7	Advice of guide (AG) 0.10
8	Structure of routine/slot (SR) 0.04

Table 15.9 WSM method matrix

Alternatives	SPHD	SJ	PK	EL	SI	CC	AG	SR
ESSDC	0.1556	0.1945	0.0156	0.119	0.18	0.023	0.1	0.04
HF	0.2	0.0278	0.0044	0.13	0.019	0.014	0.0143	0.0108
SCM	0.1556	0.1945	0.03	0.1108	0.099	0.07	0.1	0.03
RSEB	0.111	0.25	0.027	0.116	0.14	0.014	0.0143	0.04
PEM	0.111	0.13875	0.0116	0.122	0.18	0.01	0.0714	0.028
SDM	0.1556	0.13875	0.028	0.125	0.14	0.014	0.1	0.0165

Table 15.10 WPM method matrix

Alternatives	SPHD	SJ	PK	EL	SI	CC	AG	SR
ESSDC	0.951	0.94	0.98	0.988	1	0.925	1	1
HF	1	0.577	0.944	1	0.673	0.893	0.823	0.949
SCM	0.951	0.94	1	0.979	0.899	1	1	0.987
RSEB	0.89	1	0.997	0.985	0.955	0.893	0.823	0.977
PEM	0.89	0.863	0.971	0.991	1	0.872	0.967	0.986
SDM	0.951	0.863	0.998	0.994	0.955	0.893	1	0.965

4. **Experience of Lecturer (EL)**: Table 15.14
5. **Subject of Interest (SI)**: Table 15.15
6. **Course Content (CC)**: Table 15.16
7. **Advice of Guide (AG)**: Table 15.17
8. **Structure of Routine/Slot (SR)**: Table 15.18

Inferences

The results obtained are presented as bar plots for a detailed interpretation of the results as below.

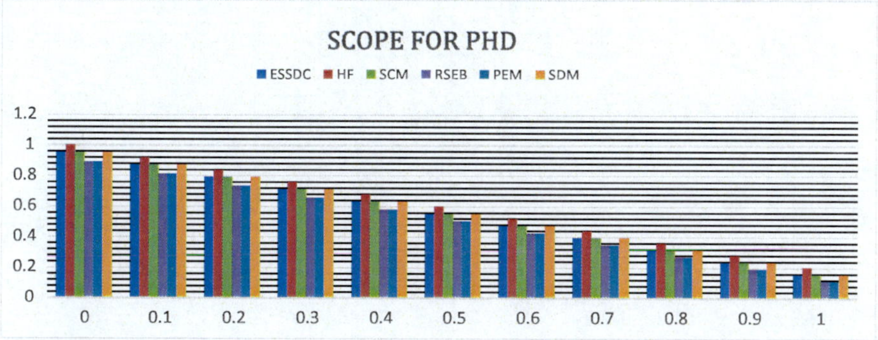

From this graph, it has been observed that if a student wants to select an elective subject for his/her aspiration for doing PHD or higher study, then from this graph he/she can easily select human factor as his elective subject, as it attains highest value for $\lambda = 0, 0.1, 0.2 \ldots 1$, compared to other elective subjects.

Accordingly all the graphs has been developed for the various attributes.

Table 15.11 WASPAS matrix for SPHD

λ	0	0.1	0.2	0.3	0.4	0.5	0.6	0.7	0.8	0.9	1
ESSDC	0.951	0.87146	0.79192	0.71238	0.63284	0.5533	0.47376	0.39422	0.31468	0.23514	0.1556
HF	1	0.92	0.84	0.76	0.68	0.6	0.52	0.44	0.36	0.28	0.2
SCM	0.951	0.87146	0.79192	0.71238	0.63284	0.5533	0.47376	0.39422	0.31468	0.23514	0.1556
RSEB	0.89	0.8121	0.7342	0.6563	0.5784	0.5005	0.4226	0.3447	0.2668	0.1889	0.111
PEM	0.89	0.8121	0.7342	0.6563	0.5784	0.5005	0.4226	0.3447	0.2668	0.1889	0.111
SDM	0.951	0.87146	0.79192	0.71238	0.63284	0.5533	0.47376	0.39422	0.31468	0.23514	0.1556

Table 15.12 WASPAS matrix for SJ

λ	0	0.1	0.2	0.3	0.4	0.5	0.6	0.7	0.8	0.9	1
ESSDC	0.94	0.86545	0.7909	0.71635	0.6418	0.56725	0.4927	0.41815	0.3436	0.269	0.1945
HF	0.577	0.52208	0.46716	0.41224	0.3573	0.3024	0.24748	0.19256	0.13764	0.083	0.0278
SCM	0.94	0.86545	0.7909	0.71635	0.6418	0.56725	0.4927	0.41815	0.3436	0.269	0.1945
RSEB	1	0.925	0.85	0.775	0.7	0.625	0.55	0.475	0.4	0.325	0.25
PEM	0.863	0.79058	0.71815	0.64573	0.5733	0.500875	0.42845	0.35603	0.2836	0.211	0.1388
SDM	0.863	0.79058	0.71815	0.64573	0.5733	0.500875	0.42845	0.35603	0.2836	0.211	0.1388

Table 15.13 WASPAS matrix for PK

λ	0	0.1	0.2	0.3	0.4	0.5	0.6	0.7	0.8	0.9	1
ESSDC	0.98	0.88356	0.78712	0.69068	0.5942	0.4978	0.40136	0.30492	0.20848	0.112	0.0156
HF	0.944	0.85004	0.75608	0.66212	0.5682	0.4742	0.38024	0.28628	0.19232	0.098	0.0044
SCM	1	0.903	0.806	0.709	0.612	0.515	0.418	0.321	0.224	0.127	0.03
RSEB	0.997	0.9	0.803	0.706	0.609	0.512	0.415	0.318	0.221	0.124	0.027
PEM	0.971	0.87506	0.77912	0.68318	0.5872	0.4913	0.39536	0.29942	0.20348	0.108	0.0116
SDM	0.998	0.901	0.804	0.707	0.61	0.513	0.416	0.319	0.222	0.125	0.028

Table 15.14 WASPAS matrix for EL

λ	0	0.1	0.2	0.3	0.4	0.5	0.6	0.7	0.8	0.9	1
ESSDC	0.988	0.9011	0.8142	0.7273	0.6404	0.5535	0.4666	0.3797	0.2928	0.206	0.119
HF	1	0.913	0.826	0.739	0.652	0.565	0.478	0.391	0.304	0.217	0.13
SCM	0.979	0.89218	0.80536	0.71854	0.6317	0.5449	0.45808	0.37126	0.28444	0.198	0.1108
RSEB	0.985	0.8981	0.8112	0.7243	0.6374	0.5505	0.4636	0.3767	0.2898	0.203	0.116
PEM	0.991	0.9041	0.8172	0.7303	0.6434	0.5565	0.4696	0.3827	0.2958	0.209	0.122
SDM	0.994	0.9071	0.8202	0.7333	0.6464	0.5595	0.4726	0.3857	0.2988	0.212	0.125

Table 15.15 WASPAS matrix for SI

λ	0	0.1	0.2	0.3	0.4	0.5	0.6	0.7	0.8	0.9	1
ESSDC	1	0.918	0.836	0.754	0.672	0.59	0.508	0.426	0.344	0.262	0.18
HF	0.673	0.6076	0.5422	0.4768	0.4114	0.346	0.2806	0.2152	0.1498	0.084	0.019
SCM	0.899	0.819	0.739	0.659	0.579	0.499	0.419	0.339	0.259	0.179	0.099
RSEB	0.955	0.8735	0.792	0.7105	0.629	0.5475	0.466	0.3845	0.303	0.222	0.14
PEM	1	0.918	0.836	0.754	0.672	0.59	0.508	0.426	0.344	0.262	0.18
SDM	0.955	0.8735	0.792	0.7105	0.629	0.5475	0.466	0.3845	0.303	0.222	0.14

Table 15.16 WASPAS matrix for CC

λ	0	0.1	0.2	0.3	0.4	0.5	0.6	0.7	0.8	0.9	1
ESSDC	0.925	0.8348	0.7446	0.6544	0.5642	0.474	0.3838	0.2936	0.2034	0.113	0.023
HF	0.893	0.8051	0.7172	0.6293	0.5414	0.4535	0.3656	0.2777	0.1898	0.102	0.014
SCM	1	0.907	0.814	0.721	0.628	0.535	0.442	0.349	0.256	0.163	0.07
RSEB	0.893	0.8051	0.7172	0.6293	0.5414	0.4535	0.3656	0.2777	0.1898	0.102	0.014
PEM	0.872	0.7858	0.6996	0.6134	0.5272	0.441	0.3548	0.2686	0.1824	0.096	0.01
SDM	0.893	0.8051	0.7172	0.6293	0.5414	0.4535	0.3656	0.2777	0.1898	0.102	0.014

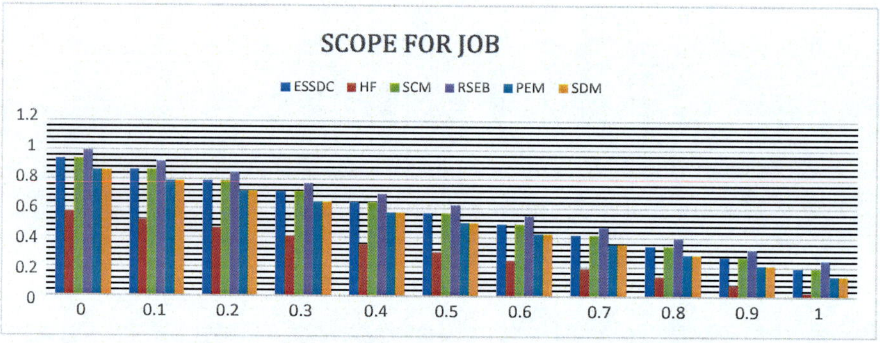

From this graph, it can be seen that if one is interested in doing elective from the perspective of job placement, then human factor subject has very less opportunity for this. An elective course on recommender system in e-business seems to offer better job placement opportunities.

Table 15.17 WASPAS matrix for AG

λ	0	0.1	0.2	0.3	0.4	0.5	0.6	0.7	0.8	0.9	1
ESSD	1	0.91	0.82	0.73	0.64	0.55	0.46	0.37	0.28	0.19	0.1
HF	0.823	0.74213	0.66126	0.58039	0.4995	0.41865	0.33778	0.25691	0.17604	0.095	0.0143
SCM	1	0.91	0.82	0.73	0.64	0.55	0.46	0.37	0.28	0.19	0.1
RSEB	0.823	0.74213	0.66126	0.58039	0.4995	0.41865	0.33778	0.25691	0.17604	0.095	0.0143
PEM	0.967	0.87744	0.78788	0.69832	0.6088	0.5192	0.42964	0.34008	0.25052	0.161	0.0714
SDM	1	0.91	0.82	0.73	0.64	0.55	0.46	0.37	0.28	0.19	0.1

Table 15.18 WASPAS matrix for SR

λ	0	0.1	0.2	0.3	0.4	0.5	0.6	0.7	0.8	0.9	1
ESSDC	1	0.904	0.808	0.712	0.616	0.52	0.424	0.328	0.232	0.136	0.04
HF	0.949	0.85518	0.76136	0.66754	0.5737	0.4799	0.38608	0.29226	0.19844	0.105	0.0108
SCM	0.987	0.8913	0.7956	0.6999	0.6042	0.5085	0.4128	0.3171	0.2214	0.126	0.03
RSEB	0.977	0.8833	0.7896	0.6959	0.6022	0.5085	0.4148	0.3211	0.2274	0.134	0.04
PEM	0.986	0.8902	0.7944	0.6986	0.6028	0.507	0.4112	0.3154	0.2196	0.124	0.028
SDM	0.965	0.87015	0.7753	0.68045	0.5856	0.49075	0.3959	0.30105	0.2062	0.111	0.0165

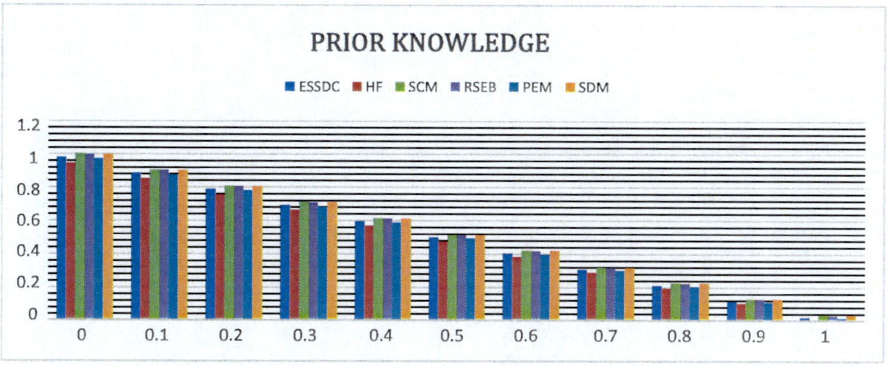
PRIOR KNOWLEDGE
ESSDC HF SCM RSEB PEM SDM
1.2
1
0.8
0.6
0.4
0.2
0
0 0.1 0.2 0.3 0.4 0.5 0.6 0.7 0.8 0.9 1

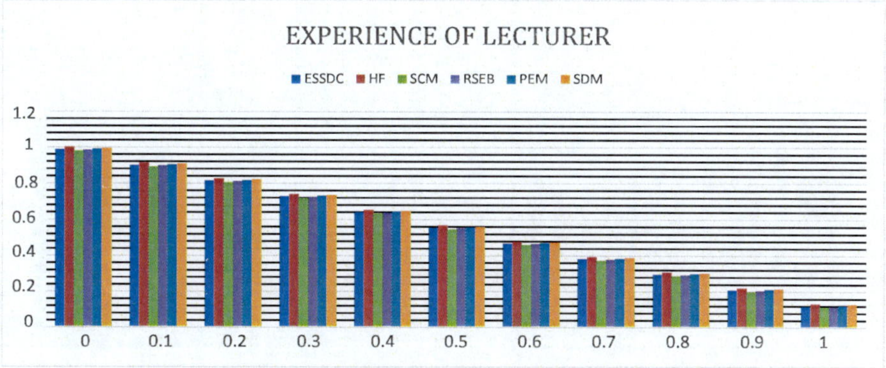
EXPERIENCE OF LECTURER
ESSDC HF SCM RSEB PEM SDM
1.2
1
0.8
0.6
0.4
0.2
0
0 0.1 0.2 0.3 0.4 0.5 0.6 0.7 0.8 0.9 1

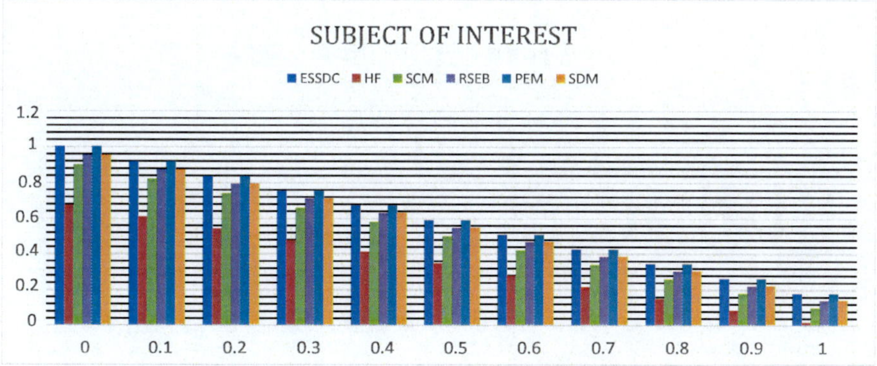
SUBJECT OF INTEREST
ESSDC HF SCM RSEB PEM SDM
1.2
1
0.8
0.6
0.4
0.2
0
0 0.1 0.2 0.3 0.4 0.5 0.6 0.7 0.8 0.9 1

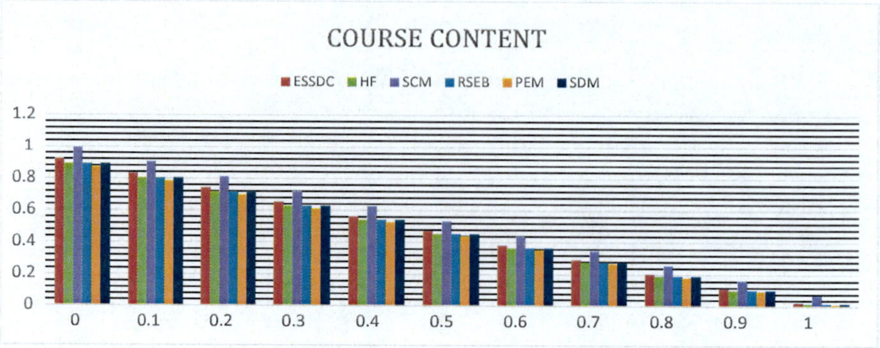
COURSE CONTENT
ESSDC
HF
SCM
RSEB
PEM
SDM
1.2
1
0.8
0.6
0.4
0.2
0
0
0.1
0.2
0.3
0.4
0.5
0.6
0.7
0.8
0.9
1

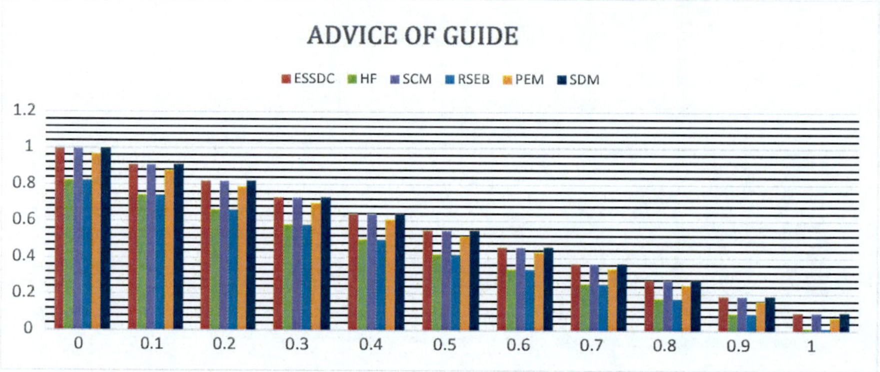
ADVICE OF GUIDE
ESSDC
HF
SCM
RSEB
PEM
SDM
1.2
1
0.8
0.6
0.4
0.2
0
0
0.1
0.2
0.3
0.4
0.5
0.6
0.7
0.8
0.9
1

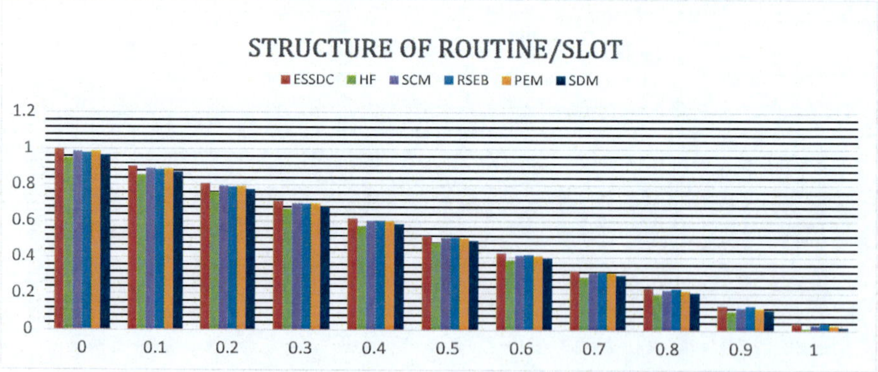
STRUCTURE OF ROUTINE/SLOT
ESSDC
HF
SCM
RSEB
PEM
SDM
1.2
1
0.8
0.6
0.4
0.2
0
0
0.1
0.2
0.3
0.4
0.5
0.6
0.7
0.8
0.9
1

Application 2: Prioritization of Lean and Agile Supply Chain Critical Success Factors for the Hotel Industry

The Council of Supply Chain Management Professionals (CSCMP) defines supply chain management as: "Supply chain management encompasses the planning and management of all activities involved in sourcing and procurement, conversion, and all logistics management activities. Importantly, it also includes coordination and collaboration with channel partners, which can be suppliers, intermediaries, third party service providers, and customers. In essence, supply chain management integrates supply and demand management within and across companies". CSCMP implies that supply chain management is successful when the goal of getting the right product to the right customer at the lowest costs is achieved. Therefore, supply chain management touches on the mixture of different supply chain activities to help maximize a company's profit and total value.

Lean Supply Chain

The basic idea of "Lean", which is basically a team-based approach to uninterrupted improvement focused on reducing or eliminating non-value added jobs or "waste" from the perspective of the client. In the recent time, the concept has been applied to the supply chain and logistics management area. Lean supply chain focuses on the use of continuous improvement that focus on abolishing all non-profit including activities across the supply chain.

Agile Supply Chain

It means "readiness to change". Agility is defined as a strategy that is more responsive in a volatile market place, where this strategy is totally demand driven. As consumer buying patterns are changing on a very rapid pace, so does the whole supply chain management changes. The fundamental drivers of agile supply chain are Speed, Cost and Efficiency. Agile supply chains are based on the sensitivity to consumer demand. Here, sensitivity refers to the ultimate consumer demand, in terms of volatility of demand.

"Leagile" Supply Chain

The products with high variety and high volume need for a collaborative approach to managing the supply chain and developing an optimal solution. Buyer and supplier involvement will be combined with other key stakeholders in the supply chain to create a leagile supply chain. This can be achieved by adopting Early Supply Chain Involvement (ESCI) bringing together the expertise of buyers, suppliers and other stakeholders to achieve a lean and agile environment. This demands reducing inventory through Just In Progress (JIP) and use of postponement strategy to cut down on waste and still be responsive to changes in demand.

XYZ Resorts International is an American property with the head office based in New York, dealing with gaming, hospitality, and entertainment. It owns and operates fifteen properties in Nevada, Mississippi, and Michigan and has 50% investments in four other properties in Dubai, Nevada, Illinois, and Macau, China. XYZ caters the customers in various sectors such as tourism, hospitality and gaming. XYZ Resorts International is a well-recognized organization in the hospitality sector having a number of hotels, casinos, entertainment centers and resorts. XYZ has been a key player in the market for 30 years. It has more than 60,000 employees of all age groups altogether. XYZ Resorts International generated about $9.8 billion in 2020, with an operating income of $1.1 billion and $3.2 as net income. The company owns a total asset of around $26 billion. Fortune magazine has named XYZ Resorts International as one of the best companies in hospitality sector.

Working towards a "Leagile" Supply Chain—Challenges in the Hotel Industry

In today's altering business setting, there is an enlarged focus on providing service to the client at the lowest likely expenses. Hotel industries, both large and small, must prioritize on how to provide comfort and facilities while keeping expenses as low as possible. Because of Covid-19 pandemic, there has been a steep reduction in corporate, group, and vacation tour count. As a result, hotel occupancy rates have fallen across the world. Most of the hotels focus on increasing cost-efficiency by altering the perks and benefits of the employees and other visible factors. Whereas in cost cutting efforts, one area of the hotel industry which is inadequately addressed is its supply chain operations and logistics. For example, MGM Resort International Hotel, which owns a dozen of properties in Las Vegas and huge internal resources, barely escaped from declaring bankruptcy in 2011. The hotel companies can improve their profitability by applying the concepts of leagile. The key challenges involved in the adoption of leagile supply are as follows:

C1. Management Training in Lean and Agile Supply Chain Principles

Management buy-in and formal training for senior and middle management in lean and agile practices is a key pre-requisite to the overhaul of strategy and adopting right supply chain practices. Strong ideological support is key to installing the ethos across the organization.

C2. Supplier Integration and Vendor Management

Developing an outreach programme with the vendors to understand their procurement processes and systems would help strengthen the relationship.

C3. Data Collection, Marketing, and Forecasting Challenges

Improved information on the customer base and their seasonal preferences is key to demand management in the hospitality industry, an important aspect to maintaining agility. Implementing an extensive feedback and customer profiling mechanism, along with investment into market research initiatives can help design and determine packages and policies that are best suited to volatile situations.

C4. Information Management Challenges

Information technology has changed the manner in which businesses interact with suppliers and customers. Today, different information systems are integrated to help forecast data, monitor inventory levels and sales trends, and in turn, companies have seen cycle time reduction, quicker order filling, inventory at the right safety stock level, and customer service improvements. Implementing a strategy of collaborative planning, forecasting, and replenishment where the hotel shares information with their logistics and supply chain partners, and jointly make decisions to improve the supply chain performance would be key to implementing a leagile strategy.

C5. Workforce Agility Challenges

The seasonal nature of demand in the hospitality industry translates into a possibility where management may find themselves either facing an oversupply of workers or an undersupply of personnel. The leagile paradigm dictates that organizations should be agile in their personnel management, and an efficient way of achieving that is empaneling a human resource firm to access workers on a contractual basis, depending on short-term requirements.

Problem Structure and Formulation

Definition of Criteria for Leagile Supply Chain

See Table 15.19.

Definition of Scale

See Table 15.20.

Design of Alternatives

See Table 15.21.

Application of WASPAS.

R Code for WASPAS and Results.

Table 15.19 Definition of criteria

No.	Criteria	Definition	Desirable value	Type of criteria	Symbol
1	Product variety	Agile supply chains are expected to cater to varied customer needs, and this requires a wide array of products to be present in the product mix	Medium	Benefit	x_1
2	Impact on service level	Leagile supply chains are differentiated by their high service levels, and this is a key success factor	High	Benefit	x_2
3	Profit margin	Profit maximization is subordinated to the capability to switch gears and meet new expectations, to ensure the organization is at the forefront of emerging markets	Moderate	Benefit	x_3
4	Inventory costs	Vendor Managed Inventory models are the prime framework in which the waste minimization requirements of the lean supply chain and the low lead time requirements of the agile supply chain are satisfied	Low (Vendor managed inventory)	Cost	x_4

(continued)

Table 15.19 (continued)

No.	Criteria	Definition	Desirable value	Type of criteria	Symbol
5	Information enrichment	Information flow and real-time updates are key to the functioning of a leagile supply chain. Information management plays a key role in achieving these objectives	Essential	Benefit	x_5
6	Forecast mechanism	An intuitive understanding of the consumer base and being prepared for surges in demand is key to the management of a leagile supply chain, and to ensure high customer satisfaction and service levels	Algorithmic/consultative/both	Cost	x_6
7	Customization	Leagile supply chains differentiate themselves by providing a high level of customization in services and processes, to best fit the needs of the consumer	High	Cost	x_7

(continued)

Table 15.19 (continued)

No.	Criteria	Definition	Desirable value	Type of criteria	Symbol
8	Rapid reconfiguration	Slack capacity and reconfiguration costs are key to the functioning of a leagile supply chain	Essential	Cost	x_8
9	Cost of implementation	Includes the monetary component of implementing a project	Low	Cost	x_9
10	Requirement of management support	Includes the level of management buy-in required for the success of a project	High	Cost	x_{10}

Table 15.20 Quantification of scale for cost and benefit attributes

Cost attributes	Scale	Benefit attributes
Very high	1	Very low
High	3	Low
Moderate	5	Moderate
Low	7	High
Very low	9	Very high

Table 15.21 Scale results for listed alternatives (Note: x1, x2, x3, x5 are benefit criteria and x4, x6, x7, x8, x9, x10 are cost criteria)

Alternative/Criteria	x_1	x_2	x_3	x_4	x_5	x_6	x_7	x_8	x_9	x_{10}
C_1	Very low	High	Low	Low	Very high	High	High	High	Moderate	Very high
C_2	High	Very high	High	Low	Very high	High	Low	Low	High	High
C_3	Very low	High	Moderate	Very low	High	Very high	Very low	Very low	Moderate	Moderate
C_4	Very low	Very high	High	Very low	Very high	High	Very low	Very low	Very high	High
C_5	Low	High	High	Low	Low	Very low	Very low	Moderate	Low	High

```
> library("MCDM")

> d <- matrix(ncol = 10, nrow = 5)

> d[1,] <- c(1,7,3,7,9,3,3,3,5,1)

> d[2,] <- c(7,9,7,7,9,3,7,7,3,3)
> d[3,] <- c(1,7,5,9,7,1,9,9,5,5)
> d[4,] <- c(1,9,7,9,9,3,9,9,1,3)
> d[5,] <- c(3,7,7,7,3,9,9,5,7,3)
> d
     [,1] [,2] [,3] [,4] [,5] [,6] [,7] [,8] [,9] [,10]
[1,]    1    7    3    7    9    3    3    3    5     1
[2,]    7    9    7    7    9    3    7    7    3     3
[3,]    1    7    5    9    7    1    9    9    5     5
[4,]    1    9    7    9    9    3    9    9    1     3
[5,]    3    7    7    7    3    9    9    5    7     3

> w <- c(0.1,0.1,0.1,0.1,0.1,0.1,0.1,0.1,0.1,0.1)
> cb <- c('max','max','max','max','max','max','max','max','max','max')
> lambda1 <- 0
> lambda2 <- 0.5
> lambda3 <- 1

> WASPAS(d,w,cb,lambda1)
  Alternatives         W Ranking
1            1 0.4258116       5
2            2 0.7253740       1
3            3 0.5875048       3
4            4 0.5768772       4
5            5 0.7013732       2

> WASPAS(d,w,cb,lambda2)
  Alternatives         W Ranking
1            1 0.4649693       5
2            2 0.7474489       1
3            3 0.6556572       3
4            4 0.6493910       4
5            5 0.7243374       2

> WASPAS(d,w,cb,lambda3)
  Alternatives         W Ranking
1            1 0.5041270       5
2            2 0.7695238       1
3            3 0.7238095       3
4            4 0.7219048       4
5            5 0.7473016       2
```

The result indicates that leagile supply chain challenge 2 receives the highest ranking with weightage (0.769) and challenge 1 is the least difficult to handle with lowest weightage (0.504).

15.4 Benefits of WASPAS

- Strong in single dimensional problem
- Ability to compensate among criteria; intuitive to decision makers; calculation is simple does not require complex computer programmes.

The weighted aggregated sum product assessment method is applied for solving MCDM problems for increasing ranking accuracy, and it has the capability to achieve the highest accuracy of estimation. This method incorporates a resistance against rank reversal of the considered alternatives. It is also found that this method has the unique capability of dealing with both single and multi-response optimization problems.

15.5 Limitations of WASPAS

- Estimates revealed do not always reflect the real situation; result obtained may not be logical.
- No solution with equal weight of decision makers.

The above-mentioned disadvantages can be overcome by incorporating fuzzy logic technique with WASPAS method.

In WASPAS, beneficial and non-beneficial criteria are treated equally, but their normalization is done separately. The normalization approach employed in WASPAS method takes into consideration only two performance values, i.e. minimum (for non-beneficial attributes) and maximum (for beneficial attributes), and does not consider all the performance values. Therefore, the normalized scores obtained in this method are not entirely realistic.

This chapter has discussed the salient features and application of WASPAS. The next chapter will introduce a new procedure step-wise weight assessment ratio analysis (SWARA) for the attribute weights determination that provides the opportunity to estimate the differences of their significances.

Selected Bibliography and Further Reading

1. Amir, F. (2011). Significance of lean, agile and leagile decoupling point in supply chain management. *Journal of Economics and Behavioural Studies, 3*(5), 287–295.
2. Blanca, A., & Martin, C. (2015). *MCDM: Multi-criteria decision making methods.* R package version 1.0. https://CRAN.R-project.org/package=MCDM.
3. Chakraborty, S., Bhattacharyya, O., Zavadskas, E.K., & Antucheviciene J. (2015). Application of WASPAS method as an optimization tool in non-traditional machining process. *Information Technology and Control*, T. 44, Nr. 1. https://doi.org/10.5755/j01.itc.44.1.7124.
4. Council of Supply Chain Management Professionals (CSCMP). (2011). *Logistics management.* Retrieved from https://cscmp.org/aboutcscmp/definitions.asp.
5. Gilaninia, S., Taleghani, M., Mousavian, S.J., Tajani, T.K., Ghoreishi, S.M., Shahidi, S.F., & Seighalani, F.Z. (2011). Comparative study of lean and agile supply chain management along with the optimal model presentation of agile supply chain management. *Kuwait Chapter of Arabian Journal of Business and Management Review, 1*(4), December 2011.
6. Gunasekaran, A. & Ngai, E.W.T. (2004). Information systems in supply chain integration and management. *European Journal of Operational Research* (159).
7. Krishnamurthy, R., & Yauch, C. A. (2007). Leagile manufacturing: a proposed corporate infrastructure. *International Journal of Operations & Production Management, 27*(6), 588–604.
8. Mistry, J. J. (2005). Supply chain management: a case study of an integrated lean and agile model. *QRAM, 2*(2).
9. Odoom C. K. (2008). Supply chain management in the hotel industry: impact on hotel performance in service delivery. In: Paper for partial fulfilment Bachelors of Science in Business Administration University of Nevada, Reno.
10. Rana, S. M. S., Osman, A., Bahari, A. B., & Solaiman, M. (2015). Determinants of supply chain performance: a strategic point of view. *International Journal of Supply Chain Management, 4*(3), 94–102, September 2015.
11. Seman, A. A., & Ibrahim N. (2015). Implementing fuzzy approach to assess the best hotel services. *ARPN Journal of Engineering and Applied Sciences, 10*(3), February 2015.
12. Vinodh, S., Sundararaj, G., & Devadasan, S. R. (2009). Total agile design system model via literature exploration. *Industrial Management and Data System, 109*(4), 570–588.
13. Zaeri M. S., Sadeghi A., Naderi A., Kalanaki A., Fasihy R., & Poyan A. (2011). Application of multi criteria decision making technique to evaluation suppliers in supply chain management. Mathematics and Computer Science Research Vol. 4 (3), pp. 100–106, March, 2011. https://www.academicjournals.org/AJMCSR ISSN 2006–9731

Chapter 16
Stepwise Weight Assessment Ratio Analysis (SWARA)

16.1 Background

Weight assessment is an important issue in many multi-attribute decision-making (MADM) problems. The popular weight assessment approaches include analytic hierarchy process (AHP), analytic network process (ANP), SWARA, etc. Stepwise Weight Assessment Ratio Analysis (SWARA) method is the most efficient method for criteria evaluation. It involves two important steps: the first is to prioritize the criteria by consulting experts, while the second is the weighting process. A step-by-step procedure for SWARA is presented in Fig. 16.1.

The step-wise weight assessment ratio analysis (SWARA) was developed by Kersuliene et al. (2010). The SWARA method involves considerably lower pair-wise comparison and easy to use compared to other popular methods such as AHP. SWARA is an algorithm primarily used for determining the relative weights of the criteria to evaluate alternatives.

16.2 Step-by-Step Procedure of SWARA

1. Every expert specifies the priority of each criterion based on his or her own implicit knowledge, information and experience.
2. The decisive factors are ranked from first to last considering total outcome. The most significant criterion is given rank 1, and the least significant criterion is given rank last.
3. Subsequently, the average value of ranks is considered to determine the overall ranks.

Figure 16.2 shows the steps involved in the evaluation procedure of SWARA.

The weights obtained by a subjective approach reflect subjective judgement of a person resulting in ranking of the alternatives for a given problem. Zavadskas (1987)

J. J. Thakkar, *Multi-Criteria Decision Making*, Studies in Systems, Decision and Control 336, https://doi.org/10.1007/978-981-33-4745-8_16

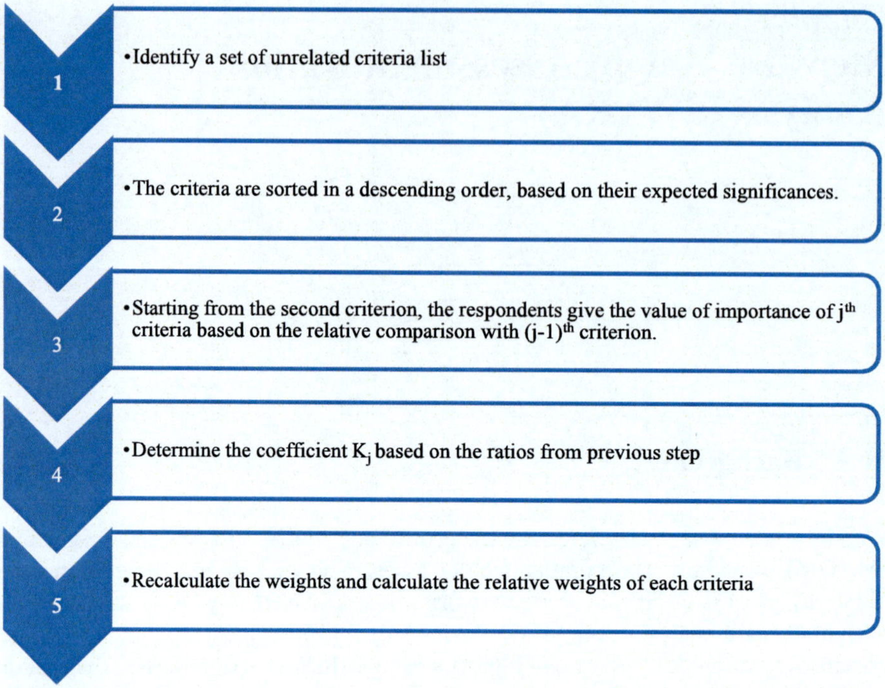

Fig. 16.1 SWARA procedure

suggested a mathematical expression for determining the integrated weight of the attributes as follows:

$$\overline{w}_j = \frac{w_j^* w_j}{\sum_1^n w_j^* w_j}; \sum_{j=1}^{n} \overline{w}_j = 1; j = 1, 2, \ldots n,$$

where w^*_j—objective weight of the j attribute

w_j—subjective weight of the j attribute

$\overline{w}_j$—integrated weight of the j attribute

16.3 Illustrative Application of SWARA for Prioritizing Supply Chain Coordination Mechanisms

The competitive factors that determine the success of today's organizations are shorter delivery times, customized products and extremely high service levels. Supply chain management (SCM) is extensively used as a tool to meet competitive priorities

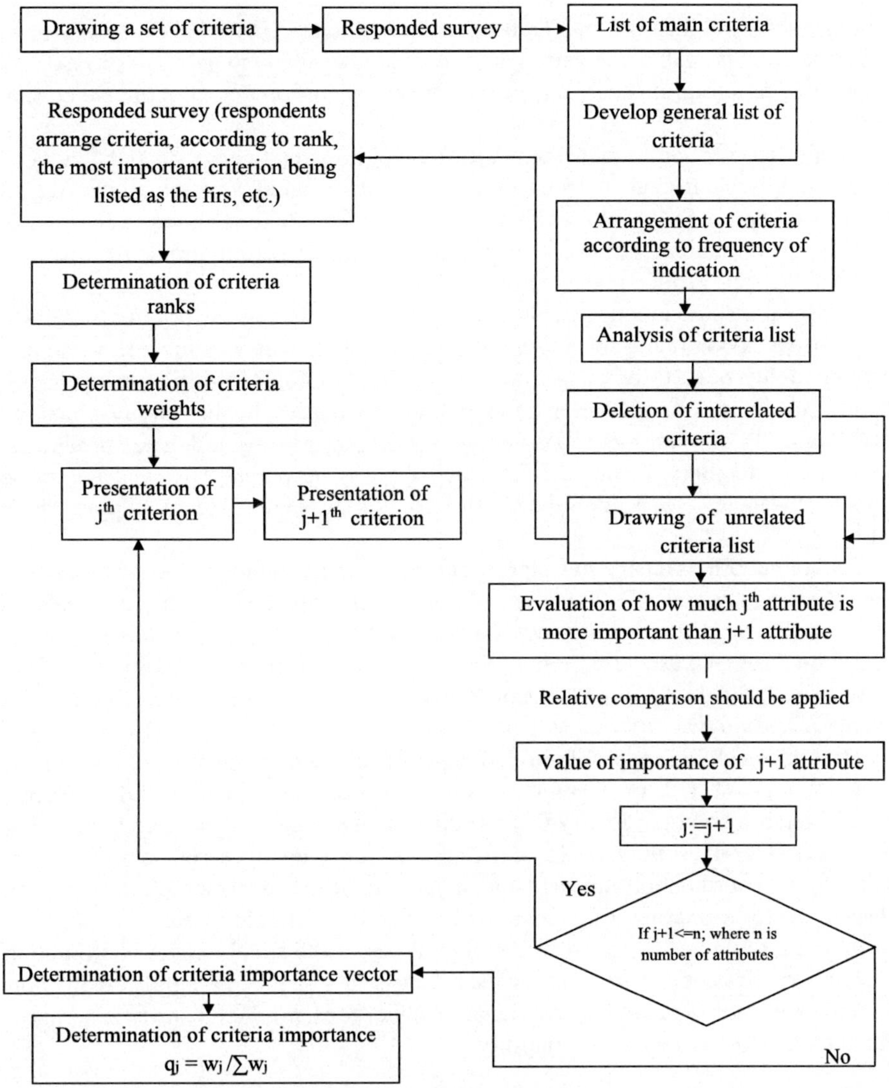

Fig. 16.2 Determining of the criteria weights based on SWARA

and providing sustainability to the businesses. The key challenge for enhancing the competitiveness through SCM is to develop business-specific priorities, evolving an alignment among key SCM drivers like facility, inventory, pricing, transportation and information. This can be achieved by establishing an appropriate coordination mechanism among supply chain partners.

In order to assess dispute resolution methods from the view point of economic, social and political, it is necessary to examine the solutions for multiple attributes. The new step-wise weight assessment ratio analysis method (SWARA) allows experts and key stakeholders such as government, industry, lawyers, dispute parties to express their opinions on significance ratio of the attributes in the process of rational decision making.

Supply chain coordination intends to integrate various activities and functions across supply chain and motivates the supply chain entities to maximize overall supply chain profitability. However, each supply chain is unique and it is necessary to rank the coordination mechanisms considering the customer requirements, financial implications and various other attributes.

A typical dispute among supply chain partners for inventory, pricing or service level demands conflicting business goals such as expediency, economic value and confidentiality need to be evaluated holistically. A manufacturing industry has to ensure an optimal coordination among multiple parties in the supply chain by developing an appropriate coordination mechanism among suppliers, producers, distributors, retailers, customers, etc, and devise mechanisms to manage those interdependencies, to accommodate the conflicting objectives of supply chain partners.

An automobile industry has significant reliance on suppliers for achieving the higher efficiency and effectiveness of supply chain operations. This demands an effective design of supply chain coordination and partnership mechanisms for costs reduction through waste elimination, continuous improvement of quality to achieve zero defects, flexibility improvement to meet the end-customer requirements, and decrease lead time at different stages of the SC.

The top five global supply chain challenges in automobile industry are—visibility, cost containment, risk management, increasing customer demands and globalization. There is a need to identify the coordination mechanisms that help in reducing the uncertainty in supply chain and improve transparency across supply chain. The various coordination mechanisms include: joint decision making, information sharing, resource sharing, implementing information technology, etc.

Figure 16.3 indicates the most common types of the supply chain coordination mechanisms. It is necessary to study the effectiveness of various supply chain coordination mechanisms with respect to market conditions; product features in-terms of demand volume, variety and variability.

16.3.1 Supply Chain Contracts

SC members coordinate by using contracts for better management of supplier–buyer relationship and risk management. The contracts specify the parameters (like quantity, price, time and quality) within which a buyer places orders and a supplier fulfils them. The objectives of SC contracts are: to increase the total SC profit, to reduce overstock/understock costs and to share the risks among the SC partners.

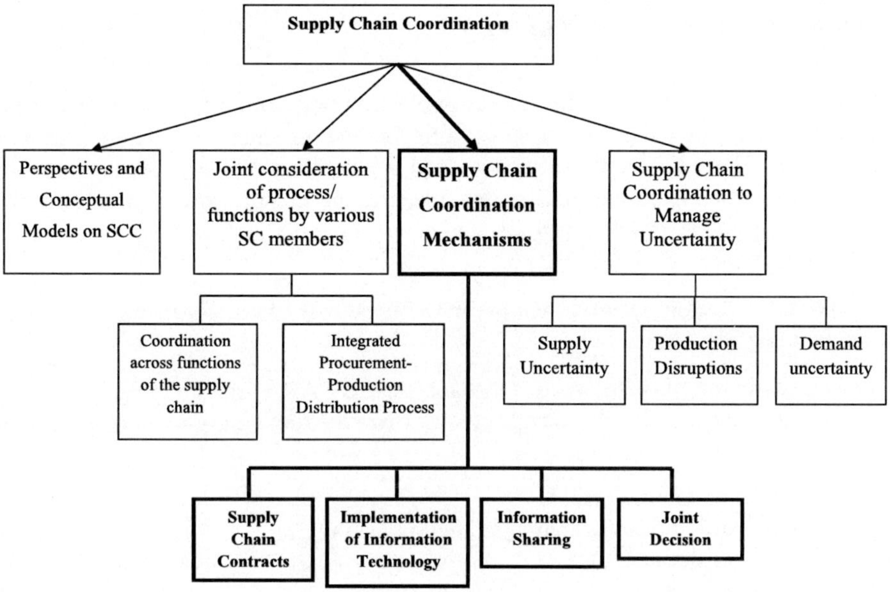

Fig. 16.3 Supply chain coordination mechanisms

16.3.2 Implementation of Information Technology (IT)

IT enhances inter-organizational coordination and improves firm performance on customer service, lead time and production costs. A seamless integration through IT across supply chain improves the real-time data sharing and hence enables better planning, tracking and estimating of the lead times. The various advancements such as Internet, EDI (electronic data interchange), ERP (enterprise resource planning), and e-business have enabled the firms to rapidly exchange products, information and funds through collaborative mechanisms.

16.3.3 Information Sharing

This demands real-time information sharing on demand, orders, inventory, POS data, etc. This enhances supply chain performance by having realistic commitments from downstream customers, reduction in inventory, offering price discounts and reducing lead time. The value of information sharing increases with improvements in service level, reduction in supplier-holding costs, reduced demand variability and length of the order cycle. The point of sales (POS) data helps the supplier to better anticipate future orders of the retailers and reduces the bullwhip effect.

16.3.4 *Joint Decision Making*

A joint decision making helps in resolving conflicts among SC members and managing futuristic uncertainties specific to demand, product variety and customer preferences. There are many factors involved in achieving coordination like human, technology, strategies, relationship, rewards, sharing of knowledge, sharing benefits, aligning goals, scheduling of frequent meetings of stakeholders for conflict resolution, and knowledge of SC concepts.

Steps to form decision matrix for prioritizing supply chain coordination mechanisms using SWARA

The step-wise weight assessment ratio analysis (SWARA) methodology has been deployed to determine criteria weights to prioritize the four below-mentioned supply chain mechanisms.

(a) Supply chain contract (C1);
(b) Implementation of information technology (IT) (C2);
(c) Information sharing (C3);
(d) Joint decision making (C4);

The steps involved in calculating the weights are as follows:

Step 1: Drawing of supply chain coordination mechanisms (criteria) list for prioritizing.

Step 2: Determination of criteria ranks by a group of experts.

(a) In case of the automobile company, the expert group has given highest priority to supply contracts as the company is very old and experienced; also, over the years, they have faced a variety of contingencies in supplier-buyer contracts. This has led to a robust supply contract model, which is considered to be a top priority for improving the business competitiveness.
(b) Implementation of information technology has gained significant acceptance with the advancements in automobile technology and stricter environmental norms. Their compliance also guarantees substantial financial waivers in case of auto industry, as the buyers and suppliers are spread all across the globe.
(c) Information sharing using various technologies such as ERP, EDI, Wi-Fi and Internet improves transparency across supply chain and helps an automobile industry to improve the performance in-terms of lead time, inventory costs and procurement costs. Hence, this attribute has received the next priority in the list.
(d) Joint decision making is important in any business model to assure a good buyer–supplier relationship, keeping interests of both the parties in mind. This extends significant advantage to the automobile industry specifically for the situations like natural calamities, accidents and unforeseen demands.

For determining criteria (C1, C2, C3, C4) weights by SWARA method, the calculations are performed by assuming the importance value for each criterion as indicated in Fig. 16.4.

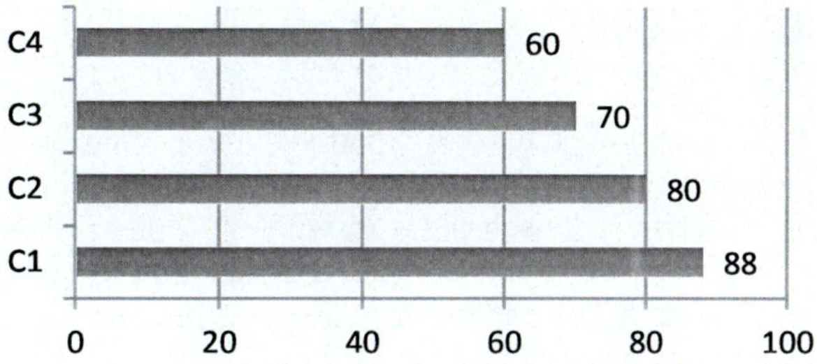

Fig. 16.4 Most important criteria in selection of supply chain coordination mechanisms

Step 3: Determination of criteria weights for each of the supply chain mechanisms (Table 16.1).

Results

The main feature of SWARA method is the possibility to estimate experts or interest groups' opinion about significance ratio of the attributes in the process of their weights determination. Here, the prioritization of attributes for four types of supply chain coordination mechanisms is conducted for automobile industry. The final decision matrix (Table 16.2) provides weights obtained for various criteria (supply chain coordination mechanisms) using SWARA methodology, along with ranking of suppliers (S1, S2, S3 and S4). This decision matrix may also contain various tangible attributes like cost, inventory, lead time, cash flow, quantity/ quality discount, etc., to enable a well-informed decision-making model. SWARA provides a decision matrix which can be subsequently solved using various MCDM approaches such as VIKOR, COPRAS, WASPAS etc.

Table 16.1 Calculation of weights as per ranks assigned by group of experts

Attributes	Comparative importance of average value S_j	Coefficient $k_j = S_j + 1$	Recalculated weight $W_j = x_{j\text{-}1}/k_j$	Weight $Q_j = w_j/\sum w_j$
$C1$		1	1	0.218
$C2$	0.08 [(88 − 80)/100]	1.08	0.925 [1/1.08]	0.201
$C3$	0.1 [(80 − 70)/100]	1.1	0.841 [0.925/1.1]	0.183
$C4$	0.1 [(70 − 60)/100]	1.1	0.765 [0.841/1.1]	0.166

Table 16.2 Final decision matrix

Attributes (weights)	$S1$	$S2$	$S3$	$S4$
$C1$ (0.218)	9	4	7	2
$C2$ (0.201)	3	8	6	5
$C3$ (0.183)	4	6	9	7
$C4$ (0.166)	7	5	3	8

16.4 Advantages of SWARA

- It accommodates the expert's opinion concerning accuracy of the weighted criteria.
- Experts can consult and work with each other as well, which leads to more concrete and accurate results compared to the other MCDM procedures.
- This method is applicable where priorities are determined based on set of policies and strategies of firms or nations.
- This method leads policy makers to make better decisions in wide-ranging situations and prioritize criteria according to the predefined goals and objectives.

16.5 Limitations of SWARA

- In certain cases ratio analysis might prove to be misleading with regard to profits.
- A single ratio may not convey much information.
- SWARA helps to determine an initial decision matrix. It is necessary to use other MCDM approach for evaluating decision matrix and deriving final priorities.

This chapter has discussed the salient features and application of MCDM technique—SWARA. The next chapter will explain the use of graph theoretic analysis (GTA). This technique is derived from the field of a combinatorial mathematics which enables a decision maker to analyse and understand the system as a whole by identifying system and sub-system up to the component level.

Selected Bibliography and Further Reading

Aghdaie, M. H., Zolfani, S. H., & Zavadskas, E. K. (2013). Decision making in machine tool selection: An integrated approach with SWARA and COPRAS-G methods. *Engineering Economics (Impact Factor: 0.87). 01/2013; 24*(1):5–17.

Arshinder, K., Kanda, A., & Deshmukh, S. G. (2011). A review on supply chain coordination: coordination mechanisms, managing uncertainty and research directions. In: T.-M. Choi and T. C. Edwin Cheng (Eds.), *Supply chain coordination under uncertainty, international handbooks on information systems*. Doi https://doi.org/10.1007/978-3-642-19257-9_3,© Berlin, Heidelberg: Springer.

Bhattacharya, S., Mukhopadhyay, D., & Giri, S. (2014). Supply chain management in Indian automotive industry : complexities, challenges and way ahead. *International Journal of Managing Value and Supply Chains (IJMVSC) 5(2) June 2014*. Doi https://doi.org/10.5121/ijmvsc.2014.5206 49.

Dehnavi, A., Aghdam, I. N., Pradhan, B., & Varzandeh, M. H. M.(2015). A new hybrid model using step-wise weight assessment ratio analysis (SWARA) technique and adaptive neuro-fuzzy inference system (ANFIS) for regional landslide hazard assessment in Iran. *CATENA, 135*, 122–148

Egri, P. (Supervisor: JozsefVancza, Ph.D.). (2008). Coordination in production networks. *Ph.D. Thesis*, Eotvos Lorand University (ELTE).

Ho, C.-F., Tai, Y. M., & Lee, M.X. (2007). Information sharing and coordination mechanism for supply chain management: a case study of the automotive industry. In: *IEEE conference on wireless communications, networking and mobile computing, 2007*. WiCom 2007

Li, X., & Wang, Q. (2007). Coordination mechanisms of supply chain systems. *European Journal of Operational Research 179*, 1–16

Keršulienė, V., Zavadskas, E. K., & Turskis, Z. (2010). Selection of rational dispute resolution method by applying new step-wise weight assessment ratio analysis (Swara). *Journal of Business Economics and Management, 11*(2): 243–258.

Kersuliene, V., & Turskis Z. (2011). Integrated fuzzy multiple criteria decision making model for architect selection. *Technological and Economic Development of Economy, 17*(4), 645–666.

Thakkar, J., Kanda, A., Deshmukh, S. G. (2007). Evaluation of buyer-supplier relationships using an integrated mathematical approach of interpretive structural modeling (ISM) and graph theoretic matrix: The case study of Indian automotive SMEs. *Journal of Manufacturing Technology Management, 19*(1), 92 – 124

Zolfani, S. H., Saparauskas, J. (2013). New application of SWARA method in prioritizing sustainability assessment indicators of energy system. *Engineering Economics, 24*(5), 408–414.

Zolfani, S. H., Aghdaie, M. H., Derakhti, A., Zavadskas, E. K., & Varzandeh, M. H. M. (2013). Decision making on business issues with foresight perspective; an application of new hybrid MCDM model in shopping mall locating. *Expert Systems with Applications 40*, 7111–7121.

Chapter 17
Graph Theoretic Analysis (GTA)

17.1 Introduction to Graph Theoretic Approach

Graph theory matrix is derived from the field of combinatorial mathematics which enables a decision maker to analyse and understand the system as a whole by identifying system and sub-system up to the component level. It is widely used to represent almost any physical situation involving discrete objects and relationship among them. A digraph is used to represent the elements and their interdependencies in terms of nodes and edges. Matrices provide a simple and very elegant means of representing digraph, as it gives one-to-one representation. The permanent of a matrix is used to characterize configurations of a system or the structure of a graph and to develop a unique representation that is independent of labelling. Graph theory is a subject of combinatorial mathematics and draws from network theory. Graph theory based mathematical formulation incorporates multi-relations among its constituent components. The framework integrates the following three representations: the digraph representation, the matrix representation and the permanent function representation. The digraph is the visual portrayal of the elements and their association which influences the quality/execution of the framework under thought. The matrix changes over the digraph into a scientific frame. The permanent function representation is a mathematical model that decides a numeric list to assess the importance of the elements chosen. Figure 17.1 shows the different stages of the graph theory-based approach in the form of a flowchart.

GTA is a systematic and logical approach which synthesizes the interrelationship among different parameters or sub-system parameters and provides a synthetic score for the entire system. It also takes care of directional relationship and interdependence among parameters. The graph theoretical methodology consists of three steps, namely digraph representation, matrix representation and permanent function

J. J. Thakkar, *Multi-Criteria Decision Making*, Studies in Systems, Decision and Control 336, https://doi.org/10.1007/978-981-33-4745-8_17

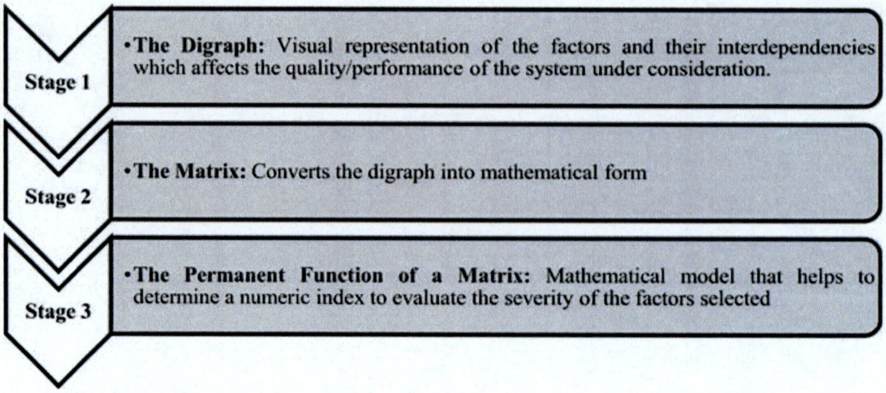

Fig. 17.1 Key components of graph theoretic approach

representation. A digraph is used to represent the structure of the system in terms of nodes and edges wherein the nodes represent the measure of characteristics and the edges correspond to the dependence of characteristics. Matrix representation is one-to-one representation of the digraph. Permanent representation is the mathematical expression of characteristics and their interdependence. Key features that highlight the uniqueness of GTA include: provides a single numerical index for all the parameters, enables conversion of qualitative factors to quantitative values, allows modelling of the interdependence of parameters and also allows visual analysis and computer processing, and also can be used for comparison of different systems. A detailed procedure of GTA is explained in Fig. 17.2.

17.2 Step-by-Step Procedure of Graph Theoretic Approach (GTA)

Step 1: The digraph representation. The digraph comprises of an arrangement of nodes $N = \{A_i\}$ with $i = 1, 2$ [...] M and an arrangement of coordinated edges $E = \{a_{ij}\}$. A node A_i speaks to the ith factor influencing the quality and edges speak to the relative significance among the factors. The quantity of nodes M considered is equivalent to the quantity of factors influencing the nature of servitization process in an automobile industry. On the off chance that a node "*i*" has the relative significance over another factor "*j*" in the investigation of value evaluation, at that point a coordinated edge or arrow is attracted from node i to node j (i.e. a_{ij}) or the other way around (Fig. 17.3).

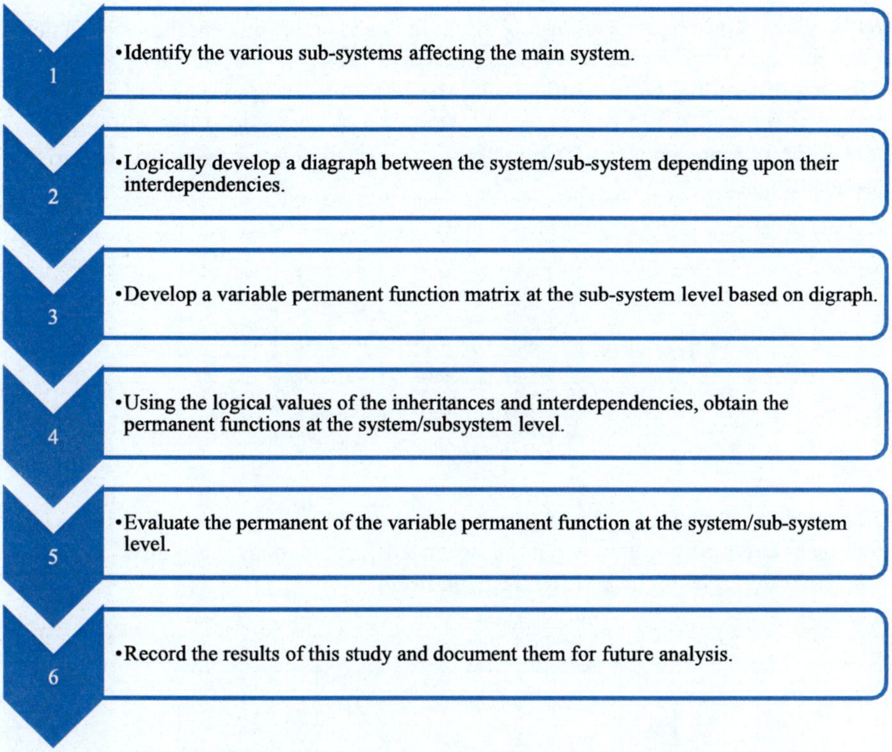

Fig. 17.2 Procedure of GTA

Fig. 17.3 A diagraph with 6 nodes

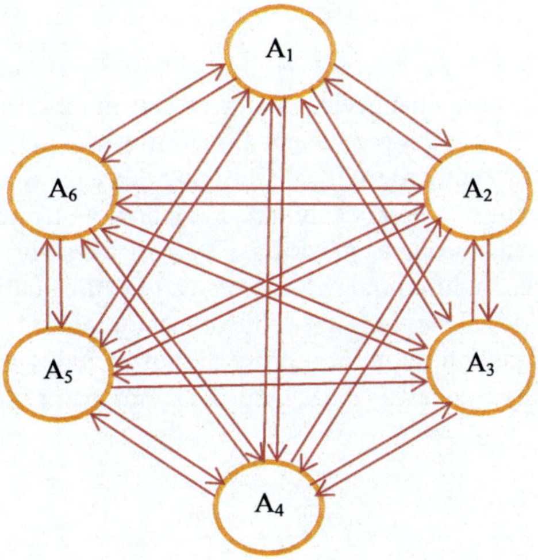

Step 2: The matrix representation. A digraph is a visual representation which helps in dissecting the framework to a constrained degree as it were. For the establishment of the expression for factors influencing the phenomenon under consideration, the digraph is represented in matrix form. Matrix representation of the digraph offers point-to-point representation. This matrix (X) is called assessment matrix or variable permanent matrix (VPM).

$$VPM = X = \begin{bmatrix} A_1 & a_{12} & a_{13} & a_{14} & a_{15} & a_{16} \\ a_{21} & A_2 & a_{23} & a_{24} & a_{25} & a_{26} \\ a_{31} & a_{32} & A_3 & a_{34} & a_{35} & a_{36} \\ a_{41} & a_{42} & a_{43} & A_4 & a_{45} & a_{46} \\ a_{51} & a_{52} & a_{53} & a_{54} & A_5 & a_{56} \\ a_{61} & a_{62} & a_{63} & a_{64} & a_{65} & A_6 \end{bmatrix} \begin{matrix} \text{Factors} \\ A_1 \\ A_2 \\ A_3 \\ A_4 \\ A_5 \\ A_6 \end{matrix} \tag{17.1}$$

In general, if there are M factors affecting the phenomenon under consideration and relative importance exists between all factors, then the matrix Y, for the considered digraph can be written as given in Eq. (17.2):

$$Y = \begin{bmatrix} X_1 & x_{12} & x_{13} & \ldots\ldots & x_{1M} \\ x_{21} & X_2 & x_{23} & \ldots\ldots & x_{2M} \\ x_{31} & x_{32} & X_3 & \ldots\ldots & x_{3M} \\ \ldots & \ldots & \ldots & \ldots\ldots & \ldots \\ x_{M1} & x_{M2} & x_{M3} & \ldots\ldots & X_M \end{bmatrix} \begin{matrix} \text{Factors} \\ X_1 \\ X_2 \\ X_3 \\ \ldots \\ X_M \end{matrix} \tag{17.2}$$

Step 3: Permanent representation. Both digraph and matrix portrayals are not novel, as they change by changing the numbering of the nodes. To build up a special portrayal, a permanent function of the matrix is required. The permanent function is a standard matrix function and is used in combinatorial mathematics. The permanent function is achieved in an analogous manner as its determinant but with all the negative signs substituted by positive signs. Utilizing this calculation, each term of the multinomial capacity can add to the quality model and none of the terms loses its criticalness because of the cancellation of terms as in the traditional matrix method. This multinomial representation includes all the information regarding the factors selected and is called variable permanent function.

The permanent function of matrix Y, containing M factors, can be written as given in Eq. (17.3). Equation (17.3) contains $M!$ terms and these terms are arranged in $M + 1$ grouping, where M is the number of elements (factors).

$$
\begin{aligned}
\operatorname{per} Y^* = {} & \prod_{1}^{M} X_i + \sum_i \sum_j \sum_k \sum_l \cdots \sum_M x_{ij} x_{ji} X_k X_l \ldots X_M \\
& + \left(\sum_i \sum_j \sum_k \sum_l \cdots \sum_M (x_{ij} x_{jk} x_{ki}) X_l X_m \ldots X_M \right. \\
& \left. + \sum_i \sum_j \sum_k \sum_l \cdots \sum_M (x_{ik} x_{kj} x_{ji}) X_l X_m \ldots X_M \right) \\
& + \left(\left\{ \sum_i \sum_j \sum_k \sum_l \cdots \sum_M (x_{ij} x_{ji})(x_{kl} x_{lk}) X_m X_n \ldots X_M \right\} \right. \\
& + \left\{ \sum_i \sum_j \sum_k \sum_l \cdots \sum_M (x_{ij} x_{jk} x_{kl} x_{li}) X_m X_n \ldots X_M \right. \\
& \left. \left. + \sum_i \sum_j \sum_k \sum_l \cdots \sum_M (x_{il} x_{lk} x_{kj} x_{ji}) X_m X_n \ldots X_M \right\} \right) \\
& + \left(\left\{ \sum_i \sum_j \sum_k \sum_l \cdots \sum_M (x_{ij} x_{ji})(x_{kl} x_{lm} x_{mk}) X_n X_o \ldots X_M \right. \right. \\
& \left. + \sum_i \sum_j \sum_k \sum_l \cdots \sum_M (x_{ij} x_{ji})(x_{kl} x_{lm} x_{mk}) X_n X_o \ldots X_M \right\} \\
& + \left\{ \sum_i \sum_j \sum_k \sum_l \cdots \sum_M (x_{ij} x_{jk} x_{kl} x_{lm} x_{mi}) X_n X_o \ldots X_M \right. \\
& \left. \left. + \sum_i \sum_j \sum_k \sum_l \cdots \sum_M (x_{im} x_{ml} x_{lk} x_{kj} x_{ji}) X_n X_o \ldots X_M \right\} \right)
\end{aligned}
\tag{17.3}
$$

17.3 Illustrative Application of Graph Theoretic Approach for Servitization in Automobile Industry

17.3.1 Background of the Problem

Automotive industries are improving their profitability through an extensive portfolio of servitization. It is necessary for them to examine the service quality of their service. We demonstrate an application of graph theory-based methodology to evaluate the factors affecting the quality in the servitization practices and proposes an index, critical service quality index or CSQI, to measure the same. A hypothetical application of GTA methodology is demonstrated for three Indian automobile companies. This ensures the validation of the proposed approach and also helps to evolve select key recommendations for improving servitization quality in the Indian automobile company. This investigation can help the automotive industry in threefold. First, it sets the basis for the assessment and classification of a set of manufacturers. Second, it helps to draw key recommendations for examined automotive industries for improving their service quality. Finally, it helps to derive an implementation framework for addressing servitization challenges. Servitization can be described as the novelty of an organization's competencies and procedures to better generate mutual value over a shift from vending product to vending PSS (Product Service System). In last three decades, automobile industry across the world has witnessed intense competition on quality and cost and is trying to leverage the competitive advantage through servitization.

17.3.2 Quality Indicators of Servitization in the Field of Automobile Sector in India

The factors affecting the service quality in automobile industry are explained as below (also see Table 17.1).

Physical Assets

Physical assets exercises a significant influence on the quality of service delivery process in an automobile product service provider. Automobile sectors providing different types of product service system require different organizational functionalities which include infrastructure, tools, machines, different facilities, spare parts, accessories, space, etc. Therefore, satisfactory space and appropriate physical assets, containing buildings, garage space, equipment, material, spare part, first aid kit, safety procedures and other ancillary facilities, should be available.

Employee and Staff

The qualified and talented workers ensure the quality of technically intensive services in automobile industry. The dedication, attitudes and relational abilities of the managers, salespersons and service delivery staffs play an essential and pivotal part in effectively running the automobile organization. The worker quality, capability and level of skill and execution ought to be satisfactory to achieve the mission and objectives for the selection of servitization strategies.

Financial Resources

There is a requirement for the vehicle specialist co-ops to be financially steady. The money-related assets ought to be satisfactory to manage not just the accomplishment of current service conveyance targets yet additionally accommodate upgrades within a reasonable time frame. There must be a provision to guarantee legitimate finance-related service and an efficient procedure. Not exclusively should there be the portion of a satisfactory spending plan for capital (non-repeating) works (counting infrastructure, machines and equipment), there must be a financial plan for operational (maintenance) work.

Governing Policies

Governing policies such as management interest in adopting new strategies, employment procedure for various staffs and managements, salary structure, opportunity for training and self-development, offers for selling, selling procedure, and decision-making capacity of different managers which directly concerns satisfaction of the customer, are all important in the context of improving the quality of servitisation.

Service Procedure

Good customer-service provider relationship requires clear interaction as well as emotional and relational support that significantly improve the loyalty of the customers towards the organisation. Different product service combinations offered by the organisation should be comprehensive to provide the customer with sufficient inputs in different types of services and their benefits. The core of the servitisation should focus on acquisition of the customer in the specific service domain and also on enhancing exposure to other value-added services.

Client's Perspective

Customers, suppliers, employees, managers, technicians, etc., are the potential stakeholders whose viewpoints reflect the quality of service given by the automobile manufacturers/sectors. How well the partner's desires are met decides the reputation of an automobile manufacturer and hence generates the demand in the market for the automobile products and services. A cursory literature survey recommends that service quality is perceived to be the fundamental precursor of customer satisfaction.

Table 17.1 shows the quality indicators and parameters for servitization process of automobile sectors.

Table 17.1 Quality indicators and parameters

Quality indicators	Parameters
Physical Assets (A_1)	Space and infrastructure available to the Automobile outlet
	Garage is well equipped with tools and machines
	Adequate lighting and comfort zones for maximum capacity of the outlet
	Proximity of service centre
	Showroom is well equipped with facilities
	Availability of spare parts and accessories
	Availability of adequate space to accommodate the vehicle under service
	Space availability for customers
	Security aspects
	First Aid availability in the outlet for employees and customers
	Safety aspects available for the employees
Worker and staff (A_2)	Strength of well-qualified employee/staff in each department
	Technicians having required qualification for service
	Employees having proper training for different operations
	Technical knowledge of the workers
	Workshop timing
	Communication skills of the employees
	Attitude of the staff towards the customer
Financial assets (A_3)	Adequate fund available for the employees
	Annual increment to the staff
	Recognition of the staff according to good work
	Funds available for purchase of modern equipment
	Funds available for maintaining the required level of spare parts and accessories
Governing policies (A_4)	Managements' interest and strategies
	Selection procedure for employees and staffs
	Flexible salary and employment for outstanding persons
	Opportunity for self-development and regular training
	Authority to head of the departments for availing needful facilities/resources in the department
	Selling procedure
	Offers for selling a product
Servicing procedure (A_5)	Service method
	Time taken in servicing
	Free service camps are arranged for the customers

(continued)

Table 17.1 (continued)

Quality indicators	Parameters
	Delivery procedure of a vehicle after service/purchase
	Offers given to the customers regarding services
	Staff-–customer interaction in service centre and outside
	Warranty period service process and benefits
	Buy-back procedure of the used vehicles
Client's perspective (A_6)	Satisfaction for service provided
	Support of staff and management in different situation
	Opportunity for understanding the benefits of service
	Attitude of customer to their service provider
	High cost of repair
	Willingness to adopt the new services introduced

17.3.3 Quality Modelling Using a Graph Theoretic Approach

To evaluate and expand the quality of servitization strategies in Indian automobile manufacturers, a suitable quality model is mandatory which can be developed by involving significant quality indicators/factors and their relations. The influence of the quality indicators/factors nominated was scrutinized using the feedback of different stakeholders such as customers, marketing employees, managers and experts from both research and production.

In the present case, digraph and matrix strategies, additionally called a graph theoretic approach (GTA), have been deployed to develop a quality model in servitization of automobile division. GTA is a precise strategy for transformation of subjective elements to quantitative esteems, and numerical displaying gives an edge to the prescribed procedure over regular strategies like cause–impact outlines, stream diagrams and so on. The fundamentally preferred standpoint of GTA is its capacity to consider various interdependent factors.

17.3.3.1 Quality Assessment of Automobile Manufacturers

The quality of servitization in automobile industry can be assessed in terms of an index to ascertain the severity of the factors selected. The numerical value obtained is known as the critical service quality index (CSQI). CSQI is computed from the permanent function of variable permanent matrix (VPMCSQI), i.e. Equation (17.3). It is a function of factors selected and their interdependencies. Hence, the CSQ index is given as:

$$\text{CSQI} = \text{per } X^* = \text{Permanent value of } VPM_{\text{CSQI}}$$

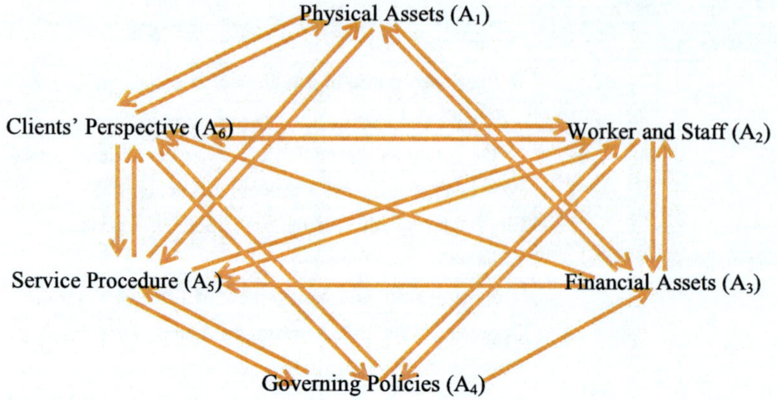

Fig. 17.4 Quality assessment diagraph (QAD)

To evaluate the CSQI, i.e. per X^*, numerical values of diagonal and off-diagonal elements in matrix (Eq. (17.1))) are required. Diagonal elements (A_i) indicate the importance or contribution (also called inheritance) of the *i*th factor to the overall CSQI, while off-diagonal elements (a_{ij}) indicate the relative importance or interdependencies among the main factors. Two different numerical scales as listed in Tables 17.2 and 17.3 have been suggested for assigning numerical values to "A_i" and "a_{ij}", respectively. In the present work, six critical elements, to be specific, physical assets (A_1), worker and staff (A_2), financial assets (A_3), governing policies (A_4), servicing procedure (A_5) and client's perspective (A_6), have been chosen that influence the nature of servitization of automobile sector. Physical assets (A_1) do not influence the worker and staff (A_2) and governing policies (A_4), yet influence the financial assets (A_3), servicing procedure (A_5) and client's perspective (A_6). Figure 17.4 indicates that there are coordinated edges from A_1 to A_3, A_5, A_6; however, there is no association between A_1 to A_2 and A_4.

It might be said that one may pick any scale, e.g. 0–1, 0–5, 1–5, 0–10, 1–10, 1–11, 0–50, 0–100, 1–100, 1–110, 0–1000, 1–1000 or some other scale for A_i and a_{ij}. In any case, the final ranking won't change, as these are relative values. It is, notwithstanding, attractive to pick a lower scale for A_i and a_{ij} to get a reasonable estimation of critical service quality index. It might be additionally specified that the scales adapted for A_i and a_{ij} can be independent of each other.

$$VPM_{\text{CSQI}} = X = \begin{bmatrix} A_1 & 4 & 3 & 0 & 5 & 1 \\ 3 & A_2 & 0 & 0 & 1 & 5 \\ 5 & 5 & A_3 & 3 & 2 & 2 \\ 4 & 0 & 2 & A_4 & 3 & 5 \\ 5 & 4 & 1 & 0 & A_5 & 3 \\ 5 & 5 & 0 & 3 & 2 & A_6 \end{bmatrix} \tag{17.4}$$

Table 17.2 Numerical values assigned to the quality indicators

Quality measure factors	Assigned value (A_i)
Exceptionally low	1
Very low	2
Low	3
Below average	4
Average	5
Above average	6
High	7
Very high	8
Exceptionally high	9

Table 17.3 Numerical values assigned to off-diagonal elements

S. No.	Qualitative measure of interdependencies	a_{ij}
1	Very strong	5
2	Strong	4
3	Medium	3
4	Weak	2
5	Very weak	1

17.3.4 Application of Graph Theoretic Model for Indian Automobile Organizations for Servitization

In view of the specialists' conclusions and a comprehensive literature survey, numerical values to the off-diagonal components have been obtained utilizing Table 17.3. The following paragraph provides the justification of the assigned numerical values in the off-diagonal elements of the matrix shown in Eq. 17.4.

Physical assets like infrastructures, tools, machineries, location of the service centre, spare parts, etc., are neither affected by the technical knowledge of the employees nor affected by the knowledge or qualification of the staff. For this reason, physical asset (A_1) factor is not connected in both directions with the workers and staff (A_2) factor. So the numerical value of the a_{12} and a_{21} is zero. Physical assets have strong influence on financial resources as if the automobile sector wants to expand the physical assets it needs finance. Henceforth, we used very strong relationship between physical asset and financial resources, i.e. a_{13} as numerical value 5 but the reverse is not true. If an automobile sector has strong financial assets, then it will first go for governing policies and management decision before strengthening physical resources. Hence, the numerical value for interrelationship a_{31} is strong, i.e. 4 not very strong. Similarly, physical assets have no relationship with governing policies of an automobile service sector but have strong relationship with servicing process. If adequate facilities are not available, then servicing process will be affected strongly. However, customer wants good service not the physical assets the garage has. So there is weak relationship between physical assets and customers viewpoint. Hence, numerical values associated with a_{15} and a_{16} are 4 and 2, respectively. Similarly, other off-diagonal values have been assigned as shown in matrix 4.

To generalize, we computed the maximum and minimum values the CSQI for both the excellent and poor service quality index for any automobile service sector. To obtain the maximum value of CSQI [Eq. (17.4)], all the diagonal elements (A_i) were assigned the numerical value of 9 (from Table 17.2), and correspondingly for obtaining the minimum value, all A_i's were assigned a value of 1 (from Table 17.2). The computed maximum and minimum values are $\text{CSQI}_{\text{max}} = 2.447514 \times 10^6$ and $\text{CSQI}_{\text{min}} = 8.6398 \times 10^4$. These computed values of maximum and minimum CSQI gave the range of the quality index of servitization in an automobile service sector. The higher the CSQI value, the better would be the quality of servitization process in a given automobile industry and the greater in the brand value.

Data collection

To assess the impact/significance of chosen factors (A_i), input of partners (i.e. representatives, directors, employees and clients) of the automobile suppliers was gathered. In view of the parameters of the quality markers (from Table 17.1), an overview instrument (Appendix) was intended to gather criticism information.

Feedback information was gathered for three automobile suppliers situated in the zone of Kharagpur in the state of West Bengal, India. Input was taken for a common service offered by the three vehicle suppliers. Participants were arbitrarily chosen from the example vehicle suppliers and the survey was regulated. Reactions from 15 service representatives, 15 managers and 10 clients were obtained. All responses received were found to be valid and accounted for:

- Response data were collected from 40 stakeholders (15 employees, 15 managers and 10 customers) for each automobile provider.
- Using feedback data, the average weight was calculated for each parameter in Tables 17.6, 17.7 and 17.8 in Appendix.
- Total weight of each of the quality indicators was calculated by adding the average weight of each parameter obtained in step-II. The different parameters were grouped under different quality indicators as listed in Table 17.4.
- A qualitative range (exceptionally low to exceptionally high) was assigned to the total weights of the quality indicators obtained in Step III. The higher the value of the weight indicates, the higher the value of importance (A_i) to the quality indicator (selected from Table 17.5).

With the end goal of task of inheritance values to the quality indicators, the scope of normal estimations of the information gathered was considered. For average value of information lying in the vicinity of 1 and 1.25, inheritance value of 1 was allotted. For average value of information lying in the vicinity of 1.25 and 1.75, inheritance value of 2 was doled out. Also, average value of information lying in the vicinity of 1.75 and 2.25 was appointed 3, those in the vicinity of 2.25 and 2.75 as 4, those in the vicinity of 2.75 and 3.25 as 5, those in the vicinity of 3.25 and 3.75 as 6, those in the vicinity of 3.75 and 4.25 as 7, those in the vicinity of 4.25 and 4.75 as 8 and

Table 17.4 Assessment of Indian automobile organizations for quality of service

Factors affecting quality of service	Company A		Company B		Company C	
	Average value	Assigned value	Average value	Assigned value	Average value	Assigned value
Physical assets (A_1)	4.83	9	3.79	7	3.47	6
Worker and staff (A_2)	4.15	7	4.26	8	4.52	8
Financial assets (A_3)	4.91	9	3.86	7	4.68	8
Governing policies (A_4)	4.05	7	3.14	5	3.84	7
Servicing procedure (A_5)	4.35	8	2.68	4	3.72	6
Client's perspective (A_6)	3.88	7	2.98	5	3.18	5

Table 17.5 CSQI for Indian automobile organizations

Category	CSQI
Minimum Value	8.6398×10^4
Company A	1.5034×10^6
Company B	6.8637×10^5
Company C	1.0099×10^6
Maximum Value	2.4475×10^6

those in the vicinity of 4.75 and 5 as 9. Table 17.4 demonstrates the average value acquired and equivalent assigned values to the quality indicators for three automobile providers.

From Fig. 17.5, it is evident that company A puts more effort on physical assets, governing policies, financial assets, servicing procedure and client's perspective, whereas company C puts more efforts on worker and staff and financial aspects. Company B is behind company A and company C in all the quality indicators, and hence, the overall CSQI for company A is lowest. Form Fig. 17.4, it is clear that for being the market leader, an organization has to put efforts to improve upon these six quality parameters. The servitization strategies are penetrating in developing countries like India to a great level. As shown in the radar diagram above, all the companies are competing with each other in all the quality characteristics.

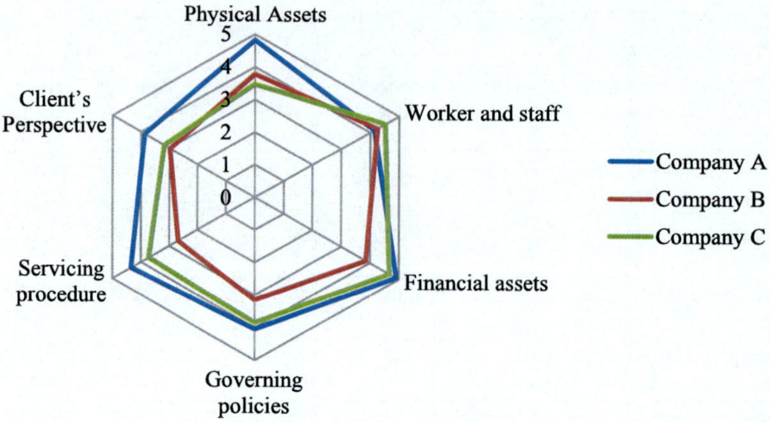

Fig. 17.5 Radar diagram of average values for the quality indicators for the three test companies

It has been evident from the analysis of case organizations that the companies receiving more weightage on client's perspectives, servicing procedure and governing policies have reported improved CSQI. The CSQI indicator has a potential to deliver critical insights for practitioners in automobile industry and extends a significant value for top management when making strategic decisions. They can design a portfolio of services by conducting benchmarking with competitors which can generate the value for company and customer both and also help the companies to understand the services which can really have small and large differences in terms of their ability to create profit.

The CSQI for three automobile providers was assessed (Table 17.5) utilizing matrix 4 and Eq. (17.3) as recorded in Table 17.8. Organization A had the most elevated CSQI $= 1.5034 \times 10^6$: it showed the larger level of satisfaction of stakeholders when contrasted with the other two organizations and subsequently a higher quality level. Organization C follows organization A and has a CSQI value as 1.0099×10^6. Organization B has the lowest CSQI value which is 6.8637×10^5. The competition between the three organizations is tough and they are close in respect of quality concerns. Hence, a remarkable product service portfolio is needed to lock the competition out. The organizations should think about the servitization integration to their product service portfolio.

17.4 Advantages of Graph Theoretic Approach

- Graph theoretic approach sometimes known as graph theoretic matrix approach (GTMA) has the ability to consider interrelationships.

- GTMA is very simple to use and could be used widely in different fields like tribology, mechanical systems, industry quality assessments, etc., where one wants to compare between various systems.
- GTMA has the ability to accommodate dynamic changes, with least mathematical complexities, and has the ability to incorporate subjective opinion of elicited experts.
- It is a versatile approach, which can consider a number of mechanisms of coordination as well as various factors and their sub-factors affecting these mechanisms.
- The information of digraphs can be computer processed since it can be expressed as a mathematical entity.
- The digraph representation is better than conventional representation like cause and effect diagram and flow chart, which provide only visual analysis but do not depict interactions among mechanisms. The digraph depicts a single numerical value for coordination index. This can help to develop comparisons among set of organizations or functions.

17.5 Limitation of Graph Theoretic Approach

- The permanent function uses qualitative values for various terms, which can be different for different experts.
- The index is industry or firm specific and hence the number of mechanisms and sub-factors should be the same for all the organizations considered for necessary comparisons.

This chapter has discussed the use of graph theoretic analysis which is based on graph theory concept. This approach helps the decision maker to develop a holistic thinking on system by identifying the system and sub-system up to component level. We have covered the salient features and applications of various MCDM techniques in the book so far. The next three chapters will help the readers to appreciate an integrated application of various MCDM techniques for improving the discriminatory power of the analysis for real-life problems. The next chapter will demonstrate an integrated application of grey AHP and grey TOPSIS.

Table 17.6 Questionnaires for customers

S. No.	Parameters	Weight
1	Adequate lighting and comfort zones for maximum capacity of the outlet	
2	Proximity of service centre	
3	Showroom is well equipped with facilities	
4	Availability of spare parts and accessories	
5	Availability of adequate space to accommodate the vehicle under service	
6	Security aspects	
7	First aid availability in the outlet for employees and customers	
8	Communication skills of the employees	
9	Attitude of the staff towards the customer	
10	Selling procedure	
11	Offers for selling a product	
12	Service method	
13	Free service camps are arranged for the customers	
14	Delivery procedure of a vehicle after service/purchase	
15	Perks given to the customers regarding services	
16	Staff–customer interaction in service centre and outside	
17	Warranty period service process and gifts	
18	Buy-back procedure of the used vehicles	
19	Satisfaction for service provided	
20	Support of staff and management in different situation	
21	Opportunity for understanding the benefits of service	
22	Readiness to pay for the service provided	
23	Willingness to adopt the new services introduced	

Appendix

The marking scale used in this research to assigning weights was as follows.

5: Excellent, 4: Good; 3: Above average; 2: Average; 1: Poor

See Tables 17.6, 17.7 and 17.8.

Table 17.7 Questionnaires for staff/employees

S. No.	Parameters	Weight
1	Space and infrastructure available to the automobile outlet	
2	Garage is well equipped with tools and machines	
3	Adequate lighting and comfort zones for maximum capacity of the outlet	
4	Showroom is well equipped with facilities	
5	Availability of spare parts and accessories	
6	Availability of adequate space to accommodate the vehicle under service	
7	Security aspects	
8	First Aid availability in the outlet for employees and customers	
9	Safety aspects available for the employees	
10	Adequate strength of well-qualified employee/staff in each section	
11	Technicians having required qualification for service	
12	Employees having proper training for different operations	
13	Technical knowledge of the workers	
14	Flexible working hours for employees	
15	Communication skills of the employees	
16	Adequate fund available for the employees	
17	Annual increment to the staff	
18	Recognition of the staff according to good work	
19	Funds available for purchase of modern equipment	
20	Funds available for maintaining the required level of spare parts and accessories	
21	Flexible salary for outstanding persons	
22	Opportunity for self-development and regular training	
23	The head of the departments is authorized for availing needful facilities/resources in their section	
24	Selling procedure	
25	Offers for selling a product	
26	Service method	
27	Free service camps are arranged for the customers	
28	Delivery procedure of a vehicle after service/purchase	
29	Perks given to the customers regarding services	
30	Staff–customer interaction in service centre and outside	
31	Attitude of customer to their service provider	

Table 17.8 Questionnaires for managers

S. No.	Parameters	Weight
1	Space and infrastructure available to the automobile outlet	
2	Garage is well equipped with tools and machines	
3	Adequate lighting and comfort zones for maximum capacity of the outlet	
4	Showroom is well equipped with facilities	
5	Availability of spare parts and accessories	
6	Adequate number of well-qualified employee/staff in each section	
7	Technicians having required qualification for service	
8	Employees having proper training for different operations	
9	Technical knowledge of the workers	
10	Communication skills of the employees	
11	Recognition of the staff according to good work	
12	Managements' interest and strategies	
13	Selection procedure for employees and staffs	
14	Flexible salary and increment for outstanding persons	
15	Opportunity for self-development and regular training	
16	The head of the departments is authorized for availing needful facilities/resources in their section	
17	Selling procedure	
18	Offers for selling a product	
19	Attitude of customer to their service provider	

Selected Bibliography and Further Reading

1. Gambhir, V., Wadhwa, N. C., & Grover, S. (2016). Quality concerns in technical education in india: a quantifiable quality enabled model. *Quality Assurance in Education, 24*(1), 2–25.
2. Gandhi, O. P., & Agrawal, V. P. (1992). FMEA—a diagraph and matrix approach. *Reliability Engineering and System Safety, 35*(2), 147–158.
3. Grover, S., Agrawal, V.P. & Khan, I.A. (2006). Role of human factors in TQM: a graph theoretic approach. Benchmarking: An International Journal, 13(4), pp. 447–468.
4. Jangra, K., Grover, S., & Aggarwal, A. (2011). Digraph and matrix method for the performance evaluation of carbide compacting die manufactured by wire EDM. *The International Journal of Advanced Manufacturing Technology, 54*(5–8), 579–591.

5. Jangra, K., Grover, S., Chan, F. T., & Aggarwal, A. (2011). Digraph and matrix method to evaluate the machinability of tungsten carbide composite with wire EDM. *The International Journal of Advanced Manufacturing Technology, 56*(9–12), 959–974.
6. Kaur, A., Kanda, A., & Deshmukh, S. G. (2006). A graph theoretic approach for supply chain coordination. *International Journal of logistics Systems and Management, 2*(4), 321–341.
7. https://doi.org/10.1080/00207543.2020.1736721.

Chapter 18
Interpretive Structural Modelling (ISM)

18.1 Introduction

The ISM procedure created by Warfield [7] is used to break down a complex issue into simpler one with the help of knowledge and wisdom of the experts. ISM is an interactive learning process which transforms unclear and poorly articulated mental models of systems into visible, well-defined models useful for many purposes. ISM facilitates a group to impose order on the complexity of the items. The method is interpretive in that the group's judgment decides whether and how items are related; it is structural in that, on the basis of the relationship, an overall structure is extracted from the complex set of items; and it is modeling in that the specific relationships and overall structure are portrayed in a digraph model [1]. The key characteristics of ISM methodology can be summarized as follows:

- It is interpretive since it determines the interrelationships among factors.
- It is a displaying system, as diagraph displays all connections and complete structure.
- It is a gathering and learning process.

A pictorial understanding of ISM method is presented in Fig. 18.1.

J. J. Thakkar, *Multi-Criteria Decision Making*, Studies in Systems, Decision and Control 336, https://doi.org/10.1007/978-981-33-4745-8_18

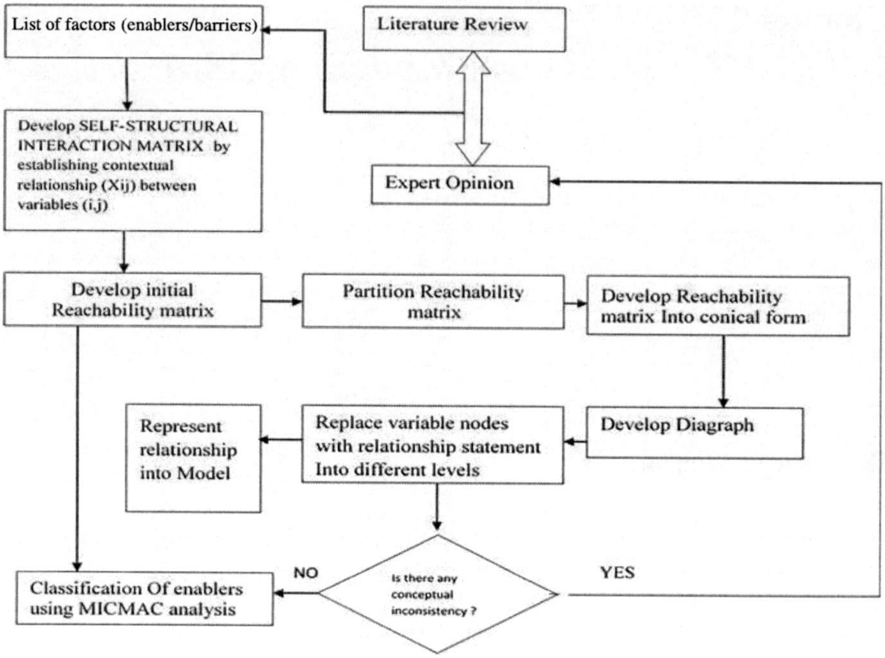

Fig. 18.1 Flow diagram of ISM

18.2 Step-by-Step Procedure of ISM Methodology

Step 1 Identify key enablers or barriers affecting the problem or phenomenon under investigation.

Step 2 A Structural self-interaction matrix (SSIM) is established for enablers, representing inter-relationship between enablers.

Step 3 Based on SSIM, initial reachability matrix is formed, and after removing all the transitivity links, final reachability is formed. Transitivity is nothing but a basic assumption for relative relationship such as if A equal to B, B equal to C then A will automatically equal to C.

Step 4 Partition the final reachability matrix to levels.

Step 5 After removing transitivity, a diagraph is drawn to represent contextual relationship.

Step 6 Convert diagraph into ISM model.

Step 7 ISM model developed in step 6 is examined for inconsistencies.

Table 18.1 Key enablers for supply chain issues in Indian SMEs in handloom industry

S. No.	Critical Issues In Handloom Industry
E1	Raw material shortage
E2	Obsolete technology of machineries
E3	Conventional product range
E4	Lack of working capital
E5	Competition from power loom and mill sector
E6	Poor textile education and skill development
E7	Lack of research and development (R and D) in cotton area
E8	Rising input costs Increased cost of Production
E9	lack of marketing
E10	Labour concerns
E11	Demand uncertainty
E12	Power shortage

18.3 An Application of ISM for Small and Medium-Scale Enterprises in Handloom and Textile Industry

This section demonstrates a step-by-step application of ISM for SMEs in handloom and textile industry.

18.3.1 Identification of Enablers

Twelve key enablers affecting handloom and textile supply chain of SMEs are identified in Table 18.1.

18.3.2 Structural Self-interaction Matrix (SSIM)

Interpretive structural modelling uses industry expert's opinion via process such as nominal group and brainstorming technique to build interrelationship between different enablers. The enablers for sickness in handloom industry were recognized through the literature review and experts' opinion. SSIM is developed (in Table 18.2) based on the contextual relationship established between enablers represented as i and j. The relationship between any two variables (in our case enablers) can be represented by four standard symbols which provides direction to the flow of relationship. The symbols used for representing these four relationships are as follows:

Table 18.2 Structural self-interaction matrix (SSIM)

S. No.	Enablers	E12	E11	E10	E9	E8	E7	E6	E5	E4	E3	E2	E1
E1	Raw material shortage	O	X	O	O	V	O	O	A	A	O	O	–
E2	Obsolete technology of machineries	O	O	O	O	X	O	O	A	A	O	–	
E3	Conventional product range	O	O	O	O	A	A	A	V	A	–		
E4	Lack of working capital	O	X	O	V	X	V	O	O	–			
E5	Competition from power loom and mill sector	O	O	X	X	A	A	A	–				
E6	Textile education and skill development is very poor	O	O	A	O	O	O	–					
E7	Lack of research and development (R and D) in cotton area	O	A	O	O	O	–						
E8	Rising input costs Increased cost of production	O	X	V	O	–							
E9	lack of marketing	O	O	O	–								
E10	Skill employment instability	O	O	–									
E11	Demand uncertainty	O	–										
E12	Power shortage	–											

V When the relation of enabler is from i to j or when enabler i is impacted by enabler j.

A When the relation of enabler is from j to i or when enabler j is impacted by enabler i.

X When relation between enablers is from both sides or when both enabler i and j is impacted by each other.

O When there is no relation between both the enablers or when both enablers I and j are not affected by each other.

Once SSIM is developed, it should further be discussed with opinion experts so that result of SSIM is validated.

18.3.3 Reachability Matrix

Representation of SSIM to 1 s and 0 s which were earlier represented as V, A, X and O is known as initial reachability matrix.

Steps to be followed for conversion of matrix represented by (i, j) are:

$$\begin{aligned} &\text{For } V\ (i,\ j) \text{ the entry becomes } 1\\ &(j,\ i) \text{ the entry becomes } 0\\ &\text{For } A\ (i,\ j) \text{ the entry becomes } 0\\ &(j,\ i) \text{ the entry becomes } 1\\ &\text{For } X\ (\mathrm{i},\ \mathrm{j}) \text{ the entry becomes } 1\\ &(j,\ i) \text{ the entry becomes } 1\\ &\text{For } O\ (i,\ j) \text{ the entry becomes } 0\\ &(j,\ i) \text{ the entry becomes } 0 \end{aligned}$$

Initial matrix is formed (in Table 18.3) after substituting the variables, and final matrix is formed after transitivity analysis (in Table 18.4).

18.3.4 Level Partitions

From the reachability matrix, the reachability set and antecedent set for each objective are determined. The reachability set consists of the element itself and other elements to which it may reach, whereas the antecedent set consists of the elements itself and the other elements, which may reach it. Then, the intersection of these sets is derived for all elements. The elements are considered as top-level elements for which the reachability and intersection sets are the same. Physically, these top-level elements of hierarchy will not reach any higher than their own level. To obtain the next level of elements, top-level elements are separated out from other elements and the same process is repeated. The whole process of partitioning is based on establishing the

Table 18.3 Initial reachability matrix

S. No.	Enablers	E1	E2	E3	E4	E5	E6	E7	E8	E9	E10	E11	E12
E1	Raw material shortage	1	0	0	0	0	0	0	1	0	0	1	0
E2	Obsolete technology of machineries	0	1	0	0	0	0	0	1	0	0	0	0
E3	Conventional product range	0	0	1	0	1	0	0	0	0	0	0	0
E4	Lack of working capital	1	1	1	1	0	0	1	1	1	0	1	0
E5	Competition from power loom and mill sector	1	1	0	0	1	0	0	0	1	1	0	0
E6	Textile education and skill development is very poor	0	0	1	0	1	1	0	0	0	0	0	0
E7	Lack of research and development (R and D) in cotton area	0	0	1	0	1	0	1	0	0	0	0	0
E8	Rising input costs Increased cost of production	0	1	1	1	1	0	0	1	0	1	1	0
E9	lack of marketing	0	0	0	0	1	0	0	0	1	0	0	0
E10	Skill employment instability	0	0	0	0	1	1	0	0	0	1	0	0
E11	Demand uncertainty	1	0	0	1	0	0	1	1	0	0	1	0
E12	Power shortage	0	0	0	0	0	0	0	0	0	0	0	1

Table 18.4 Transitivity and final matrix (Note: 1* shows transitivity)

S. No.	Enablers	E1	E2	E3	E4	E5	E6	E7	E8	E9	E10	E11	E12
E1	Raw material shortage	1	1*	1*	1*	1*	0	1*	1	0	1*	1	0
E2	Obsolete technology of machineries	0	1	1*	1*	1*	0	0	1	0	1*	1*	0
E3	Conventional product range	1*	1*	1	0	1	0	0	0	1*	1*	0	0
E4	Lack of working capital	1	1	1	1	1*	0	1	1	1	1*	1	0
E5	Competition from power loom and mill sector	1	1	0	0	1	1*	0	1*	1	1	1*	0
E6	Textile education and skill development is very poor	1*	1*	1	0	1	1	0	0	1*	1*	0	0
E7	Lack of research and development (R and D) in cotton area	1*	1*	1	0	1	0	1	0	1*	1*	0	0
E8	Rising input costs Increased cost of production	1*	1	1	1	1	1*	1*	1	1*	1	1	0
E9	lack of marketing	1*	1*	0	0	1	0	0	0	1	1*	0	0
E10	Skill employment instability	1*	1*	1*	0	1	1	0	0	1*	1*	0	0
E11	Demand uncertainty	1	1*	1*	1	1*	0	1	1	1*	1*	1	0
E12	Power shortage	0	0	0	0	0	0	0	0	0	0	0	1

precedence relationships and arranging the elements in topological order. Finally, the reachability matrix is converted into the canonical (lower triangular) format by arranging the elements according to their levels.

A level partitioning matrix is presented in Table 18.5.

18.3.5 Conical Level

When enablers of common level are grouped together in rows and columns, then it is termed as conical level (as presented in Table 18.6).

1. Drive power: When all power in rows is summed up.
2. Dependence power: When all power in columns is summed up.

These two then give the highest ranks to the enablers by calculating number of 1 s of all rows and columns.

18.3.6 Formation of Diagraph

The diagraph is obtained when we remove the transitivity which represents the enablers and their interdependencies, and hence, it becomes the visual demonstration of interdependence of enablers.

18.3.7 Formation of ISM Model

From the canonical matrix form of reachability matrix (Table 18.6), the structural model is generated by means of vertices or nodes and lines of edges. If there is a relationship between element i and j, this is shown by an arrow which points from i to j. This graph is called a direct graph or digraph. Next, the elements descriptions are written in the digraph to call it the ISM. The ISM model for handloom and textile industry supply chain enablers is shown in Fig. 18.2. ISM representation helps to appreciate hierarchy of enablers and their interrelationships. This enables the decision makers to plan the initiatives in the appropriate direction.

Table 18.5 Level partitioning

S. No.	Enabler	Reachability set (column)	Antecedent set (row)	Intersection set	Iteration no. and level
Iteration I					
E1	Raw material shortage	1,3,4,5,6,7,8,9,10,11	1,2,3,4,5,7,8,10,11	1,3,4,5,7,8,10,11	
E2	Obsolete technology of machineries	1,2,3,4,5,6,7,8,9,10,11	2,3,4,5,8,10,11	2,3,4,5,8,10,11	
E3	Conventional product range	1,2,3,4,6,7,8,10,11	1,2,3,5,9,10	1,2,3,10	
E4	Lack of working capital	1,2,4,8,11	1,2,3,4,5,7,8,9,10,11	1,2,4,8,11	II
E5	Competition from power loom and mill sector	1,2,3,4,5,6,7,8,9,10,11	1,2,5,6,8,9,10,11	1,2,5,6,8,9,10,11	
E6	Textile education and skill development is very poor	5,6,8,10	1,2,3,5,6,9,10	5,6,10	
E7	Lack of research and development (R and D) in cotton area	1,4,7,8,11	1,2,3,5,7,9,10	1,7	
E8	Rising input costs Increased cost of production	1,2,4,5,8,11	1,2,3,4,5,6,7,8,9,10,11	1,2,4,5,8,11	II
E9	lack of marketing	3,4,5,6,7,8,9,10,11	1,2,5,9,10	5,9,10	
E10	Skill employment instability	1,2,3,4,5,6,7,8,9,10,11	1,2,3,5,6,9,10	1,2,3,5,6,9,10	
E11	Demand uncertainty	1,2,4,5,8,11	1,2,3,4,5,7,8,9,10,11	1,2,4,5,8,11	II
E12	Power shortage	12	12		I

(continued)

Table 18.5 (continued)

S. No.	Enabler	Reachability set (column)	Antecedent set (row)	Intersection set	Iteration no. and level
Iteration II					
E1	Raw material shortage	3,6,7,9,10	3,7,10	3,7,10	
E2	Obsolete technology of machineries	3,6,7,9,10	3,10	3,10	
E3	Conventional product range	3,6,7,10	3,9,10	3,10	
E5	Competition from power loom and mill sector	3,6,7,9,10	6,9,10	6,9,10	
E6	Textile education and skill development is very poor	6,10	3,6,9,10	6,10	III
E7	Lack of research and development (R and D) in cotton area	4,7	3,7,9,10	7	
E9	lack of marketing	3,6,7,9,10	9,10	9,10	
E10	Skill employment instability	3,6,7,9,10	3,6,9,10	3,6,9,10	
Iteration III					
E1	Raw material shortage	3,7,9	3,7	3,7	
E2	Obsolete technology of machineries	3,7,9	3	3	
E3	Conventional product range	3,7	3,9	3	
E5	Competition from power loom and mill sector	3,7,9	9	9	

(continued)

Table 18.5 (continued)

S. No.	Enabler	Reachability set (column)	Antecedent set (row)	Intersection set	Iteration no. and level
E7	Lack of research and development (R and D) in cotton area	7	3,7,9	7	IV
E9	lack of marketing	3,7,9	9	9	
E10	Skill employment instability	3,7,9	3,9	3,9	
Iteration IV					
E1	Raw material shortage	3,9	3	3	VI
E2	Obsolete technology of machineries	3,9	3	3	VI
E3	Conventional product range	3	3,9	3	V
E5	Competition from power loom and mill sector	3,9	9	9	VI
E9	lack of marketing	3,9	9	9	VI
E10	Skill employment instability	3,9	3,9	3,9	V

18.4 Advantages of ISM

(i) It helps to present a complex problem into a graphical form. It incorporates experts' subjective judgments and their knowledge base in the most systematic manner.

(ii) It helps in transforming the vague and unclear mental models to well-defined model which helps in theory building.

(iii) It helps to identify the structure and provides rationale for decision making. The outstanding features of the technique include (1) incorporating experts' subjective judgments and their knowledge base in the most systematic manner,

Table 18.6 Dependence and driving power

S. No.	Enablers	E1	E2	E3	E4	E5	E6	E7	E8	E9	E10	E11	E12	Driving power
E1	Raw material shortage	1	1	1	1	1	0	1	1	0	1	1	0	9
E2	Obsolete technology of machineries	0	1	1	1	1	0	0	1	0	1	1	0	7
E3	Conventional product range	1	1	1	0	1	0	0	0	1	1	0	0	6
E4	Lack of working capital	1	1	1	1	1	0	1	1	1	1	1	0	10
E5	Competition from power loom and mill sector	1	1	0	0	1	1	0	1	1	1	1	0	8
E6	Textile education and skill development is very poor	1	1	1	0	1	1	0	0	1	1	0	0	7
E7	Lack of research and development (R and D) in cotton area	1	1	1	0	1	0	1	0	1	1	0	0	7
E8	Rising input costs Increased cost of Production	1	1	1	1	1	1	1	1	1	1	1	0	11
E9	lack of marketing	1	1	0	0	1	0	0	0	1	1	0	0	5
E10	Skill employment instability	1	1	1	0	1	1	0	0	1	1	0	0	7
E11	Demand uncertainty	1	1	1	1	1	0	1	1	1	1	1	0	10
E12	Power shortage	0	0	0	0	0	0	0	0	0	0	0	1	1
	Dependence power	10	11	9	5	11	4	5	6	9	11	6	1	

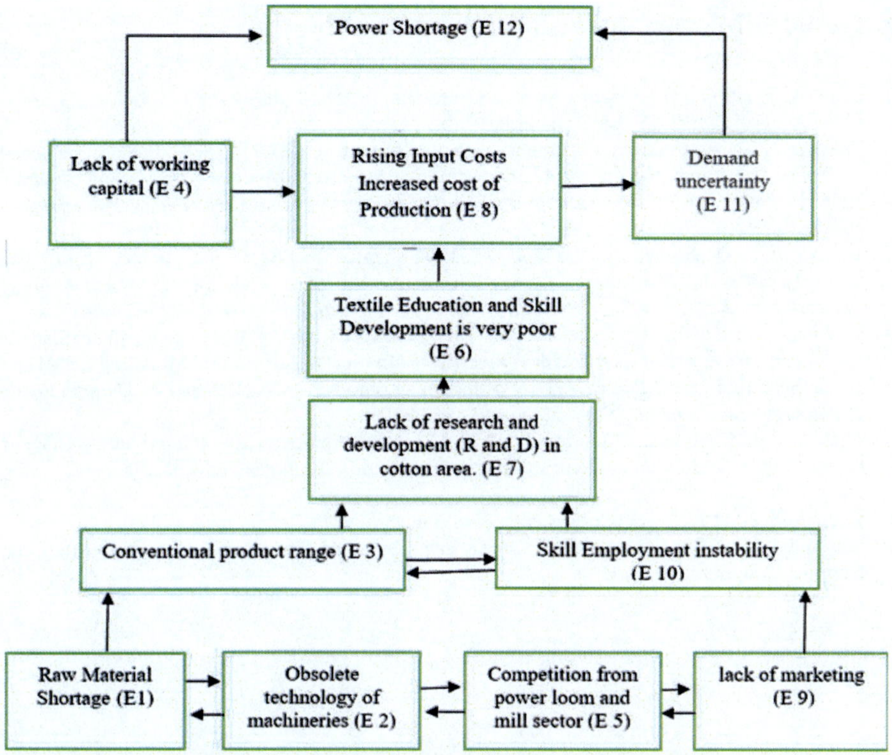

Fig. 18.2 ISM model for handloom and textile industry supply chain enablers

(2) providing ample opportunity for revision of judgments, and (3) computational efforts involved are far less for criteria ranging from 10 to 15 numbers and can be used as a handy tool in real-life applications.

18.5 Limitations of ISM

(i) It can only be instrumentalized with the help of domain experts who have developed significant insights into the problem area through a long experience.
(ii) Large number of variables are difficult to handle as it makes ISM interpretation complex. The subsequent chapters represent integrated applications of MCDM techniques.

Selected Bibliography and Further Reading

1. Sage, A. P. (1977). *Interpretive Structural Modeling: Methodology for Large-Scale Systems (91–164).* New York, NY: McGraw-Hill.
2. Thakkar, J. J., Deshmukh, S. G., Gupta, A. D., & Shankar, R. (2007). Development of balanced scorecard: An integrated approach of Interpretive Structural Modeling (ISM) and Analytic Network Process (ANP). *International Journal of Productivity and Performance Management, 56*(1), 25–59.
3. Thakkar, J. J., Deshmukh, S. G., Gupta, A. D., & Shankar, R. (2005). Selection of Third-Party Logistics (3PL): A hybrid approach using Interpretive Structural Modelling (ISM) and Analytic Network Process (ANP). *Supply Chain Forum: An International Journal, 6*(1), 32–46.
4. Thakkar, J. J., Kanda, A., & Deshmukh, S. G. (2008). Evaluation of buyer-supplier relationships using an integrated mathematical approach of Interpretive Structural Modeling (ISM) and graph theoretic matrix: The case study of Indian automotive SMEs. *Journal of Manufacturing Technology Management, 19*(1), 92–124.
5. Thakkar, J. J., Kanda, A., & Deshmukh, S. G. (2008). Interpretive structural modeling (ISM) of IT-enablers for Indian manufacturing SMEs. *Information Management & Computer Security, 16*(2), 113–136.
6. Warfield, J. N. (1976). *Societal systems: Planning, policy, and complexity.* Wiley.
7. Warfield, J. N. (1974). *Structuring complex systems,* Battelle Monograph No 4, Battelle Memorial Institute, Columbus, Ohio, USA.

Chapter 19
Application of Integrated Approach of Grey AHP and Grey TOPSIS

19.1 Background

In this chapter, we discuss the integration of multi-criteria decision-making models, namely grey analytical hierarchy process (AHP) and grey technique for order preference by similarity to the ideal solution (TOPSIS). These approaches integrated for calculating a comprehensive risk index for the food supply chain. The integrated approach is applied for assessing the risks in food supply chain in Indian context. The necessary inputs are collected from the experts in the domain of food supply chain. Through this example, we illustrate how an integrated methodology involving AHP and TOPSIS can be applied to quantify risk and generate critical insights.

To accommodate uncertainty involved in experts' opinions, there is a growing interest in using the fuzzy logic with MCDM approaches. Another approach that helps in considering uncertainty in the inputs is the grey systems theory. The major disadvantage of fuzzy sets is that it can be difficult to model discrete nature of the data, specifically, if there is insufficient information. In such cases, where the level of uncertainty is high and it is difficult to identify membership functions, grey systems theory is often a better choice.

The supply chain of the food distribution involves a complex network with a number of stages and nodes involved in the entire procurement and distribution process. It is exposed to various risks such as space unavailability, rise in prices of labour and petroleum products etc. In addition, various natural disasters and non-availability of various infrastructures like transportation induces an inefficiency in the food supply chain. This demands detailed understanding and investigation of the risks involved. The existing food supply chain in India is highly inefficient because of frequent occurence of various risks at various stages in the value chain. Inability to timely mitigate these risks leads to supplier disruptions, delays in transportation and deterioration of the stored foodgrains thus leading to quality and safety concerns. The Food Corporation of India (FCI) aims to undertake procurement and transport of grains from surplus states to deficit states in an efficient way that reduces the cost and enhances the safety of foodgrains.

J. J. Thakkar, *Multi-Criteria Decision Making*, Studies in Systems, Decision and Control 336, https://doi.org/10.1007/978-981-33-4745-8_19

This chapter demonstrates the application of the combined grey AHP and grey TOPSIS approach to quantify the major risks involved in foodgrains supply chain in India.

19.2 Development of Risk Quantification Framework

The approach that is demonstrated in this example involves integration of grey AHP and grey TOPSIS. The objective is to develop an overall comprehensive risk index that explains the current state of risk management. AHP is used to calculate the weights for various risk factors associated with the food supply chain in India. The weights calculated using grey AHP form the inputs to the next step that involves computation of the overall risk index using grey TOPSIS. AHP and TOPSIS combined with grey systems theory are applied to accommodate the imprecise nature of the inputs obtained from experts and mitigate other issues such as human bias and ambiguity in judgements.

19.2.1 Mathematical Notation

The following notation is used to describe grey number

E : The set of decision - makers
P, Q: Criterion set of indices
$\otimes x_{pq}^{e}$: Grey number for an evaluator e, which will estimate the impact of criterion p on criterion q,
where $e \in E, p \in P, q \in Q$
N_{pq}^{e} : The overall normalized crisp value for an evaluator e, where $e \in E, p \in P, q \in Q$
Y_{pq}^{e} : The final crisp value for an evaluator e, where $e \in E, p \in P, q \in Q$

19.2.2 Grey Approach

The major advantage of grey approach is the ability to produce probable results even when only a small amount of input data is available. The conversion of grey numbers to crisp numbers involves a three-step procedure called modified converting fuzzy data into crisp Scores (CFCS) presented below:-

For a grey number $\otimes x^e_{pq} = [\underline{\otimes} x^e_{pq}, \overline{\otimes} x^e_{pq}]s$ the crisp values can be obtained using the following three stages:

Stage 1. Normalization

$$\underline{\otimes}\overline{x}^e_{pq} = \left(\underline{\otimes} x^e_{pq} - \min_q \underline{\otimes} x^e_{pq}\right) / \Delta^{\max}_{\min}$$

$$\overline{\otimes}\overline{x}^e_{pq} = \left(\overline{\otimes} x^e_{pq} - \min_q \underline{\otimes} x^e_{pq}\right) / \Delta^{\max}_{\min}, \text{ where } \Delta^{\max}_{\min} = \max_q \overline{\otimes} x^e_{pq} - \min_q \underline{\otimes} x^e_{pq} \tag{19.1}$$

Stage 2. Determination of the overall normalized crisp value

$$N^e_{pq} = \frac{\underline{\otimes}\overline{x}^e_{pq}\left(1 - \underline{\otimes}\overline{x}^e_{pq}\right) + \left(\overline{\otimes}\overline{x}^e_{pq}\right)^2}{1 + \overline{\otimes}\overline{x}^e_{pq} - \underline{\otimes}\overline{x}^e_{pq}} \tag{19.2}$$

Stage 3. Calculation of the final crisp value

$$Y^e_{pq} = \min_q \underline{\otimes} x^e_{pq} + N^e_{pq} \Delta^{\max}_{\min} \tag{19.3}$$

19.2.3 AHP Method

The AHP method addresses the problem in three parts. The first part is the broader problem or issue that needs to be resolved, the second part is the consideration of the alternate solutions that are available to solve the problem. The third part deals with the criteria used to evaluate the alternative solutions. Table 19.1 is used to convert subjective opinions of the experts to an objective numerical value which can be easily compared.

Based on the opinions received from the experts, a pairwise comparison matrix (PCM) is developed as shown in Figure 19.1. The diagonal elements are always taken as "1" as it represents the comparison of each element with itself. For other elements, if element in ith row jth column is represented as a_{ij}, then the element in jth row ith

Table 19.1 Greyscale for the importance weight of the evaluators

Linguistic terms	Grey numbers	Normal value
Equal importance	[1]	1
Little importance	[1, 3]	2
Strong importance	[3, 5]	4
Very strong importance	[5, 7]	6
Extreme importance	[7,9]	8

Figure 19.1 AHP pairwise comparison matrix

$$\begin{bmatrix} 1 & a_{12} & a_{13} & a_{1j} & .. & a_{1n} \\ 1/a_{21} & 1 & a_{23} & a_{2j} & .. & .. \\ 1/a_{13} & 1/a_{12} & 1 & 1/a_{12} & .. & .. \\ 1/a_{1j} & 1/a_{2j} & 1/a_{3j} & 1 & .. & a_{jn} \\ .. & .. & .. & .. & 1 & .. \\ 1/a_{1n} & .. & .. & 1/a_{jn} & .. & 1 \end{bmatrix}$$

column becomes $1/a_{ij}$. The estimated eigenvalues provide the priority values for the elements.

The value of each alternative and weight of each criterion are multiplied which are then added up to get the overall score for each alternative. The value thus obtained is used to rank all the alternatives. Consistency index is calculated to determine the consistency of judgement. The consistency index is compared with random judgement, and if the ratio is below 0.1, the results are accepted. A higher ratio indicates that the judgement values are close to random and need to be reviewed. In this example, the AHP procedure is only used to obtain weights for each criterion. These weights are used as input for the TOPSIS procedure.

19.2.4 TOPSIS Technique

Hwang and Yoon [2] proposed TOPSIS for inquiring into the multi-criteria decision-making environment. The method follows the notion that the chosen alternative solution has the shortest distance from the positive-ideal solution and the farthest from the negative-ideal solution. The procedure for TOPSIS is as follows:

1. Estimate the normalized decision matrix $\left(R_{ij}\right)$. The value of R_{ij} is measured as

$$R_{ij} = y_{ij} / \sqrt{\sum_{j=1}^{J} y_{ij}^2},$$

   where $j = 1, \ldots, J;\ i = 1, \ldots, N$
2. The weighted normalized decision matrix $\left(D_{ij}\right)$ is estimated using following equation. $D_{ij} = w_i R_{ij}$,
   where $j = 1, \ldots, J;\ i = 1, \ldots, N$ and w_i is the weight of the criterion *i*.
3. Estimate the ideal and negative-ideal solution.

$$A^* = \{D_1^*, \ldots, D_n^*\}$$
$$= \left\{\left(\max_j D_{ij} \middle| i \in I'\right), \left(\min_j D_{ij} \middle| i \in I''\right)\right\}$$

and

$$A^- = \{D_1^-, \ldots, D_n^-\}$$
$$= \left\{\left(\min_j D_{ij} \middle| i \in I'\right), \left(\max_j D_{ij} \middle| i \in I''\right)\right\}.$$

where I' is linked with benefit criteria, and I'' is linked to cost criteria.

4. Estimate the separation of each alternative from the ideal solution, which is given as

$$S_j^* = \sqrt{\sum_{i=1}^{n} \left(v_{ij} - v_i^*\right)^2}, \quad j = 1, \ldots, J.$$

Similarly, the separation from the negative-ideal solution is provided as

$$S_j^- = \sqrt{\sum_{i=1}^{n} \left(v_{ij} - v_i^-\right)^2}, \quad j = 1, \ldots, J.$$

5. Calculate the score $\left(C_j^*\right)$ of the risk type to the ideal solution. The score of the alternative a_j on A^* is defined as

$$C_j^* = \frac{S_j^-}{S_j^* + S_j^-}, \quad j = 1, \ldots, J.$$

6. Based on the scores, rank the different criteria.

19.3 Illustrative Application of an Integrated Grey AHP and Grey TOPSIS for Food Supply Chain Risk Assessment

An integrated application of Grey AHP and Grey TOPSIS is demonstrated for quantification of risks for the foodgrains supply chain. The necessary inputs for this application are collected from the executives of Food Corporation of India and academic experts in foodgrains supply chain. The steps involved in the application of an integrated approach are explained as below.

Level I: Collection of necessary data and recognition of common risks.

The various categories of risks involved in the food supply chain were identified through the literature and comments of experts. The risks were categorized into four major categories which were further divided into various sub-risks.

Level II: Evaluating risks using grey AHP and grey TOPSIS approach.

After the set of risks and sub-risks were identified and categorized, a comprehensive analysis of the inputs obtained from industrial and academic experts was carried out. The experts were asked to inspect the relationships between different pairs of risks and sub-risks. The judgements provided by these experts were analysed using an integrated approach of grey AHP and grey TOPSIS. The complete procedure involved in the analysis is described below:

Step 1: For each evaluator, find out the crisp direct-relation matrix. The procedure of this step is outlined below.

Step 1a: Estimate a grey pairwise comparison scale for each component. For AHP calculation, a 5-level scale is used with the following scale items: [1, 1] = equal importance, [1, 3] = little importance, [3, 5] = strong importance, [5, 7] = very strong importance and [7,9] = extreme importance. For TOPSIS calculation, the risk is measured based on five different parameters: categories of risk, the probability of occurrence, their impact on foodgrains supply chain performance, increase in activity duration and increase in activity cost due to the given risk. The parameter "categories of risk" is divided into five linguistic expressions, namely strategic (S), mid-strategic-tactical (ST), tactical (T), mid-tactical-operational (TO) and operational (O). The parameter "probability of occurrence and impact of risk" is divided into low (L), medium (M), high (H), very high (VH) and extreme (E). The parameter "Increase in activity duration and cost due to a given risk" is divided into no influence (NI), low influence (LI), medium influence (MI), high influence (HI) and very high influence (VHI). The grey linguistic scales for the assessments of respondents are defined in Table 19.2.

Step 1b: A direct-relation matrix $(\otimes_{ij}^{k})_{m\times n}$ is given to each evaluator separately for comparison of risks in the food supply chain in India.

Step 1c: The grey direct-relation matrix as explained in Eqs. (19.1)–(19.3) is used to acquire crisp values. Now, considering the work experience and position of experts, we allocate weights for each evaluator. The detailed work background of the evaluators is provided below.

Manager 1 has an experience of more than 25 years and is responsible for taking decisions about the various operations and for making the strategy for procurement, storage and movement, sales and finance operations in FCI. Manager 2 is a junior engineer with more than five years of experience. He is mainly responsible for construction and maintenance operations in FCI.

Academic expert 1 is a faculty who has been researching since 2013 in food supply chain, food grains transportation planning and inventory management. Academic experts 2 and 3 are research scholars, who are working on a sponsored project on sustainable food supply chain.

Table 19.2 Grey linguistic scale for the respondents' assessments for (a) the categories of risk; (b) the parameter of probability of occurrence and impact of risk; (c) the parameter of increase in activity duration and cost due to given risks

(a) Linguistic scale for the categories of risk		(b) Linguistic scale for the parameter of probability of occurrence and impact of risk		(c) Linguistic scale for the parameter of increase in activity duration and cost due to given risks	
Linguistic terms	Grey number	Linguistic terms	Grey number	Linguistic terms	Grey number
Strategic (S)	[0,0.2]	Low (L)	[0,0.2]	No influence (NI)	[0, 0]
Mid-strategic-tactical (ST)	[0.2,04]	Medium (M)	[0.2,04]	Low influence (LI)	[0, 0.2]
Tactical (T)	[0.4,0.6]	High (H)	[0.4,0.6]	Medium influence (MI)	[0.2, 0.5]
Mid-tactical-operational (TO)	[0.6,0.8]	Very high (VH)	[0.6,0.8]	High influence (HI)	[0.5, 0.8]
Operational (O)	[0.8,1]	Extreme (E)	[0.8,1]	Very high influence (VHI)	[0.8, 1]

Table 19.3 Greyscale weights of importance for evaluators

Evaluator	Linguistic	Greyscale
Manager 1	Extreme	[0.8,1]
Manager 2	Very high	[0.6,0.8]
Academic Expert 1	Very high	[0.6,0.8]
Academic Expert 2	High	[0.4,0.6]
Academic Expert 3	Medium	[0.2,0,4]

Step 2: Using the weights assigned to the evaluators in Table 19.3 and their comments, next step is to find the significance of all criteria. To find the weights of the different decision criteria, AHP method is used. Table 19.4 represents grey pairwise comparison for risk type "supply risks" (i.e. R1 to R5). Based on the expert advice, the pairwise comparison matrix is derived and solved using grey extent analysis and AHP method. The weight for the remaining risk types (i.e. R6 to R16) is calculated in the similar manner as shown in Table 19.5.

Step 3: Assigning the score using grey TOPSIS.

In this step, the score to the criteria is assigned using grey TOPSIS. The weight calculated in the previous step using AHP is utilized to find the score for all criteria. The linguistic expressions given by the experts are presented in Table 19.6. As discussed earlier, each risk is measured against different parameters using the inputs received from the experts. With the help of Table 19.2, the initial decision matrix is obtained for all experts. In the next step, grey values are converted into crisp values with the help of Eqs. (19.1)–(19.3). These scores are then multiplied by weights assigned to the similar risk. This will be called as the first-level risks or weighted

Table 19.4 Grey pairwise comparison for risk type "supply risks"

	Sub-Risks	R1	R2	R3	R4	R5	Weights
R1	Farmer's inability of supply	[1]	[5, 7]	[3, 5]	[0.142,0.2]	[0.142,0.2]	0.357
R2	Non-availability of the procurement centre	[0.142,0.2]	[1]	[1, 3]	[0.111,0.142]	[0.333,1]	0.173
R3	Poor quality of the supplied food grains	[0.2,0.333]	[0.333,1]	[1]	[0.2,0.111]	[1, 3]	0.227
R4	Communication failure	[5, 7]	[7,9]	[3, 5]	[1]	[3, 5]	0.217
R5	Natural disaster	[5, 7]	[1, 3]	[0.333,1]	[0.2,0.333]	[1]	0.386

Table 19.5 Weight assigning to the remaining risk types

	Sub-risk	Weighs
R6	Storage warehouse	0.280
R7	Poor handling—loading/unloading at different locations	0.191
R8	Poor packaging and preservation	0.272
R9	In transit loss	0.302
R10	Timely availability of the vehicles	0.315
R11	Malfunctioning in PDS	0.594
R12	Theft or risks from outside elements	0.322
R13	Technological risk	0.139
R14	Labour strikes	0.304
R15	Sudden shoot in demand	0.658
R16	Forecasting error	0.702

normalized decision matrix. The weighted normalized decision matrix is given in Table 19.6. The intermediate results are presented in Tables 19.7, 19.8, 19.9, 19.10 and 19.11. Table 19.12 displays the score determined for each risk using grey TOPSIS method for five common parameters: categories of risks, the probability of occurrence, their impact on food supply chain performance, an increase in activity duration and activity cost due to given risk.

From Table 19.12, we can infer that *Technological risk* presents the maximum risk in case of a food supply chain followed by *poor quality of food grains. Non-availability of the procurement centre, storage problem at the warehouse* and *inappropriate handling* are the other key risks that influence the overall risk index value.

Table 19.6 Input from the industrial experts

Risks	Categories of risks	Probability of incident of the risk	Impact of the risk	Increase in activity duration due to given risks	Increase in activity cost due to given risks
Farmer's inability of supply	[0.6,0.8]	[0.2,04]	[0,0.2]	[0.5, 0.75]	[0.25,0.5]
Non-availability of the procurement centre (mechanism)	[0.6,0.8]	[0.6,0.8]	[0.2,04]	[0, 0.25]	[0.75, 1]
Poor quality of the supplied food grains	[0.2,04]	[0.2,04]	[0,0.2]	[0.25,0.5]	[0.5, 0.75]
Communication failure	[0.6,0.8]	[0,0.2]	[0.6,0.8]	[0.75, 1]	[0, 0.25]
Natural disaster	[0.2,04]	[0,0.2]	[0.2,04]	[0, 0.25]	[0.75, 1]
Storage warehouse	[0.4,0.6]	[0.4,0.6]	[0.2,04]	[0.5, 0.75]	[0.25,0.5]
Handling–loading/unloading at different locations	[0.6,0.8]	[0.4,0.6]	[0.6,0.8]	[0.75, 1]	[0.25,0.5]
Packaging and preservation	[0.8,1]	[0,0.2]	[0.4,0.6]	[0, 0]	[0.75, 1]
In transit loss	[0.8,1]	[0.4,0.6]	[0.8,1]	[0.75, 1]	[0.5, 0.75]
Timely availability of the vehicles (Rails)	[0.6,0.8]	[0,0.2]	[0.8,1]	[0.25,0.5]	[0.75, 1]
Malfunctioning in PDS	[0.4,0.6]	[0.6,0.8]	[0.6,0.8]	[0.75, 1]	[0, 0]
Theft or risks from outside elements	[0.4,0.6]	[0.2,04]	[0.4,0.6]	[0.5, 0.75]	[0, 0.25]
Technological risk	[0.6,0.8]	[0,0.2]	[0.4,0.6]	[0, 0.25]	[0.75, 1]
Labour strikes	[0.8,1]	[0.8,1]	[0.6,0.8]	[0.25,0.5]	[0.5, 0.75]
Sudden shoot in demand	[0.6,0.8]	[0,0.2]	[0.8,1]	[0.75, 1]	[0.5, 0.75]
Forecasting errors	[0,0.2]	[0.2,04]	[0.8,1]	[0.5, 0.75]	[0.75, 1]

While an application demonstrated in this chapter integrates the use of grey AHP and grey TOPSIS, it is possible to modify the solution procedure using other fuzzy, crisp and grey approaches. In addition, grey AHP can be substituted with grey ANP for the determination of weights, which would additionally accommodate the interrelationship between different criteria while computing the weights.

This chapter has demonstrated an integrated application of grey AHP and grey TOPSIS. The next chapter is developed to demonstrate an integrated application of fuzzy FMEA and VIKOR.

Table 19.7 Calculation of the crisp value

Risks	Categories of risks	Probability of incident of the risk	Impact of the risk	Increase in activity duration due to given risks	Increase in activity cost due to given risks
Farmer's inability of supply	0.733	0.267	0.03	0.65	0.35
Non- availability of the procurement centre (mechanism)	0.733	0.733	0.267	0.05	0.95
Poor quality of the supplied food grains	0.267	0.267	0.03	0.35	0.65
Communication failure	0.733	0.03	0.733	0.95	0.05
Natural disaster	0.267	0.03	0.267	0.05	0.95
Storage warehouse	0.5	0.5	0.267	0.65	0.35
Handling–loading/unloading at different locations	0.733	0.5	0.733	0.95	0.35
Packaging and preservation	0.967	0.03	0.5	0.05	0.95
In transit loss	0.967	0.5	0.967	0.95	0.65
Timely availability of the vehicles (Rails)	0.733	0.03	0.967	0.35	0.95
Malfunctioning in PDS	0.5	0.733	0.733	0.95	0.05
Theft or risks from outside elements	0.5	0.267	0.5	0.65	0.05
Technological risk	0.733	0.03	0.5	0.05	0.95
Labour strikes	0.967	0.967	0.733	0.35	0.65
Sudden shoot in demand	0.733	0.03	0.967	0.95	0.65
Forecasting errors	0.03	0.267	0.967	0.65	0.95

Table 19.8 Normalized decision matrix

Risks	Categories of risks	Probability of incident of the risk	Impact of the risk	Increase in activity duration due to given risks	Increase in activity cost due to given risks
Farmer's inability of supply	0.073	0.052	0.003	0.076	0.037
Non-availability of the procurement centre (mechanism)	0.073	0.141	0.029	0.006	0.100
Poor quality of the supplied food grains	0.026	0.052	0.003	0.041	0.068
Communication failure	0.073	0.006	0.080	0.110	0.005
Natural disaster	0.026	0.006	0.029	0.006	0.100
Storage warehouse	0.050	0.097	0.029	0.076	0.037
Handling–loading/unloading at different locations	0.073	0.097	0.080	0.110	0.037
Packaging and preservation	0.096	0.006	0.055	0.006	0.100
In transit loss	0.096	0.097	0.106	0.110	0.068
Timely availability of the vehicles (Rails)	0.073	0.006	0.106	0.041	0.100
Malfunctioning in PDS	0.050	0.141	0.080	0.110	0.005
Theft or risks from outside elements	0.050	0.052	0.055	0.076	0.005
Technological risk	0.073	0.006	0.055	0.006	0.100
Labour strikes	0.096	0.187	0.080	0.041	0.068
Sudden shoot in demand	0.073	0.006	0.106	0.110	0.068
Forecasting errors	0.003	0.052	0.106	0.076	0.100

Table 19.9 Obtain weighted decision matrix

Risks	Categories of risks	Probability of incident of the risk	Impact of the risk	Increase in activity duration due to given risks	Increase in activity cost due to given risks
Farmer's inability of supply	0.026	0.018	0.001	0.027	0.013
Non-availability of the procurement centre (mechanism)	0.013	0.024	0.005	0.001	0.017
Poor quality of the supplied food grains	0.006	0.012	0.001	0.009	0.016
Communication failure	0.016	0.001	0.017	0.024	0.001
Natural disaster	0.010	0.002	0.011	0.002	0.039
Storage warehouse	0.010	0.019	0.006	0.015	0.007
Handling–loading/unloading at different locations	0.010	0.013	0.011	0.015	0.005
Packaging and preservation	0.019	0.001	0.011	0.001	0.020
In transit loss	0.058	0.058	0.058	0.058	0.058
Timely availability of the vehicles (Rails)	0.017	0.001	0.025	0.010	0.024
Malfunctioning in PDS	0.022	0.062	0.035	0.049	0.002
Theft or risks from outside elements	0.011	0.012	0.013	0.017	0.001
Technological risk	0.007	0.001	0.005	0.001	0.010
Labour strikes	0.022	0.043	0.018	0.009	0.016
Sudden shoot in demand	0.036	0.003	0.052	0.054	0.034
Forecasting errors	0.002	0.026	0.054	0.038	0.051

Table 19.10 Obtain the positive- and negative-ideal solution

A+	0.002	0.001	0.001	0.001	0.001
A−	0.058	0.062	0.058	0.058	0.058

Table 19.11 Separation measure

Risks	Positive-ideal solution	Negative-ideal solution
Farmer's inability of supply	0.042	0.096
Non- availability of the procurement centre (mechanism)	0.031	0.106
Poor quality of the supplied food grains	0.021	0.113
Communication failure	0.032	0.108
Natural disaster	0.040	0.108
Storage warehouse	0.027	0.107
Handling–loading/unloading at different locations	0.024	0.108
Packaging and preservation	0.028	0.111
In transit loss	0.128	0.004
Timely availability of the vehicles (Rails)	0.038	0.101
Malfunctioning in PDS	0.088	0.071
Theft or risks from outside elements	0.026	0.108
Technological risk	0.012	0.122
Labour strikes	0.053	0.087
Sudden shoot in demand	0.088	0.069
Forecasting errors	0.086	0.071

Table 19.12 Relative closeness to ideal solution—final risk index value

Risks	Risk index value
Farmer's inability of supply	0.697
Non-availability of the procurement centre (mechanism)	0.773
Poor quality of the supplied food grains	0.846
Communication failure	0.771
Natural disaster	0.730
Storage warehouse	0.801
Handling–loading/unloading at different locations	0.820
Packaging and preservation	0.800
In transit loss	0.030
Timely availability of the vehicles (Rails)	0.728
Malfunctioning in PDS	0.449
Theft or risks from outside elements	0.809
Technological risk	0.913
Labour strikes	0.622
Sudden shoot in demand	0.437
Forecasting errors	0.453

Selected Bibliography and Further Reading

1. Chen, C.-T. (2000). Extensions of the TOPSIS for group decision-making under fuzzy environment. *Fuzzy Sets and Systems, 114*(1), 1–9.
2. Hwang, C.L., & Yoon, K. (1981). *Multiple Attribute Decision Making: Methods and Applications*. Springer-Verlag, New York.
3. Rathore, R., Thakkar, J. J., & Jha, J. K. (2017). A quantitative risk assessment methodology and evaluation of food supply chain. *the International Journal of Logistics Management, 28*(4), 1272–1293. https://doi.org/10.1108/IJLM-08-2016-0198.
4. Saaty, T. L. (1977). A scaling method for priorities in hierarchical structures. *Journal of Mathematical Psychology, 15*(3), 234–281.
5. Saaty, T. L. (1990). How to make a decision: The analytic hierarchy process. *European Journal of Operational Research, 48*(1), 9–26.
6. Saaty, T.L. (2013). Analytic hierarchy process. In *Encyclopedia of operations research and management science* (pp. 52–64). Boston, MA: Springer.
7. Vaidya, O. S., & Kumar, S. (2006). Analytic hierarchy process: An overview of applications. *European Journal of Operational Research, 169*(1), 1–29.
8. Vargas, L. G. (1990). An overview of the analytic hierarchy process and its applications. *European Journal of Operational Research, 48*(1), 2–8.
9. Zahedi, F. (1986). The analytic hierarchy process—a survey of the method and its applications. *interfaces, 16*(4), 96–108.

Chapter 20
An Application of Integrated Approach of Fuzzy FMEA and VIKOR

20.1 Background

Foodgrains supply chain (FSC) is affected by select critical and significant issues such as insufficient storage capacity, wastage and corruption. It is essential to have better procurement, transportation and storage facility to enhance the efficiency of foodgrains supply chain in India. It is difficult to forecast the uncertainty and associated risks because the traditional FSC involves a large number of stages and nodes. In general, there are several risks which enhance the probability of the uncertainties associated with unavailability of the procurement centres, poor quality of the supplied foodgrains, communication failure, natural disaster, in-transit loss, malfunctioning, forecasting error, timely availability of the vehicles and technological risk. In India, the rise in population is increasing the demand of the foodgrains. To address these issues, it is important to identify and evaluate the risks involved in the FSC.

For examining the potential failures and risks, failure mode and effects analysis (FMEA) is one of the important and suitable continuous improvement tools. FMEA is an effective technique which can be used to prioritize the risk using the risk priority number (RPN). RPN is calculated by multiplying the probability of occurrence (P), severity (S) and detection (D) of the potential risks. Despite growing interest in examining the priorities of the issues, the traditional FMEA approach is not adequate because of several limitations. Typically, this includes:

1. We may obtain the same RPN value by using the several combinations of P, S and D. But the hidden risk inferences may be completely different.
2. All three factors considered to calculate the RPN have equal importance. A relative importance need to be considered.
3. One of the key limitations of the traditional FMEA is the involvement of crisp number while are obtained from the experts. This leads to biasness and ambiguity of the final outcomes.

J. J. Thakkar, *Multi-Criteria Decision Making*, Studies in Systems, Decision and Control 336, https://doi.org/10.1007/978-981-33-4745-8_20

To overcome this limitation, and address the problems more accurately and adequately, a wide range of mathematical methods have been used. MCDM techniques use decision matrix which can help to compare and rank alternatives. VIKOR method with its unique features can help to overcome select limitations involved in the use of FMEA. Therefore, an integrated application reported here has employed the use of Fuzzy VIKOR as a replacement of traditional RPN.

An integrated application of Fuzzy FMEA and VIKOR reported here attempts to identify and evaluate the risks of food supply chain for developing countries like India. The integrated approach accommodates the subjective inputs of industrial experts. FMEA helps to identify the probable failures or risk modes, effects and causes related to the respective elements. It evaluates both quantitative and qualitative measures. The VIKOR method is used to solve decision problems with conflicting and noncommensurable (different units) criteria, assuming that compromise is acceptable for conflict resolution, the decision maker wants a solution that is the closest to the ideal, and the alternatives are evaluated according to all established criteria. An integrated application of Fuzzy FMEA and VIKOR for risk assessment in Indian food supply chain is presented in the next section.

20.2 Data Collection

The data has been collected from five experts having significant experience and knowledge of food supply chain in India. The experts were familiar with the FSC and vulnerability of the system and were able to provide their feedback on FSC risks for three parameters namely Occurrence (P), Severity (S) and Detection (D). The detailed explanation of each risk is provided in the FMEA evaluation sheet and experts were familiarized with the various risks. The experts conveyed their ratings of P, S and D according to the definition given in Table 20.1.

Table. 20.1 Description of ratings of Probability of Occurrence (P), Severity (S) and Detection (D)

Rating	Linguistic variable	Membership value for risk factors	Membership value for weigh of the criteria
1	Low	(1,2,2,3)	(0,0.1,0.3)
3	Medium	(3,4,5,6)	(0.1,0.3,0.5)
5	High	(6,7,7,8)	(0.3,0.5,0.7)
7	Very high	(7,8,8,9)	(0.7,0.9,1)
9	Extreme	(8,9,10,10)	(0.9,1,1)

Probability of occurrence (P): How often a given risk event occurs
Severity of occurrence (S): What is degree of harshness, if the particular risk is occurred
Detection rating (D): Detectability of risk (whether we are able to detect the given risk)

20.3 Proposed Risk Quantification and Estimation Methodology

This section involves the procedure to be followed to evaluate the risks in FSC.

The issues of risk estimation with FMEA approach could be acknowledged as group MCDM problem. Let us consider group MCDM has Z decision makers $DM_z(z = 1, 2, \ldots, Z)$, k alternatives $A_i(i = 1, 2, \ldots, k)$ and m is decision criteria (risk factor) $D_j(j = 1, 2, \ldots, m)$. By using the VIKOR approach, alternatives are estimated with respect to the criteria (m). The following steps are involved in VIKOR method for calculating the risk factor.

1. Collect the expert opinion to obtain the fuzzy weight of the risk factor, and the fuzzy ratings of the alternatives. Based on the input, construct the fuzzy decision matrix.
   Let us consider the fuzzy rating of the zth decision maker is $\tilde{a}_{ij}^{z} = (a_{ij1}^z, a_{ij2}^z, a_{ij3}^z, a_{ij4}^z)$ and weight of zth decision maker is $\tilde{w}_j^{z} = w_{j1}^z, w_{j2}^z, w_{j3}^z, w_{j4}^z$.
   The aggregated fuzzy ratings ($\tilde{a}_{ij}$) can be found as:

$$\tilde{a}_{ij} = (a_{ij1}, a_{ij2}, a_{ij3}, a_{ij4})$$

   Where

$$a_{ij1} = \min_z a_{ij1}^z, \quad a_{ij2} = \frac{1}{Z}\sum_{z=1}^{Z} a_{ij2}^z, \quad a_{ij3} = \frac{1}{Z}\sum_{z=1}^{Z} a_{ij3}^z, \quad a_{ij4} = \max_z a_{ij4}^z.$$

   The aggregated fuzzy weights ($\tilde{w}_j$) can be found as.

$$\tilde{w}_j = (w_{j1}, w_{j2}, w_{j3}, w_{j4})$$

   Where

$$w_{j1} = \min_z w_{j1}^z, \quad w_{j2} = \frac{1}{Z}\sum_{z=1}^{Z} w_{j2}^z, \quad w_{j3} = \frac{1}{Z}\sum_{z=1}^{Z} w_{j3}^z, \quad w_{j4} = \max_z w_{j4}^z.$$

   The value of $\tilde{a}_{ij} = (a_{ij1}, a_{ij2}, a_{ij3}, a_{ij4})$ and $\tilde{w}_j = (w_{j1}, w_{j2}, w_{j3}, w_{j4})$ is considered as the linguistic variables which defined in the trapezoidal fuzzy numbers.

2. Defuzzify the values of $\tilde{a}_{ij}$ and $\tilde{w}_j$ by using the following formula. In this step, conversion of a fuzzy number into crisp value is done

$$\text{deffuz}(a_{ij}) = \frac{a_{ij1}a_{ij2} + a_{ij3}a_{ij4} + \frac{1}{3}(a_{ij4} - a_{ij3})^2 - \frac{1}{3}(a_{ij2} - a_{ij1})^2}{-a_{ij1} - a_{ij2} + a_{ij3} + a_{ij4}}$$

$$\text{deffuz}(w_j) = \frac{w_{j1}w_{j2} + w_{j3}w_{j4} + \frac{1}{3}(w_{j4} - w_{j3})^2 - \frac{1}{3}(w_{j2} - w_{j1})^2}{-w_{j1} - w_{j2} + w_{j3} + w_{j4}}$$

3. Determine the best f_j^* and worst value f_j^- values of all risk factor, $j = 1, 2, \ldots, m$. For finding out the best and worst value:

$$f_j^* = \left\{ \begin{array}{ll} \max\limits_j a_{ij}, & \text{for benefit criteria} \\ \min\limits_j a_{ij} & \text{for cost criteria} \end{array} \right\}, \quad i = 1, 2, \ldots, k$$

$$f_j^- = \left\{ \begin{array}{ll} \min\limits_j a_{ij}, & \text{for benefit criteria} \\ \max\limits_j a_{ij} & \text{for cost criteria} \end{array} \right\}, \quad i = 1, 2, \ldots, k$$

4. Use the following equation to determine the values S_i and R_i, where $i = 1, 2, \ldots, k$

$$d_{ij} = \frac{d\left(f_j^*, a_{ij}\right)}{d(f_j^*, f_j^-)}$$

$$S_i = \sum_{j=1}^{m} w_j d_{ij}$$

$$R_i = \max\left(w_j, d_{ij}\right)$$

5. Calculate the value of Q_i, where $i = 1, 2, \ldots, k$. Use the following equation for finding the Q_i

$$Q_i = v\left[\frac{S_i - S^*}{S^- - S^*}\right] + (1 - v)\left[\frac{R_i - R^*}{R^- - R^*}\right]$$

where $S^* = \min_i(S_i)$, $S^- = \max_i(S_i)$, $R^* = \min_i(R_i)$, and $R^- = \max_i(R_i)$, and the value of $v = 0.5$. Here v is the weight for the strategy of maximum group utility and $1 - v$ is termed as weight of individual regret.
6. Rank the risk factor based on the value of Q_i, S_i and R_i
7. Recommend the compromise solution: The alternative A^1, which is considered as the best solution, is termed as best solution, if the following conditions are tested.

$\langle C_1 \rangle$: Acceptable advangate: The alternative can be considered as acceptable advantage if, $Q(A^2) - Q(A^1) \geq \frac{1}{k-1}$, where $Q(A^1)$ is presented as the best solution from the study and $Q(A^2)$ is alternative with second position in the ranking list.

$\langle C_2 \rangle$: Acceptable stability: The alternative A^1 should be the best solution while doing the raking by S_i and/or R_i. For $v = 0.5$, this defined as "voting by consensus" and for $v > 0.5$ voting by majority and for $v < 0.5$ is explained as "with veto".

If the one of the conditions mentioned above is not satisfied, then set of compromised solution is recommended which is explaining below.

1. Alternative A^1 and A^2 as compromised solution, if only the condition $\langle C_2 \rangle$ is not fulfilled.
2. Alternatives $A^1, A^2, \ldots, A^K$ as compromised solution, if only the condition $\langle C_1 \rangle$ is not fulfilled. A^K is determined by the relation $Q(A^K) - Q(A^1) < \frac{1}{k-1}$, for maximum K (the positions of these alternatives are "in closeness").

20.4 Illustrative Application of an Integrated Fuzzy FMEA and VIKOR for Food Supply Chain Risk Assessment

This section evaluates the risks for an Indian food supply chain. The inputs are collected based on the personal interview of managers of FCI. The managers are employed at a different locations and different positions such as assistant general manager, depot manager general manager and graduate trainee officer. The following steps are involved in the risk estimation for FSC.

Step 1 Based on the available literature and expert comments, risks are categorized into different categories.

Step 2 Five experts are interviewed to estimate the most severe risks factor. The factors P, S and D are explained based on the historical evidences. Inputs are collected in the form of a questionnaire, which is provided to all the experts.

Step 3 The membership value for the risk factors and weights of the criteria are shown in Table 20.1. The third column of Table 20.2 represents the weight assigned to the five experts for different categories of P, S and D. The input of the experts is shown in the subsequent column of Table 20.2.

Step 4 The linguistic values of Table 20.2 are converted into the trapezoidal fuzzy number. The aggregated fuzzy values of risk mode and weight of the criteria are obtained which is used to find out the fuzzy weigh of each risk. The aggregated fuzzy ratings are shown in Table 20.3.

Step 5 Crisps values for fuzzy values of risk mode and weight of the criteria are calculated, and it is presented in Table 20.4.

Step 6 Best value and worst value of all risk factors are determined from Table 20.6 which is given below.

Table. 20.2 Expert inputs for risk parameter

Expert	Parameter	Weights	Risk mode													
			1	2	3	4	5	6	7	8	9	10	11	12	13	14
Expert 1	P	E	M	H	L	M	M	M	M	L	M	M	M	L	L	M
	S	VH	H	VH	L	H	H	H	M	M	H	H	H	M	M	H
	D	VH	H	H	H	H	H	H	H	M	H	H	M	H	M	M
Expert 2	P	M	M	VH	M	M	M	VH	M	L	M	M	M	VH	L	H
	S	H	H	VH	L	VH	L	H	L	H	VH	H	E	H	M	VH
	D	M	H	H	H	H	H	M	H	M	H	VH	M	H	M	M
Expert 3	P	VH	M	M	M	L	VH	M	M	VH	H	M	H	M	M	L
	S	VH	L	L	H	M	VH	L	VH	VH	H	M	H	M	M	M
	D	H	H	H	VH	VH	H	H	H	H	L	VH	M	VH	M	M
Expert 4	P	M	M	H	M	L	L	M	L	H	L	H	M	L	M	L
	S	H	H	L	L	VH	M	H	L	L	M	L	M	H	L	L
	D	M	H	H	VH	VH	VH	M	H	M	H	M	H	H	H	H
Expert 5	P	H	VH	VH	E	H	L	L	H	L	M	M	L	M	M	M
	S	VH	L	M	M	L	L	M	H	M	VH	E	VH	M	M	VH
	D	H	H	L	M	H	E	E	VH	VH	H	H	M	H	M	M

Table. 20.3 Aggregated fuzzy value of risks modes and weight of the criteria

Risk Mode	P	S	D
R1	(3,6,5.6,9)	(1,5,5,8)	(6,7,7,8)
R2	(3,6.8,7,9)	(1,4,4.4,9)	(1,6,6,8)
R3	(1,4.6,5.4,10)	(1,4.8,5,9)	(3,6.8,7,9)
R4	(1,3.8,4.2,8)	(1,5.8,6,9)	(6,7.4,7.4,9)
R5	(1,4,4.4,9)	(1,4.6,4.8,9)	(6,7.6,7.8,10)
R6	(1,4.4,5,9)	(1,5.4,5.6,8)	(3,6.2,6.8,10)
R7	(1,4.2,4.8,8)	(1,4.6,4.8,9)	(6,7.2,7.2,9)
R8	(1,4.2,4.8,9)	(1,5,5.4,9)	(3,5.4,6,9)
R9	(1,4.2,4.8,8)	(3,6.8,7,9)	(1,6,6,8)
R10	(3,4.6,5.4,8)	(1,5.8,6.2,10)	(3,6.8,7,9)
R11	(1,4.2,4.8,8)	(3,7,7.4,10)	(3,4.6,5.4,8)
R12	(1,4,4.4,9)	(3,5.2,5.8,8)	(6,7.2,7.2,9)
R13	(1,3.2,3.8,6)	(1,3.6,4.4,6)	(3,4.6,5.4,8)
R14	(1,3.8,4.2,8)	(1,5.8,6,9)	(3,4.6,5.4,8)
Weight	(0.1,0.6,1)	(0.3,0.74,1)	(0.1,0.5,1)

Table. 20.4 Crisps values for fuzzy values of risk mode and weight of the criteria

Risk mode	P	S	D
R1	5.974	2.238	5.467
R2	5.310	2.792	1.857
R3	2.320	2.356	5.310
R4	2.459	1.966	7.467
R5	2.554	2.434	7.930
R6	2.500	2.009	4.658
R7	2.237	2.434	6.390
R8	2.380	2.189	4.891
R9	2.237	5.310	1.857
R10	4.931	2.035	5.310
R11	2.237	4.773	4.931
R12	2.554	5.250	6.356
R13	2.321	2.172	4.931
R14	2.459	1.966	4.931
Weight (Wj)	0.53	0.61	0.55

$$f_O^* = 2.237;\ f_S^* = 1.966;\ f_D^* = 1.857$$
$$f_O^- = 5.974;\ f_S^- = 5.310;\ f_D^- = 7.930$$

Step 7 For all risk modes, the value of S, R and Q is estimated as Table 20.5.

Table. 20.5 The values of S, R and Q for all failure modes

	Risk modes													
	R1	**R2**	R3	R4	R5	R6	R7	R8	R9	R10	R11	R12	R13	R14
S	0.907	0.586	0.396	0.540	0.680	0.299	0.496	0.336	0.610	0.707	0.790	0.951	0.328	0.310
R	0.530	0.436	0.313	0.508	0.550	0.254	0.410	0.275	0.610	0.382	0.512	0.599	0.278	0.278
Q	0.786	0.442	0.144	0.511	0.663	0.130	0.347	0.051	0.700	0.448	0.684	0.979	0.051	0.039

Table. 20.6 The ranking of the failure modes by S, R and Q in decreasing order

	Risk modes													
	R1	R2	R3	R4	R5	R6	R7	R8	R9	R10	R11	R12	R13	R14
S	2	7	10	8	5	14	9	11	6	4	3	1	12	13
R	4	7	10	6	3	14	8	13	1	9	5	2	11	11
Q	2	8	10	6	5	11	9	12	3	7	4	1	12	14

Step 8 Ranking of all risk modes based on S, R and Q is shown in Table 20.6.

This chapter has demonstrated an integrated application of FMEA-Fuzzy VIKOR for estimating the risks in food supply chain in India. FMEA is one of the well-established methodology for risk estimation. Three different criteria P, S and D are used in the traditional FMEA approach. For a complex system, such as FSC, FMEA analysis gets affected because of factors such as uncertainty, biasness and ambiguity involved in the subjective opinions received from the experts. An integrated approach of FMEA and Fuzzy VIKOR helps to overcome such difficulties.

This chapter has demonstrated an integrated application of Fuzzy FMEA and VIKOR. The next chapter will help the readers to appreciate an integrated use of DEMATEL and ANP, called as DANP.

Selected Bibliography

1. Afshari, H., Issa, M. H., & Radwan, A. (2016). Using failure mode and effects analysis to evaluate barriers to the greening of existing buildings using the leadership in energy and environmental design rating system. *Journal of Cleaner Production, 127,* 195–203.
2. Chin, K. S., Wang, Y. M., Poon, G. K. K., & Yang, J. B. (2009). Failure mode and effects analysis by data envelopment analysis. *Decision Support Systems* (Vol. 48, no 1, pp. 246–256). Elsevier B.V.
3. Faezy Razi, F., Danaei, A., Ehsani, A., & Dolati, C. (2013). An FMEA analysis using grey theory and grey rough sets. *Decision Science Letters, 2*(4), 299–304.
4. Garcia, P. A. A., Schirru, R., Frutuoso e Melo, P. F. (2005) A fuzzy data envelopment analysis approach for FMEA. *Progress in Nuclear Energy, 46*(3–4), 359–373.
5. Kutlu, A. C., & Ekmekçioğlu, M. (2012). Fuzzy failure modes and effects analysis by using fuzzy TOPSIS-based fuzzy AHP. *Expert Systems with Applications, 39*(1), 61–67.
6. Liu, H.-C., Liu, L., Liu, N., & Mao, L.-X. (2012). Risk evaluation in failure mode and effects analysis with extended VIKOR method under fuzzy environment. *Expert Systems with Applications, 39*(17), 12926–12934.
7. Abdelgawad, M., Fayek, A. R., & ASCE, A. M. (2010). Risk management in the construction industry using combined fuzzy FMEA and fuzzy AHP. *Journal of Construction Engineering and Management, 136*(9), 1028–1036.
8. Mohsen, O., & Fereshteh, N. (2017). An extended VIKOR method based on entropy measure for the failure modes risk assessment—a case study of the geothermal power plant (GPP). *Safety Science, 92,* 160–172.
9. Safari, H., Faraji, Z., & Majidian, S. (2016). Identifying and evaluating enterprise architecture risks using FMEA and fuzzy VIKOR. *Journal of Intelligent Manufacturing, 27*(2), 475–486.
10. Wang, Y. M., Chin, K. S., Poon, G. K. K., & Yang, J. B. (2009). Risk evaluation in failure mode and effects analysis using fuzzy weighted geometric mean. *Expert Systems with Applications, 36*(2), 1195–1207.

Chapter 21
An Integrated DEMATEL-ANP (DANP) MCDM Approach for Quantifying the Supply Chain Risk: A Case of Indian Petroleum Supply Chain

This chapter demonstrates an application of an integrated approach of DEMATEL-ANP (DANP) to quantify the overall supply chain risk and hence assessing the priority values of individual risk and sub-risk drivers. A case of Indian Petroleum SC is considered as a viable study in order to demonstrate the efficacy of the proposed method.

21.1 Proposed Integrated Framework of DEMATEL and ANP (DANP)

Quantifying risk index as a measure of overall risk based on the various drivers of risk and also sub-risks is a multi-criteria decision making (MCDM) problem. Risk and sub-risk drivers are more often subjective in nature. Measuring these factors usually requires expertise and knowledge in the field in which risk assessment needs to be done. The subjective opinions of these experts need to be converted in to a quantifiable objective risk index. This chapter demonstrates an integrated application of DEMATEL-ANP to quantify and establish the interrelationships between various criteria. The various supply chain risk and sub risk drivers exhibit complex relationships among them which plays an important role in quantifying the risk index. Consider the example of a petroleum supply chain, there is a higher degree of interaction between the *Alliance risk* and the *Social risk*. Alliance risk refers to risk involved in a project in case of joint venture between a domestic and a foreign company, while social risk arises out of differences in culture between both the foreign and domestic company.

Once the interrelationships among the risk drivers are identified using DEMATEL, ANP is applied to assess the importance of each individual driver of risk and sub-risk in relation to the complete SC risk. Subsequently, the global priority values for each of the criterion and the overall risk index are calculated using the ANP method. An integrated methodology of DEMATEL-ANP is presented in Fig. 21.1.

J. J. Thakkar, *Multi-Criteria Decision Making*, Studies in Systems, Decision and Control 336, https://doi.org/10.1007/978-981-33-4745-8_21

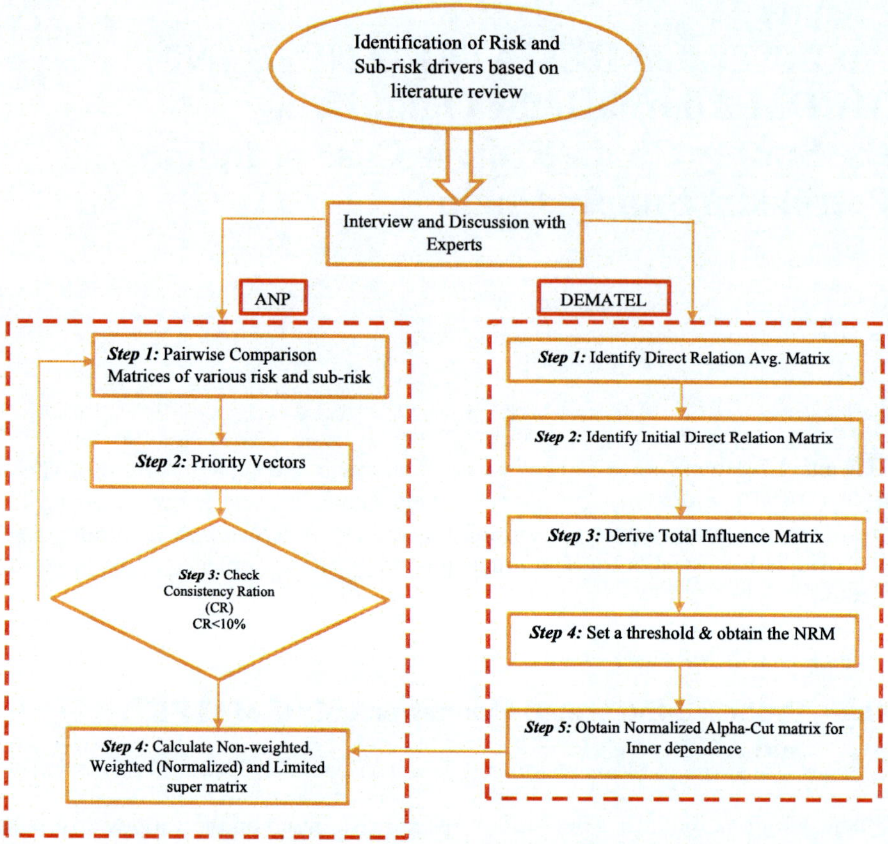

Fig. 21.1 Integrated ANP-DEMATEL risk quantification framework

21.2 Application of an Integrated DEMATEL-ANP (DANP) Approach for Indian Petroleum Supply Chain

India ranks third in overall consumption of crude oil and other petroleum products in the world, with almost of 4.5% of the global demand arising in India. The various risk factors and also the sub-risk factors were identified through an extensive review of the existing literature in the domain of petroleum supply chain. The relevance of the identified factors is checked from the experts in the domain of petroleum supply chain. Recognizing the major risk and sub-risk drivers (as shown in Table 21.1) of the Indian petroleum supply chain will help in identifying the barriers and can facillitate the government in adopting necessary policy decisions to avoid or mitigate these risk events.

The *risk drivers* are originated from the overall *SC risk index* and their mutual relationships are indicated by *inter-relationship*. Further, the *sub-risk drivers* are

Table 21.1 Hierarchical classification of risk and sub-risk drivers in Indian Petroleum supply chain

Goal	Risk driver	Sub-risk driver	Explanation/example in Indian context
Overall risk index in Indian Petroleum supply chain	1. Environment	1.1 Natural disaster	*Gujarat* Earthquake; *The Indian Oc*ean Tsunami; Cyclone *Phailin, Hud Hud*
		1.2 Man-made disaster	*Bhopal* gas leak, *Mumbai* terror attacks
		1.3 Legal/political	The Legal (or Political) policies, which affect firm's performance like minimum wage, pricing guidelines
		1.4 Social	Labour Skills and availability, Labour strikes
		1.5 Economic crisis	Economic Downturn and financial crisis experienced in 1991
	2. Upstream	2.1 Inventory	Shortage of raw materials (resources) like industrial gas and coal bed
		2.2 Crude supply	Disruption in crude supply
		2.3 Delay	Inbound Delay of crude oil due to maritime logistics
	3. Process	3.1 Manufacturing competitiveness	Operational Innovation, manufacturing technology, R&D
		3.2 IT system failure	Failure of Information (communication) system, cyber-attacks
		3.3 Quality	(Inferior) Quality of the petroleum products
		3.4 Machine breakdown	Machine/Tool/Power failures
		3.5 Capacity	Under/Over utilization of plant capacity
	4. Downstream	4.1 Demand	Demand fluctuation of the petroleum products
		4.2 Transportation/logistics	Accidents, adverse weather condition, vandalism during transportation
		4.3 Forecast	Inaccurate Forecast, poor prediction of customer demand

(continued)

Table 21.1 (continued)

Goal	Risk driver	Sub-risk driver	Explanation/example in Indian context
		4.4 Customer taste change	The slow and gradual shift of the customer preference towards the alternative resources like solar and biodiesel

originated from their corresponding *risk drivers* and they *influence* the overall risk index of the SC. The relationship between the sub-risk drivers for a particular risk driver are indicated by the *inner dependency.*

Step-by-Step Application of an integrated DEMATEL-ANP (DANP) MCDM approach for quantifying the supply chain risk: A case of Indian petroleum supply chain.

Phase 1: Applying ANP (Ignoring the interdependence between factors).

Initially, in this phase the prioritization of the various risk and sub-risk drivers was performed ignoring the interrelationship among these factors. This involved the application of ANP to calculate the priority values.

Step 1: Model construction and Problem structuring.

Real-life decision-making problems are not easy to construct hierarchically due to the presence of interdependence among elements at different hierarchical levels. Some elements from a higher level may interact with an element from a lower level. ANP methodology is an extension of AHP which takes into account the interdependence among criteria. Here, a hierarchical model is constructed which takes into account the dependency between the risk and the corresponding sub-risk drivers. In such a model, the overall risk index depends only on the sub-risk drivers.

Step 2: Pairwise comparison matrices and Priority Vectors.

Once the hierarchical model is constructed, the next step involves pairwise comparison of elements with each other for a control criterion. The sub-elements of each element are also compared among themselves to compute their contribution to the overall score. This process is accomplished using the questionnaires administered to the experts in the decision-making panel. Their responses for each comparison are recorded using the Saaty's nine-point scale. If the score of ith element in comparison with jth element is given by a_{ij}, then the corresponding score for jth element in comparison with ith element is $1/a_{ij}$.

Step 1: Construction of Pairwise comparison matrix

The pairwise comparison matrix (PCM) for comparing the risk drivers is shown in Table 21.2. The PCM is used as an input to the AHP for calculating the *priority values* in the subsequent steps.

Table 21.2 Pairwise comparison matrix for the risk drivers

Overall risk	Environment	Upstream	Process	Downstream
Environment	1.00	4.00	2.00	3.00
Upstream	0.25	1.00	0.33	0.50
Process	0.50	3.00	1.00	2.00
Downstream	0.33	2.00	0.50	1.00

The next step involves calculation of the eigenvalues (or the Priority Vectors) along with check for consistency.

Step 2. Consistency Check

The consistency ratio of the PCM is checked in order to ensure the consistency and mutual reliability of the responses among the decision criteria.

Formula for calculating, CR = CI/RI

The detailed step of calculating CI can be found from Saaty (2000).

The CR calculated for the above table is 0.0097 (<0.1), which is satisfactory.

Step 3. Normalization of the PCM

The PCM is further normalized to make it column stochastic, so that each column adds up to one. The normalization is performed by dividing each element of the PCM with their corresponding column summation. (Normalized matrix is shown in Table 21.3).

Step 4. Priority Values

The priority values of the decision criteria are obtained by averaging the normalized matrix across the rows. The ranks are calculated in Table 21.4 according to "higher the better" principle.

Table 21.3 Normalized matrix for the risk drivers

Overall risk	Environment	Upstream	Process	Downstream
Environment	0.481	0.40	0.522	0.462
Upstream	0.120	0.10	0.086	0.076
Process	0.240	0.30	0.261	0.308
Downstream	0.159	0.20	0.130	0.154

Table 21.4 Priority values and ranks

Risk drivers	PV	AHP rank
Environment	0.466	1
Upstream	0.096	4
Process	0.277	2
Downstream	0.161	3

Table 21.4 represents both the rank and priority value of each alternative. We observe that the risk factor "Environmental risk" is the highest ranked factor with contribution of almost 47% to the overall risk. Similarly, the second and third rank in the priority is "*Process risk*" and "*Downstream risk*" with sores of 28% and 16%, respectively.

Importance of Sub-Risk Drivers

Similarly, the priority values for each of the sub risk driver are computed from the PCM of the corresponding risk driver at the higher level in the hierarchy. Tables 21.5, 21.6, 21.7 and 21.8 summarize the results of the PCM for each set of sub-risk drivers. The detailed PCM for the sub-risk drivers is provided in the Appendices 21.1, 21.2, 21.3, 21.4 and 21.5.

From Table 21.5, we can infer that *Natural disasters* contribute the most towards *environmental risk* with 36%, followed by the factors such as *Legal/Political* (26%) and *Economic crisis* (21%).

Similarly, the priorities of sub-risk drivers related to "Upstream" are provided in Table 21.6. In case of *upstream* driver, supply of crude oil seems to be the most important sub-risk with 56% contribution. This is understandable because availability of crude oil is very important for its processing downstream (refining and distribution). Delay in availability is another major sub-risk that affects *Upstream* risk factor. (Table 21.7)

Table 21.7 provides the priorities of sub-risk drivers for the "process related risks". *Quality* is the sub-risk driver that receives the highest rank with 44% priority value. The major reasons are the hazardous nature of petroleum products and its chemical properties that makes it necessary to focus on quality to avoid accidents and adulteration during transport. The next important factor is *IT system failure* which has a priority value of 27%.

From Table 21.8, we can observe that in case of the *downstream* risk driver, *Transportation/Logistics* is the sub-risk driver that account for the highest priority

Table 21.5 PV of the Sub-risk drivers of environment

Risk driver	Sub-risk drivers	PV
Environment	Natural disaster	0.36
	Manmade disaster	0.10
	Legal/political	0.26
	Social	0.07
	Economic crisis	0.21

Table 21.6 PV of the Sub-risk drivers of upstream

Risk driver	Sub-risk drivers	PV
Upstream	Inventory	0.12
	Crude supply	0.56
	Delay	0.32

Table 21.7 PV of the Sub-risk drivers of process

Risk driver	Sub-risk drivers	PV
Process	Manufacturing competitiveness	0.04
	IT system failure	0.27
	Quality	0.44
	M/c Breakdown	0.17
	Capacity	0.09

Table 21.8 PV of the Sub-risk drivers of downstream

Risk driver	Sub-risk drivers	PV
Downstream	Demand	0.15
	Transportation/logistics	0.47
	Forecast	0.29
	Customer taste change	0.09

value with a contribution of about 47%. Distribution and logistics are major cost factors accounting for significant portion of the overall costs. The next important sub-risk under the *Downstream* risk is *Forecast,* as errors in forecasting can significanlty impact the overall production planning. Variations in demand which usually occur due to availability and incentivisation of alternative fuels are the next important sub-risk with 15% priority value.

Phase 2: DEMATEL

In the second phase, DEMATEL method has been employed to establish the inter-relationship between various risk and sub-risk drivers. The steps of DEMATEL are explained in brief, followed by the calculation in the following section.

Step1: Construction of the Direct-Relation Matrix (DRM)

The DEMATEL procedure starts with the construction of the DRM (denoted by *A*) (shown in Table 21.9) below, it basically represents the overall influence of one criterion over another.

The DRM forms the input for the subsequent steps of the DEMATEL procedure. The standard 0–4 (0 represents no influence and 4 represents high influence) DEMATEL scale is followed to collect the data and construct the above matrix.

Table 21.9 Direct-relation matrix

Goal	Environment	Upstream	Process	Downstream
Environment	0.00	3.00	2.00	2.00
Upstream	0.00	0.00	1.00	3.00
Process	0.00	2.00	0.00	4.00
Downstream	1.00	2.00	1.00	0.00

Step 2: Development of a Normalized Direct-Relation Matrix

The next step involves computation of the *normalized direct relation matrix* (X) from the *direct relation matrix* (A) using the following expression.

$$X = k \cdot A$$

where

$$k = \min\left(\frac{1}{\max_{1 \le i \le n}\left(\sum_{j=1}^{n} a_{ij}\right)}, \frac{1}{\max_{1 \le j \le n}\left(\sum_{i=1}^{n} a_{ij}\right)}\right)$$

The *normalized direct-relation matrix* (X) is presented in Table 21.10.

Step 3: Development of Total Influence Matrix

The total influence matrix (T) is calculated using the following expression where I denotes the identity matrix.

$$T = X(I - X)^{-1}$$

Total Relation Matrix (T) is shown in Table 21.11:

Step 4: Computation of Row and Column Sum and Modelling of Casual map

The summation of rows and columns of the total-relation matrix (T) is calculated using the following expressions and represented as vectors D and R, respectively.

Table 21.10 Normalized direct-relation matrix

Goal	Environment	Upstream	Process	Downstream
Environment	0.00	0.333	0.222	0.222
Upstream	0.00	0.00	0.111	0.333
Process	0.00	0.222	0.00	0.444
Downstream	0.11	0.222	0.111	0.000

Table 21.11 Total relation matrix

Goal	Environment	Upstream	Process	Downstream
Environment	0.0653	0.5669	0.3651	0.5879
Upstream	0.0544	0.1706	0.1966	0.4897
Process	0.0737	0.4247	0.1373	0.6634
Downstream	0.1387	0.3703	0.2106	0.2479

$$\boldsymbol{D_i} = \left[\sum_{j=1}^{n} \boldsymbol{t_{ij}}\right]_{n\times1} = [\boldsymbol{t_i}]_{n\times1}, \boldsymbol{i} = 1, 2, 3\ldots, \boldsymbol{n}$$

$$\boldsymbol{R_i} = \left[\sum_{j=1}^{n} \boldsymbol{t_{ij}}\right]_{1\times n} = [\boldsymbol{t_i}]_{n\times1}, \boldsymbol{j} = 1, 2, 3\ldots, \boldsymbol{n}$$

D_i (summation of ith row in matrix T) indicates the total (both direct and indirect) effects caused by ith criterion over the rest criteria. Similarly, R_j (summation of jth column in matrix T) shows the total (both direct and indirect) effects exerted by jth criterion over the rest criteria (Table 21.12).

A causal graph is plotted on a two dimensional plane and all the criteria are positioned on it. The horizontal axis vector $(D_k + R_k)$ called "*prominence*" is calculated by the summation of D and R while $k = i = j = 1$ which indicates how much importance the criterion carries. On the other hand, the vertical axis $(D_k + R_k)$ called "*relation*" is computed by the subtraction of R from D, which classifies the criteria into a cause group and an effect group. Usually, the criterion belongs to the *cause* group, when $(D_k + R_k)$ is *positive*. Otherwise, the criterion belongs to the *effect* group, if $(D_k + R_k)$ is *negative* (Table 21.13).

Step 5: Selection of threshold level (α) (Fig. 21.2)

A threshold value is usually chosen to separate out the minor effects and only capture the more important relationships between the different decision criteria. This research work follows Rajesh and Ravi (6) for calculating the threshold value (α) by averaging all the elements in matrix T, as shown in equation [where n is the total number criteria].

Table 21.12 Calculation of Row sum (D) and Column sum (R)

Goal	Environment	Upstream	Process	Downstream	*D*
Environment	0.0653	0.5669	0.3651	0.5879	**1.5852**
Upstream	0.0544	0.1706	0.1966	0.4897	**0.9113**
Process	0.0737	0.4247	0.1373	0.6634	**1.2991**
Downstream	0.1387	0.3703	0.2106	0.2479	**0.9675**
R	**0.3321**	**1.5325**	**0.9096**	**1.9889**	

Table 21.13 Calculation of D + R and D−R scores

Risk drivers	*D*	*R*	*D + R*	*D−R*
Environment	1.5852	0.3321	1.9173	1.2531
Upstream	0.9113	1.5325	2.4438	-0.6212
Process	1.2991	0.9096	2.2087	0.3895
Downstream	0.9675	1.9889	2.9564	-1.0214

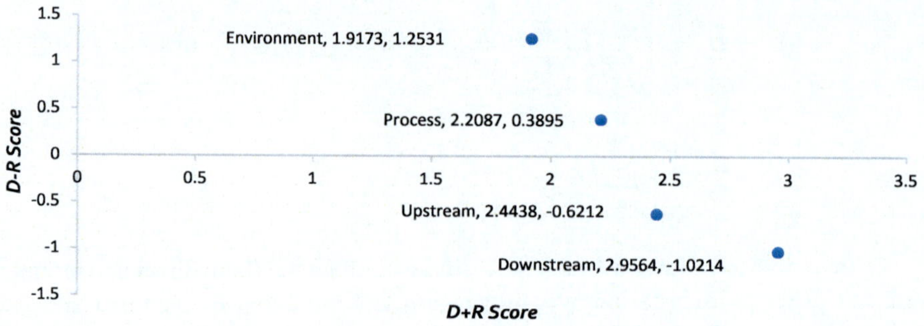

Fig. 21.2 Casual plot of Risk drivers

$$\alpha = \frac{\sum_{i=1}^{n} \sum_{j=1}^{n} [t_{ij}]}{n^2}$$

The value of the elements in matrix T which are less than the threshold value (α) basically signifies the *minor effect* (*interrelationship*) and hence are eliminated in the further step.

The threshold value ($\boldsymbol{\alpha}$) calculated for the matrix, T (**Table**Table 11) = 0.2*97.*

The effects greater than or equal to the calculated threshold value (α), i.e. 0.29**7** are shown in the new *diagraph or Network Relation Map* (NRM) in the next step.

Step 6: Obtain α-cut total influence matrix (T_α) and plot NRM

The values that are below the calculated threshold value $\boldsymbol{\alpha}$ in the matrix T are set to zero, as these values that the interrelationship is not significant. The new matrix thus generated after the values lower than $\boldsymbol{\alpha}$ is filtered out is called the called the $\boldsymbol{\alpha}$-cut total influence matrix, $\boldsymbol{T_\alpha}$.

where if $t_{ij} < \alpha$, then $t_{ij}^{\alpha} = 0$ else $t_{ij}^{\alpha} = t_{ij}$.

where = $[t_{ij}]$, called the total influence matrix.

$\boldsymbol{T_\alpha} = \left[t_{ij}^{\alpha}\right]$, called the α-cut total influence matrix (shown in Table 21.14).

The values of the matrix show the degree of influence that each risk drivers, (in rows, *i*) has on each risk drivers (in column, *j*). for instance, a score of 0.5*669* represents that *Environmental* risk has a very strong influence (or impact) on *Upstream* risk, which is slightly more than that of the downstream risk (**0.5*879***). It can be

Table 21.14: α-cut total influence matrix

Goal	Environment	Upstream	Process	Downstream
Environment	0	0.5669	0.3651	0.5879
Upstream	0	0	0	0.4897
Process	0	0.4247	0	0.6634
Downstream	0	0.3703	0	0

further observed that, all the elements in the first column (corresponds to the *Environmental* risk) are zeros, which means that the *Environmental* risk is not influenced (or caused/triggered) by any of the risk driver.

The normalized transpose of the α-cut total influence matrix $\left(\boldsymbol{T}_s'\right)$ is used in the ANP supermatrix. It basically captures the interrelationship between the decision criteria. (shown below) (Table 21.15).

The values obtained from the alpha cut total influence matrix (Table 21.14) are used to draw a network relationship map (NRM). The arrows represent the *degree of influence* that one criterion has on the other. For instance, no arrow is pointing towards the Environment driver, which means no other criteria have a significant influence on the *environment*. The degree of influence less than alpha (0.2**97**) is discarded in order to avoid the complexities in capturing the interaction effect. The NRMs for the other risk sub-drivers within their respective risk drivers are given in the appendix (Appendix 21.9, 21.10, 21.11 and 21.12) (Fig. 21.3).

Table 21.15 Normalized and transposed α-cut total influence matrix

Goal	Environment	Upstream	Process	Downstream
Environment	0	0	0	0
Upstream	0.373	0	0.3903	1
Process	0.2402	0	0	0
Downstream	0.3868	1	0.6097	0

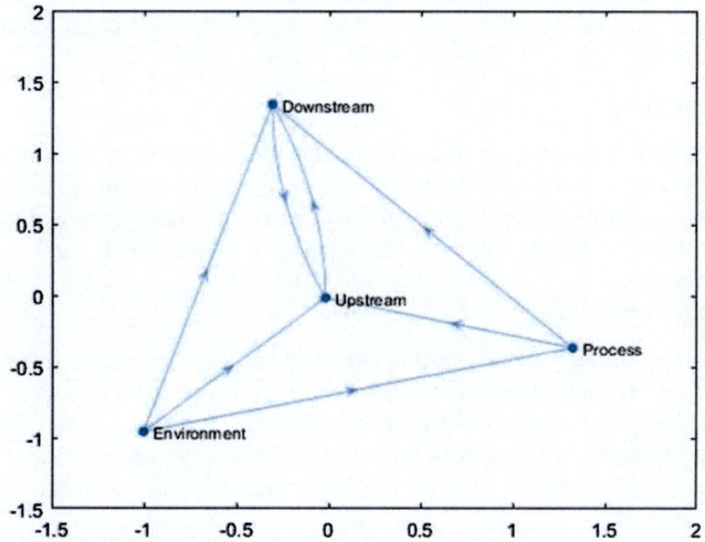

Fig. 21.3 Network relation map for the risk drivers

Table 21.16 Global weights and rankings of risk drivers

Risk drivers	Global PV	Rank
Environment	0.084368247	**3**
Upstream	0.102449108	**2**
Process	0.060322496	**4**
Downstream	0.11511348	**1**

Phase 3: Integrated DEMATEL-ANP

Limit Matrix and final scores of risk and sub-risk drivers

A supermatrix is constructed by utilizing all the Priority Vectors (obtained from pairwise comparison) and Alpha cut total relationship matrices (obtained from DEMATEL). The supermatrix is normalized in order to make it column stochastic and finally a limit matrix is calculated by raising the normalized supermatrix to a power of 32., basically represents the limiting (converging) values of each indicator after taking all the interrelationship and degree of importance into consideration in a long term basis.

The limiting values (global priorities) and the local priorities of the various risk drivers are summarized in Table 21.16.

The results (ranks of the risk drivers) obtained from the above Table 21.16 can be compared to Table 21.4, it can be further observed that there is some significant change in the weights (PVs) and hence ranks of the various risk drivers. The *Downstream* criteria appear to be the most important risk driver followed by *Upstream* and *Environment* after taking all the mutual interrelationship and causal relations into consideration.

Overall Risk Score

The overall risk score can be determined from the first row of the limit matrix, which is equal to 0.3***6255*** (or ***36.25%***). This index signifies that the overall level of petroleum supply chain risk under the considered scenario is **36%**. It will help the supply chain managers and the DM to further strategize their plans to mitigate (manage) the risk.

Summary of Key Points:

1. The case example provided in this chapter provides a complete framework to quantify the overall SC risk in relation to various operational measures.
2. Quantification of such risk at each operational level helps better understand the risk involved at the supply chain level. This approach helps in capturing operational effectiveness in terms of the measures such as quality, customer satisfaction, etc.

3. Prioritization of the various risk and sub-risk factors is necessary to manage the overall supply chain risk, since it is impossible to tackle all the problems simultaneously. Identifying the most important set of factors helps the management to focus on a narrow set of risks that are likely to bring significant advantage for the organization.

Appendices

(Tables 21.17, 21.18, 21.19, 21.20, 21.21, 21.22, 21.23, 21.24) (Figs. 21.4, 21.5, 21.6, 21.7).

Table 21.17 Pairwise comparison matrix of the sub-risk drivers w.r.t "environment"

Environment	Natural disaster	Manmade disaster	Legal/political	Social	Economic crisis
Natural disaster	1.00	3.00	2.00	4.00	2.00
Manmade disaster	0.33	1.00	0.33	2.00	0.25
Legal/political	0.50	3.00	1.00	3.00	2.00
Social	0.25	0.50	0.33	1.00	0.33
Economic crisis	0.50	4.00	0.50	3.00	1.00

Table 21.18 Pairwise comparison matrix of the sub-risk drivers w.r.t "upstream"

Upstream	Inventory	Crude supply	Delay
Inventory	1.00	0.25	0.33
Crude supply	4.00	1.00	2.00
Delay	3.00	0.50	1.00

Table 21.19 Pairwise comparison matrix of the sub-risk drivers w.r.t "process"

Process	Manufacturing competitiveness	IT system failure	Quality	M/c breakdown	Capacity
Manufacturing competitiveness	1.0000	0.1429	0.1250	0.2500	0.3333
IT system failure	7.0000	1.0000	0.5000	2.0000	3.0000
Quality	8.0000	2.0000	1.0000	3.0000	5.0000
M/c breakdown	4.0000	0.5000	0.3333	1.0000	3.0000
Capacity	3.0000	0.3333	0.2000	0.3333	1.0000

Table 21.20 Pairwise comparison matrix of the sub-risk drivers w.r.t "downstream"

Downstream	Demand	Transportation/logistics	Forecast	Customer taste change
Demand	1.0000	0.2500	0.3333	3.0000
Transportation/logistics	4.0000	1.0000	2.0000	4.0000
Forecast	3.0000	0.5000	1.0000	3.0000
Customer taste change	0.3333	0.2500	0.3333	1.0000

Table 21.21 Average direct influence of the sub-risk drivers within "environment"

Environment	Natural disaster	Manmade disaster	Legal/political	Social	Economic crisis
Natural disaster	0.00	1.00	4.00	2.00	3.00
Manmade disaster	0.00	0.00	4.00	3.00	4.00
Legal/political	0.00	3.00	0.00	4.00	3.00
Social	0.00	3.00	3.00	0.00	2.00
Economic crisis	0.00	1.00	4.00	3.00	0.00

Table 21.22 Average direct influence of the sub-risk drivers within "upstream"

Upstream	Inventory	Crude supply	Delay
Inventory	0.00	4.00	3.00
Crude supply	3.00	0.00	1.00
Delay	3.00	4.00	0.00

Table 21.23 Average direct influence of the sub-risk drivers within "process"

Process	Manufacturing competitiveness	IT system failure	Quality	M/c breakdown	Capacity
Manufacturing competitiveness	0.0000	1.0000	4.0000	4.0000	2.0000
IT system failure	1.0000	0.0000	1.0000	2.0000	1.0000
Quality	3.0000	1.0000	0.0000	2.0000	2.0000
M/c breakdown	3.0000	3.0000	4.0000	0.0000	2.0000
Capacity	2.0000	1.0000	2.0000	2.0000	0.0000

Table 21.24 Average direct influence of the sub-risk drivers within "downstream"

Downstream	Demand	Transportation/logistics	Forecast	Customer taste change
Demand	0.0000	3.0000	4.0000	4.0000
Transportation/logistics	1.0000	0.0000	2.0000	1.0000
Forecast	2.0000	3.0000	0.0000	2.0000
Customer taste change	4.0000	3.0000	3.0000	0.0000

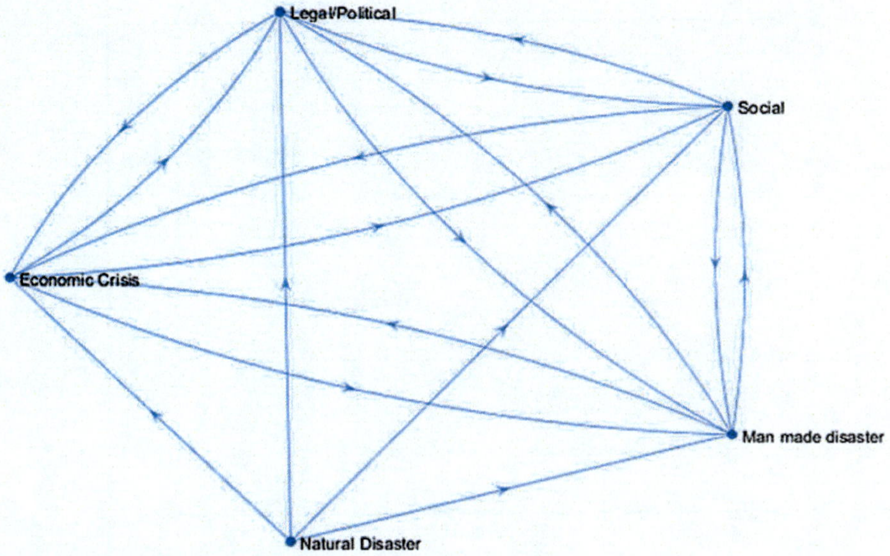

21.4 Network relation map for "environment" risk

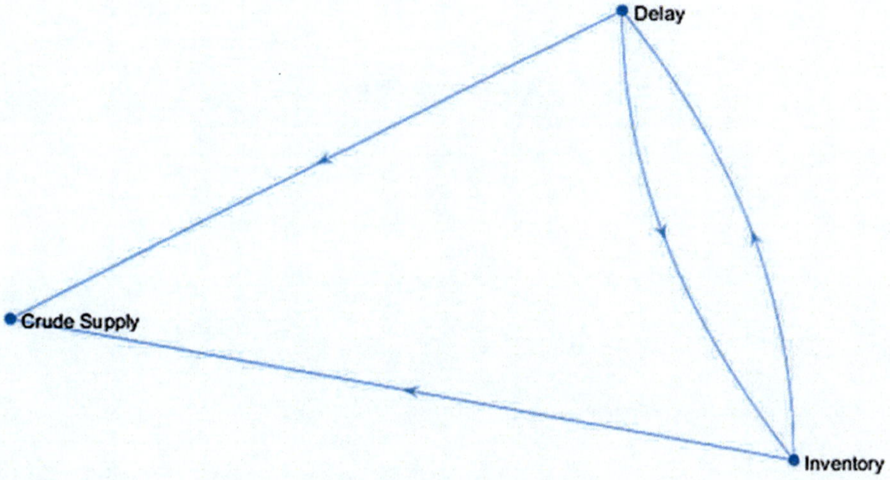

Fig. 21.5 Network relation map for "upstream" risk

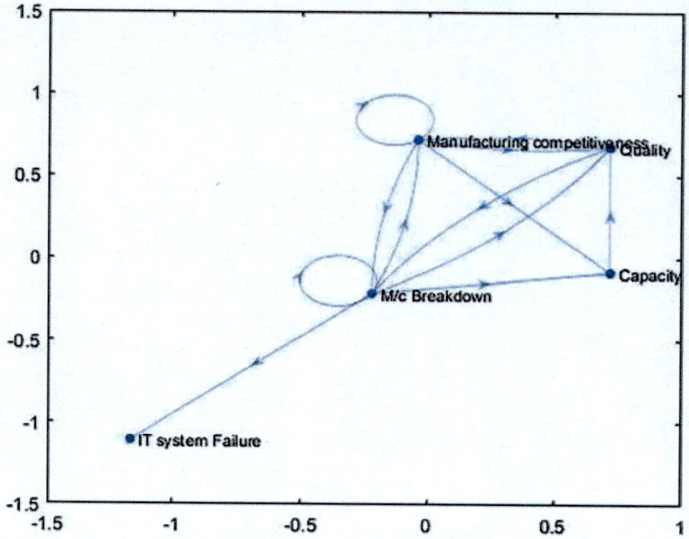

Fig. 21.6 Network relation map for "process" risk

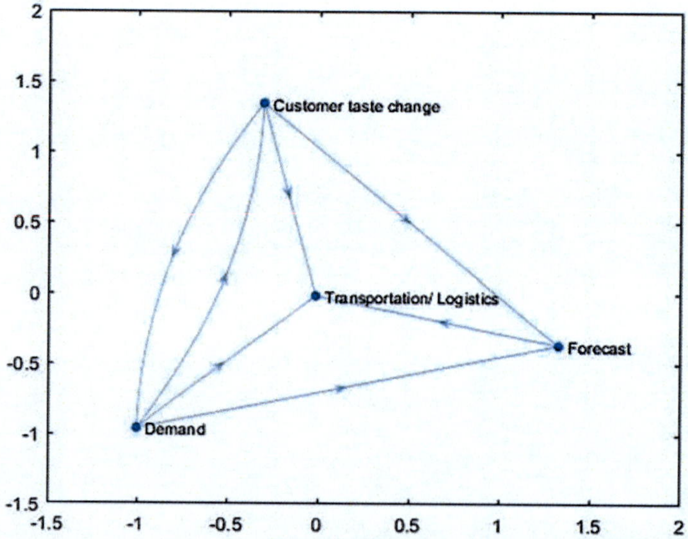

Fig. 21.7 Network relation map for "downstream" risk

Selected Bibliography and Further Reading

1. Department of Industrial Policy and Promotion, Ministry of Petroleum & Natural Gas (2017). *Oil & Gas Sector Achievements Report*. New Delhi.
2. EIA, US Energy Information Administration. (2017). *Country analysis brief: India*. Retrieved from: https://www.eia.gov/beta/international/analysis.cfm?iso=IND
3. Gandhi, S., Mangla, S. K., Kumar, P., & Kumar, D. (2016). A combined approach using AHP and DEMATEL for evaluating success factors in implementation of green supply chain management in Indian manufacturing industries. *International Journal of Logistics Research and Applications, 5567,* 1–25. https://doi.org/10.1080/13675567.2016.1164126.
4. Gölcük, I., & Baykasoğlu, A. (2016). An analysis of DEMATEL approaches for criteria interaction handling within ANP. *Expert Systems with Applications, 46,* 346–366.
5. https://doi.org/10.1016/j.eswa.2015.10.041
6. Rajesh, R., & Ravi, V. (2015). Modeling enablers of supply chain risk mitigation in electronic supply chains: A Grey-DEMATEL approach. *Computers and Industrial Engineering, 87,* 126–139. https://doi.org/10.1016/j.cie.2015.04.028.
7. https://doi.org/10.1108/JMTM-10-2017-0218

Glossary

G1: Generic Terminology Used in MCDM

S. No.	Terminology	Definition
1	Alternatives	It is a set of finite choices available with the decision maker which need to be prioritized and ranked for the better decision making
2	Multiple attributes	A real-life problem has set of alternatives which need to be prioritized for set of criteria and sub-criteria. The criteria and sub-criteria which are known as attributes may be represented in the hierarchical form for a structured decision making
3	Conflict among attributes	Attributes represent different dimensions of the alternatives such as prioritizing various manufacturing unities for set of attributes—productivity, wastages, employee turn-over ratio, etc. There is likely possibility that attributes are conflicting in nature. This means improvement in one attribute may have negative impact on other attributes
4	Incommensurable units	The attributes in the multi-criteria decision-making problem have different measurement units. For example, cost is measured in monetary unit, risk is represented as qualitative term (high or low), payback period in years, etc. This incommensurability of units in MCDM problems makes it difficult to solve
5	Decision weights	This is the weightage or importance assigned to attributes based on prior experience or group-based consensus. The weights need to be normalized to add up to one
6	Decision matrix	It is most convenient to represent MADM problem in the matrix format. *A* decision matrix **A** is an ($M \times N$) matrix in which element a_{ij} indicates the performance of alternative A_i when it is evaluated in terms of decision criterion C_j (for $i = 1, 2, 3, \ldots, M$, and $j = 1, 2, 3, \ldots, N$). It is necessary that the decision maker determines the weights of the decision criteria (denoted as W_j, for $j = 1, 2, 3, \ldots, N$) based on prior experience or group-based consensus

(continued)

J. J. Thakkar, *Multi-Criteria Decision Making*, Studies in Systems, Decision and Control 336, https://doi.org/10.1007/978-981-33-4745-8

(continued)

S. No.	Terminology	Definition
7	Completeness	This ranks all the possible alternatives in the order of preferences by considering the influence of various criteria considered in MADM problem
8	Transitivity	If action a_1 is preferred to a_2, and action a_2 is preferred to a_3, then a_1 is preferred to a_3
9	Concordance index	The amount of evidence to support the conclusion that alternative A_j outranks, or dominates, alternative Ak (quantification of positive arguments)
10	Discordance index	Counterpart of the concordance index
11	Ideal solution	A decision maker usually expects to maximize the value of benefit criteria (e.g., profit) among the alternatives. Similarly, it is expected that cost criteria need to be minimized. The most preferred alternative is considered as the ideal solution in MCDM problem
12	Negative-ideal solution	The alternative which is least preferred is considered as the negative-ideal solution

G2: Key Terminology Used in Different MCDM Techniques

S. No.	Term	Definition
Analytic Hierarchy Process (AHP) Hierarchical prioritization of criteria and alternatives		
1.	Hierarchies	A hierarchy is a stratified system which consists of an overall goal, a set of alternatives for reaching the goal and a set of criteria and sub-criteria that relate the alternatives to the goal
2.	Pairwise comparison	It is a process of comparing various alternatives or criteria with each other to understand their relative importance for a given problem
3.	Consistency index (CI)	Deviation or degree of consistency which can be calculated by the following formula $\mathrm{CI} = \frac{\lambda\max - n}{n-1}$ where n = size of comparison matrix
4.	Consistency ratio (CR)	Consistency ratio measures the level of inconsistency $\mathrm{CR} = \frac{\mathrm{CI}}{\mathrm{RI}}$ • Where RI is the average value of CI for random matrices using the Saaty scale obtained by Forman • Matrix is considered as consistent one if $\mathrm{CR} < 0.1$

(continued)

(continued)

S. No.	Term	Definition
Analytic Network Process (ANP) Causality-based prioritization of criteria and alternatives		
1.	Pairwise comparison matrices and priority vectors	It incorporates in interdependencies which are not considered in AHP for prioritizing the set of alternatives with respect to pre-defined criteria. The influence of each element on the other elements can be represented by an eigenvector
2.	Priority vector	It can be derived as an estimate of the relative importance associated with the elements (or clusters) being compared by solving the following equation $A * w = \lambda_{max} * w$ where A is the matrix of pairwise comparison W is eigenvector λ_{max} is the largest eigenvalue of A
3.	Supermatrix formation	• The supermatrix concept is similar to the Markov chain process • It represents a partitioned matrix, where each matrix segment represents a relationship between two clusters in a system
TOPSIS (Technique for Order of Preference by Similarity to Ideal Solution) Nearness to ideal solution-based ranking of alternatives		
1.	Decision or evaluation matrix	An evaluation matrix consisting of m alternatives and n criteria, with the intersection of each alternative and criteria given as x_{ij}, We therefore have a matrix $(x_{ij})_{m \times n}$
2.	Normalized decision matrix	It is a transformed matrix which is dimensionless in nature. This facilitates the comparison of the various criteria on an equal platform irrespective of their individual units
3.	Positive-ideal and negative-ideal solutions	The solution which maximizes the benefit criteria and minimizes the cost criteria is considered as positive-ideal solution. Opposite to this, a negative-ideal solution maximizes the cost criteria and minimizes the benefit criteria

(continued)

(continued)

S. No.	Term	Definition
DEMATEL		
Impact relations map for criteria and alternatives		
1	Average matrix	Each respondent was asked to evaluate the direct influence between any two factors by an integer score ranging from 0, 1, 2 and 3, representing "no influence", "low influence", "medium influence" and "high influence", respectively
2	Normalized initial direct-relation matrix (D)	In order to transform it into a dimensionless matrix, which allows the comparison of the various criteria
3	Total relation matrix (T)	The total relation matrix T is defined as $T = D(I - D)^{-1}$, where I is the identity matrix
ELECTRE (The ELimination Et Choix Traduisant la REalit´e)		
Coefficients of importance-based ranking of alternatives		
1.	Weighted normalized	Values allow the comparison of corresponding normalized values for different data sets in a way that eliminates the effects of certain gross influences $v_{ij} = w_j * x_{ij}$ w_j = attribute weights x_{ij} = performance measure
2.	Concordance set	If alternative is better than or equal to other elements of pair, it is considered under concordance set and defined by C $C(p, q) = \{j, v_{pj} \geq v_{qj}\}$
3.	Concordance matrix	Concordance matrix is the matrix generated by adding the values of weights of concordance set elements $C_{pq} = \sum_{j^*} w_{j^*}$ j^* = concordance set
4.	Discordance set	If alternative is worse than the other elements of the pair for relevant criteria, it is considered under discordance set and defined by D $D(p, q) = \{j, v_{pj} < v_{qj}\}$
5.	Discordance matrix	Discordance matrix is prepared by dividing discordance set member values to total value of whole set $D_{pq} = \frac{\left(\sum_{j0} \lvert v_{pj0} - v_{qj0} \rvert\right)}{\left(\sum_{j} \lvert v_{pj} - v_{qj} \rvert\right)}$

(continued)

(continued)

S. No.	Term	Definition
6.	Credibility value	The only values that are sufficiently close to most suitable value are considered
LINMAP Calculates distances between the evaluation values and the optimal one		
1.	Normalized probabilistic linguistic	Each normalized probabilistic linguistic evaluation vector of each alternative. We use it to replace the distance in the classical LINMAP
2.	The consistency index	A linear programming model is then built to derive the weight of each criterion
3.	Comparative analyses	Comparative analyses between the alternatives
SWARA (Step-wise Weight Assessment Ratio Analysis) Enables logical calculation of weights and relative importance of criteria		
1.	Relative weights of criteria	The evaluation criteria should be sorted in descending order based on their expected significances
2.	Ratio SJ value	The respondent must indicate the relative importance of criterion *j* with respect to the previous criterion (*j*–1), starting from the second criterion
3.	Overall rank	The overall ranks to the group of experts are determined according to the mediocre value of ranks $w_j = \frac{q_j}{\sum_{k=1}^{n} q_k}$ w_j = relative weight q_j = recalculated weight of jth criterion n = number of the criteria
PROMETHEE The method is based on preference function approach		
1.	Preference degree	Computes for each pair of possible decisions and for each criterion, the value of the preference degree
2.		Aggregate the preference degrees of all criteria for each pair of possible decisions

(continued)

(continued)

S. No.	Term	Definition
3.	Outranking flows	The ranking of the possible decisions consists in computing the outranking flows $\emptyset^{+}(a)=\frac{1}{n-1}=\sum_{x\in A}\pi(a,x)$ $\emptyset^{-}(a)=\frac{1}{n-1}=\sum_{x\in A}\pi(x,a)$ $\emptyset^{+}(a)=$ positive outranking flow $\emptyset^{-}(a)=$ negative outranking flow $n=$ number of possible decisions $\pi(a,b)=$ global preference index $a,b=$ decisions $A=$ set of possible decisions
4.	Net outranking	Using the outranking flows to establish a complete ranking between the possible decisions. The ranking is based on the net outranking flows $\emptyset(a)=\emptyset^{+}(a)-\emptyset^{-}(a)$ $\emptyset(a)=$ net outranking flow
MOORA (Multi-objective optimization on the basis of ratio analysis method) MOORA refers to a ratio system in which each response of an alternative on an objective is compared to a denominator, which is representative for all alternatives concerning that objective		
1.	Pertinent evaluation	Identify the pertinent evaluation attributes and represent the available information in the form of a decision matrix
2.		Square root of the sum of squares of each alternative per attribute $x_{ij}^{*}=x_{ij}/\sqrt{\left[\sum_{i=1}^{m}x_{ij}^{2}\right]}$ $i=1,\ldots,m \quad j=1,\ldots,n$ $x_{ij}=$ performance measure $m=$ number of alternative $n=$ number of attribute

(continued)

(continued)

S. No.	Term	Definition
3.	Normalized values	These normalized performances are added in case of maximization (for beneficial attributes) and subtracted in case of minimization (for non-beneficial attributes) $y_i = \sum_{j=1}^{g} x_{ij}^* - \sum_{j=g+1}^{n} x_{ij}^*$ y_i = normalized assessment value g = number of attributes to be maximized $(n - g)$ = number of attributes to be minimized
4.	Ordinal ranking	An ordinal ranking of y_i shows the final preference $y_i = \sum_{j=1}^{g} w_j x_{ij}^* - \sum_{j=g+1}^{n} w_j x_{ij}^*$ w_j = attribute weights
COPRAS (The preference ranking method of complex proportional assessment method) Considers maximizing and minimizing criteria separately to select the best alternative based on both the ideal and the anti-ideal solutions		
1.	Weighted normalized decision matrix,	Product of normalized performance value of alternative to the weight of particular criteria $R = [r_{ij}]_{m*n} = x_{ij} / \sum_{i=1}^{m} x_{ij}$, $i = 1, \ldots, m, \quad j = 1, \ldots, n$ r_{ij} = normalized performance value x_{ij} = assesment value i = alternative, j = criterion
2.		The sums of weighted normalized values are calculated for both the beneficial and non-beneficial criteria $= r_{ij} w_j$ w_j = weight of criterion $\sum_{i=1}^{m} y_{ij} = w_i$
3.	Relative significances	Determination of the relative significances of the alternatives $S_{+ij/-ij} = \sum_{j=1}^{n} y_{+ij/-ij}$ $Q_i = S_{+i} + \frac{S_{-i} \cdot \sum_{i=1}^{m} S_{-i}}{S_{-i} \cdot \sum_{i=1}^{m} (S_{-min}/S_{-i})}$

(continued)

(continued)

S. No.	Term	Definition
4.	Utility values	Utility values of the candidate alternatives range from 0 to 100% The greater the value, the higher is the priority of the alternative $U_i = \frac{Q_i}{Q\max}.100\%$ $Q_{max} =$ maximum relative significance value
WASPAS (Weighted aggregates sum product assessment) Integrates two well-known MCDM approaches: weighted sum (WS) method and weighted product (WP) method		
1.	Optimal performance rating	Determine the optimal performance rating for each criterion
2.	Relative importance on WS method	the relative importance of i-th alternative, based on weighted sum method $Q_i^{(1)} = \sum_{j=1}^{n} w_j r_{ij}$ $Q_i^{(1)} =$ relative importance based on WS method $r_{ij} =$ normalized performance rating
3.	Relative importance on WP method	the relative importance of i-th alternative, based on WP method $Q_i^{(2)} = \prod_{j=1}^{n} r_{ij}^{w_j}$ $Q_i^{(2)} =$ relative importance based on WP method
4.	Total relative importance	The total relative importance, or more precisely the joint generalized criterion of weighted aggregation of additive and multiplicative methods $Q_i = 0.5 * Q_i^{(1)} + 0.5 * Q_i^{(2)}$

Appendix A
Key Research Published in MCDM Domain During 2014–2019 by the Faculty of Department of Industrial and Systems Engineering, IIT Kharagpur

Faculty	Problem	Application of technique	Year of publication	Journal
J J Thakkar	Development of balanced scorecard for health care using interpretive structural modelling and analytic network process	ISM and ANP	2014	Journal of Advances in Management Research
	Sustainable supply chain practices: an empirical investigation on Indian automobile industry	SEM	2016	Production Planning and Control
	Development of composite sustainable supply chain performance index for the automobile industry	Fuzzy AHP, liberatore score and S/N ratio approach	2015	International Journal of Sustainable Engineering
	A comparative assessment of the performance of select higher education institutes in India	DEA–AHP approach	2016	Quality Assurance in Education

(continued)

J. J. Thakkar, *Multi-Criteria Decision Making*, Studies in Systems, Decision and Control 336, https://doi.org/10.1007/978-981-33-4745-8

(continued)

Faculty	Problem	Application of technique	Year of publication	Journal
	An investigation on lean-green implementation practices in Indian SMEs using analytical hierarchy process (AHP) approach	AHP	2016	Journal of Cleaner Production
	Selection of green suppliers based on GSCM practices: Using fuzzy MCDM approach in an electronics company	Fuzzy AHP and Fuzzy TOPSIS	2017	IGI Global
	Schedule and cost overrun analysis for R&D projects using ANP and system dynamics	System Dynamics and ANP	2017	International Journal of Quality and Reliability Management
	A quantitative risk assessment methodology and evaluation of food supply chain	Grey AHP and Grey TOPSIS	2017	The International Journal of Logistics Management
	An integrated approach for lead time reduction of military aircraft major overhaul: A case of ABC company	AHP, theory of constraint (TOC) and simulation for $M/M/1/K$ queuing environment	2018	International Journal of Quality and Reliability Management
	Ranking of drivers for integrated lean-green manufacturing for Indian manufacturing SMEs	TOPSIS and SAW	2018	Journal of Cleaner Production
	A quantitative framework for lean and green assessment of supply chain performance	DEMATEL and AHP	2018	International Journal of Productivity and Performance Management

(continued)

(continued)

Faculty	Problem	Application of technique	Year of publication	Journal
	Interdependence analysis of lean-green implementation challenges: a case of Indian SMEs	Interpretive structural modelling (ISM) and interpretive ranking process (IRP)	2018	Journal of Manufacturing Technology Management
	A hybrid approach for quantifying supply chain risk and prioritizing the risk drivers: A case of Indian petroleum supply chain	DEMATEL and AHP	2018	Journal of Manufacturing Technology Management
	Analysis of supply chain complexity drivers for Indian mining equipment manufacturing companies combining SAP-LAP and AHP	SAP-LAP and AHP	2018	Resources Policy
	Risk management framework for outsourcing in the defence sector: a case from India	FMEA	2018	International Journal of Production Research
	The distribution strategy selection for an e-tailer using a hybrid DANP VIKOR MCDM model	DANP, VIKOR	2019	Benchmarking: An International Journal
J Maiti	Risk analysis using FMEA: fuzzy similarity value and possibility theory-based approach	Fuzzy FMEA	2014	Expert Systems with Applications
	Human error identification and risk prioritization in overhead crane operations using HTA, SHERPA and fuzzy VIKOR method	HTA, SHERPA and fuzzy VIKOR	2015	Expert Systems with Applications

(continued)

(continued)

Faculty	Problem	Application of technique	Year of publication	Journal
	Supplier selection in uncertain environment: a fuzzy MCDM approach	DEMATEL, ANP and fuzzy VIKOR	2017	Proceedings of the First International Conference on Intelligent Computing and Communication (Book)
M K Tiwari	Green supply chain performance measurement using fuzzy ANP-based balanced scorecard: a collaborative decision-making approach	A fuzzy analytic network process (ANP)-based green-balanced scorecard	2013	Production Planning and Control
	Multi-objective optimization approach to product-planning in quality function deployment incorporated with fuzzy ANP	QFD and fuzzy ANP	2014	Applications of Multi-Criteria and Game Theory Approaches
	Recycler selection using fuzzy AHP by considering sustainability	Fuzzy AHP	2014	Advanced Materials Research
	Novel fuzzy hybrid multi-criteria group decision-making approaches for the strategic supplier selection problem	Fuzzy consensus-based possibility measure and fuzzy TOPSIS method	2015	Expert Systems with Applications
	Interpretive structural modelling-analytic network process integrated framework for evaluating sustainable supply chain management alternatives	ISM and ANP	2016	Applied Mathematical Modelling
B Mahanty	Fuzzy risk assessment for electro-optical target tracker	Technology index (TI) and fuzzy failure mode effect analysis (FMEA)	2016	International Journal of Quality and Reliability Management

(continued)

(continued)

Faculty	Problem	Application of technique	Year of publication	Journal
S P Sarmah	Supply chain performance measurement for third-party logistics	Supply Chain Operations Reference (SCOR), fuzzy AHP, TOPSIS	2014	Benchmarking: An International Journal
	Multi-criteria classification of spare parts inventories—a web-based approach	Fuzzy rule-based approach	2015	Journal of Quality in Maintenance Engineering
	A methodology for risk assessment and formulation of mitigation strategies for trucking Industry	Modified FMEA	2016	International Journal of Performability Engineering
	Risk evaluation and mitigation of sustainable road freight transport operation: a case of trucking industry	Fuzzy FMEA	2019	International Journal of Production Research
Mamta Jenamani	Sustainability in supply chain through e-procurement—an assessment framework based on DANP and liberatore score	DANP and liberatore score	2015	IEEE Systems Journal
	Risk assessment of outsourcing e-procurement services: integrating SWOT analysis with a modified ANP-based fuzzy inference system	ANP	2016	Production Planning and Control

Appendix B
List of Software in MCDM

Name of software	Key features	Supported MCDM techniques
Ahoona	• Enables decision making for real life problems	WSM
Altova MetaTeam	• Reduces the amount of time needed for users to learn database software	WSM
ChemDecide	• A software package containing a decision structuring tool	AHP, ELECTRE III and MARE
Criterium Decision Plus	• Help manage complex decisions in business and government	AHP, SMART
Definite 3.1	Facillitates decision making for a finite set of alternatives	SMART, AHP, ELECTRE 2, regime method, graphical analysis and extensive sensitivity analysis
Diviz	Helps to design, execute and share complex MCDA/M algorithms and experiments	SMART, AHP, ELECTRE, ANP, ELECTRE, MOORA
D-Sight	• Visual and interactive tool for multi-criteria decision aid problems	PROMETHEE methods and multi-attribute utility theory
D-Sight	• Facillitates better group and business decisions	PROMETHEE, UTILITY
Decision Lens	• Enterprise prioritization and resource optimization solution for the public sector	AHP, ANP
Expert Choice	Decision-making software for linear hierarchical structure	AHP
MCDA Package for R	Supporting the multi-criteria decision aiding process	SMART, AHP, ELECTRE, ANP, ELECTRE, MOORA

(continued)

J. J. Thakkar, *Multi-Criteria Decision Making*, Studies in Systems, Decision and Control 336, https://doi.org/10.1007/978-981-33-4745-8

(continued)

Name of software	Key features	Supported MCDM techniques
PriEsT	Decision-making tool to calculate the pairwise comparison	Analytic hierarchy process (AHP)
Super Decisions	• Decision-making tool to calculate the pairwise comparison	AHP, ANP
• 1000 Minds	Software for multi-criteria decision-making, prioritisation and resource allocation	PAPRIKA

Appendix C
Use of Excel in MCDM for Machine Selection Problem

Key Criteria:

- Budget (C_1),
- On-time delivery (C_2),
- Performance history (C_3),
- Technical capability (C_4),

Alternatives:

- $Machine_1$
- $Machine_2$
- $Machine_3$

Important terminology used in calculation:

1. **Weights**: These estimates relative importance of criteria.
    - Each attribute is given certain points on 0–10 scale (10 very good, 1 low)
    - Example

Criteria	Weight
Budget	4
On-time delivery	2
Performance history	6
Technical capability	8

2. **Decision makers (DM)**: These are the experts who are assigned to with the task of weighting each attribute (there can be 'n' number of decision makers).
3. **Decision matrix**: A table that is used to objectively make decision about making selection from range of options. Decision makers rate each attribute of each alternative.

J. J. Thakkar, *Multi-Criteria Decision Making*, Studies in Systems, Decision and Control 336, https://doi.org/10.1007/978-981-33-4745-8

- Example: For $Machine_1$

Criteria	Decision maker			Attributes weights
	DM_1	DM_1	DM_1	
Budget	10	8	9	$= \frac{10+8+9}{3} = 9$
On-time delivery	7	8	6	$= \frac{7+8+6}{3} = 7$
Performance history	5	8	7	$= \frac{5+8+7}{3} = 6$
Technical capability	9	5	7	$= \frac{9+5+7}{3} = 7$

Similarly,
For $Machine_2$

Criteria	Decision maker			Attributes weights
	DM_1	DM_1	DM_1	
Budget	9	6	9	$= \frac{9+6+9}{3} = 8$
On-time delivery	8	9	4	$= \frac{8+9+4}{3} = 7$
Performance history	7	10	10	$= \frac{7+10+10}{3} = 9$
Technical capability	5	7	6	$= \frac{5+7+6}{3} = 6$

- For $Machine_3$

Criteria	Decision maker			Attributes weights
	DM_1	DM_1	DM_1	
Budget	9	4	8	$= \frac{9+4+8}{3} = 7$
On-time delivery	7	9	8	$= \frac{7+9+8}{3} = 8$
Performance history	6	5	7	$= \frac{6+5+7}{3} = 6$
Technical capability	7	6	5	$= \frac{7+6+5}{3} = 6$

Decision matrix

Criteria	$Machine_1$	$Machine_2$	$Machine_3$
Budget	9	8	7
On-time delivery	7	7	8
Performance history	6	9	6
Technical capability	7	6	6

- For solving this problem, **TOPSIS** approach is adapted. **TOPSIS** selects the alternative that is the closest to the ideal solutions and farthest from negative-ideal solution.

Steps to be followed in excel

Step 1—Standardize the decision matrix.

- This step transforms various attribute dimensions into non-dimensional attributes, which allow comparisons across criteria.
- For standardizing, each column of decision matrix is divided by root of sum of square of respective

Criteria	$Machine_1$	$Machine_2$	$Machine_3$	
Budget	9	8	7	13.928
On time delivery	7	7	8	12.728
Performance history	6	9	6	12.369
Technical capability	7	6	6	11.000

(Note: First calculate the sum of squares by using the function SUMSQ of each row using SQRT to calculate the final value)

Criteria	$Machine_1$	$Machine_2$	$Machine_3$
Budget	9/13.928	8	7
On time delivery	7	7	8
Performance history	6	9	6
Technical capability	7	6	6

Criteria	Machine$_1$	Machine$_2$	Machine$_3$
Budget	0.65		
On time delivery			
Performance history			
Technical capability			

Similarly,

Criteria	Machine$_1$	Machine$_2$	Machine$_3$
Budget	0.65	0.57	0.50
On time delivery	0.55	0.55	0.63
Performance history	0.49	0.73	0.49
Technical capability	0.64	0.55	0.55

Step 2—Construct weighted standardized decision matrix by multiplying attributes weight to each rating.

Criteria	Machine$_1$	Machine$_2$	Machine$_3$
Budget	0.65	0.57	0.50
On time delivery	0.55	0.55	0.63
Performance history	0.49	0.73	0.49
Technical capability	0.64	0.55	0.55

Criteria	Weight
Budget	4
On time delivery	2
Performance history	6
Technical capability	8

Criteria	Machine$_1$	Machine$_2$	Machine$_3$
Budget	2.60		
On time delivery			
Performance history			
Technical capability			

Criteria	Machine$_1$	Machine$_2$	Machine$_3$
Budget	0.65	0.57	0.50
On time delivery	0.55	0.55	0.63
Performance history	0.49	0.73	0.49
Technical capability	0.64	0.55	0.55

Criteria	Weight
Budget	4
On time delivery	2
Performance history	6
Technical capability	8

Criteria	Machine$_1$	Machine$_2$	Machine$_3$
Budget	2.60	2.28	
On time delivery			
Performance history			
Technical capability			

Similarly,

Criteria	Machine$_1$	Machine$_2$	Machine$_3$
Budget	2.60	2.28	2.00
On time delivery	1.10	1.10	1.26
Performance history	2.94	4.38	2.94
Technical capability	5.12	4.40	4.40

Step 3—Determine ideal solution and negative-ideal solution.

A set of maximum values for each criterion is ideal solution.

Criteria	Machine$_1$	Machine$_2$	Machine$_3$	
Budget	2.60	2.28	2.00	=MAX(E26:G26)
Oil time delivery	1.10	1.10	1.26	
Performance history	2.94	4.38	2.94	
Technical capability	5.12	4.40	4.40	

Criteria	$Machine_1$	$Machine_2$	$Machine_3$	
Budget	2.60	2.28	2.00	2.60
On time delivery	1.10	1.10	1.26	1.26
Performance history	2.94	4.38	2.94	4.38
Technical capability	5.12	4.40	4.40	5.12

Ideal solution = {2.6, 1.26, 4.38, 5.12}
A set of minimum values for each criterion is negative-ideal solution.

Criteria	$Machine_1$	$Machine_2$	$Machine_3$	
Budget	2.60	2.28	2.00	=MIN(E32:G32)
On time delivery	1.10	1.10	1.26	
Performance history	2.94	4.38	2.94	
Technical capability	5.12	4.40	4.40	

Criteria	$Machine_1$	$Machine_2$	$Machine_3$	
Budget	2.60	2.28	2.00	2.00
On time delivery	1.10	1.10	1.26	1.10
Performance history	2.94	4.38	2.94	2.94
Technical capability	5.12	4.40	4.40	4.40

Negative-ideal solution = {2.0, 1.10, 2.94, 4.4}

Step 4—Determine separation from ideal solution. S_i^*

Criteria	$Machine_1$	$Machine_2$	$Machine_3$
Budget	$(2.6–2.6)^2$	$(2.28–2.6)^2$	$(2.00–2.6)^2$
On-time delivery	$(1.10–1.26)^2$	$(1.10–1.26)^2$	$(1.26–1.26)^2$
Performance history	$(2.94–4.38)^2$	$(4.38–4.38)^2$	$(2.94–4.38)^2$
Technical capability	$(5.12–5.12)^2$	$(4.40–5.12)^2$	$(4.40–5.12)^2$

Criteria	Machine $_1$	Machine$_2$	Machine$_3$
Budget	0	0.10	0.36
On-time delivery	0.02	0.02	0
Performance history	2.07	0	2.07
Technical capability	0	0.51	0

$$(0 + 0.02 + 2.07 + 0)^{1/2} = 0.74$$

Similarly,

$$S_i^* = [1.44 \qquad 0.79 \qquad 1.55]$$

Step 5—Determine separation from negative-ideal solution.

Criteria	Machine$_1$	Machine$_2$	Machine$_3$
Budget	$(2.6–2.0)^2$	$(2.28–2.0)^2$	$(2.00–2.0)^2$
On-time delivery	$(1.10–1.1)^2$	$(1.10–1.1)^2$	$(1.26–1.1)^2$
Performance history	$(2.94–2.94)^2$	$(4.38–2.94)^2$	$(2.94–2.94)^2$
Technical capability	$(5.12–4.4)^2$	$(4.40–4.4)^2$	$(4.40–4.4)^2$

Criteria	Machine$_1$	Machine$_2$	Machine$_3$
Budget	0.36	0.07	0
On-time delivery	0	0	0.02
Performance history	0	2.07	0
Technical capability	0.51	0	0

$$(0.36 + 0 + 0 + 0.51)^{1/2} = 0.93$$

$$S_i' = [0.93 \qquad 1.46 \qquad 0.14$$

Step 6—Determine relative closeness to ideal solution.

Criteria	$Machine_1$	$Machine_2$	$Machine_3$
S_i^*	1.44	0.79	1.55
S_i'	0.93	1.46	0.14
$(S_i^* + S_i')$	2.37	2.25	1.69
$S_i'/(S_i^* + S_i')$	0.93/2.37	1.46/2.25	0.14/1.69
Technical capability	0.39	0.65	0.08

↑ Max.

The results obtained in Step 6 indicate the following priority preferences.

$Machine_2$ > $Machine_1$ > $Machine_3$